Rural Aquaculture

Edited by

P. Edwards

Aquaculture and Aquatic Resources Management
School of Environment
Resources and Development
Asian Institute of Technology
Klong Luang
Thailand

D.C. Little

Institute of Aquaculture
University of Stirling
Stirling
UK

and

H. Demaine

Aquaculture and Aquatic Resources Management
School of Environment
Resources and Development
Asian Institute of Technology
Klong Luang
Thailand

CABI *Publishing*

CABI *Publishing* is a division of CAB *International*

CABI Publishing
CAB International
Wallingford
Oxon OX10 8DE
UK

CABI Publishing
10 E 40th Street
Suite 3203
New York, NY 10016
USA

Tel: +44 (0)1491 832111
Fax: +44 (0)1491 833508
Email: cabi@cabi.org
Web site: www.cabi-publishing.org

Tel: +1 212 481 7018
Fax: +1 212 686 7993
Email: cabi-nao@cabi.org

A catalogue record for this book is available from the British Library, London, UK

Library of Congress Cataloging-in-Publication Data
Rural aquaculture / edited by P. Edwards, D. Little, and H. Demaine.
    p. cm.
  Papers presented at a conference held in November 1998 in Chiang Mai, Thailand.
  Includes bibliographical references (p.).
  ISBN 0-85199-565-9 (alk. paper)
    1. Aquaculture--Congresses. 2. Aquaculture--South Asia--Congresses. 3. Aquaculture--Asia, Southeastern--Congresses. I. Edwards, Peter, 1943- II. Little, D. (David) III. Demaine, Harvey.

  SH117.S64 R87 2001
  639.8'095--dc21

                                       2001043048

ISBN 0 85199 565 9

Printed and bound in the UK by Biddles Ltd, Guildford and King's Lynn, from copy supplied by the authors.

Contents

SPECIALISED AND INTENSIVE SYSTEMS

SEED

Contributors

Abedin, J., GOLDA Project, CARE–Bangladesh, GPO Box 226, Dhaka 1000, Bangladesh

Barman, B.K., Northwest Fisheries Extension Project, Parbatipur, Dinajpur, Bangladesh

Be, N.V., Department of Environment and Natural Resources Management, College of Agriculture, Cantho University, Cantho, Vietnam

Beveridge, M.C.M., Institute of Aquaculture, University of Stirling, Stirling FK9 4LA, UK

Bimbao, M.A., International Center for Living Aquatic Resources Management (ICLARM), PO Box 500, 10670 Penang, Malaysia

Brugere, C., Institute of Aquaculture, University of Stirling, Stirling FK9 4LA, UK

Chapman, G., GOLDA Project, CARE–Bangladesh, GPO Box 226, Dhaka 1000, Bangladesh

Clarke, G.J.C., University of Wales Swansea, Swansea SA2 8PP, UK

Cong, N.V., Department of Environment and Natural Resources Management, College of Agriculture, Cantho University, Cantho, Vietnam

Cuong, N.X., Research Institute for Aquaculture No. 1, Dinh Bang, Tu Son, Bac Ninh, Vietnam

Demaine, H., Aquaculture and Aquatic Resources Management, School of Environment, Resources and Development, Asian Institute of Technology, PO Box 4, Klong Luang, Pathumthani 12120, Thailand

Dey, M.M., International Center for Living Aquatic Resources Management (ICLARM), GPO Box 500, 10670 Penang, Malaysia

Dung, D.T., Department of Environment and Natural Resources Management, College of Agriculture, Cantho University, Cantho, Vietnam

Edwards, P., Aquaculture and Aquatic Resources Management, School of Environment, Resources and Development, Asian Institute of Technology, PO Box 4, Klong Luang, Pathumthani 12120, Thailand

Felsing, M., Institute of Aquaculture, University of Stirling, Stirling FK9 4LA, UK

Giang, T.T., Faculty of Fisheries, University of Agriculture and Forestry, Ho Chi Minh City, Vietnam

Gowing, J., Centre for Land Use and Water Resources Research, Porter Building, University of Newcastle, Newcastle upon Tyne NE1 7RU, UK

Gregory, R., Aquaculture and Aquatic Resources Management, School of Environment, Resources and Development, Asian Institute of Technology, PO Box 4, Klong Luang, Pathumthani 12120, Thailand

Gupta, M.V., International Center for Living Aquatic Resources Management (ICLARM), GPO Box 500, 10670 Penang, Malaysia

Guttman, H., Aquaculture and Aquatic Resources Management, School of Environment, Resources and Development, Asian Institute of Technology, PO Box 4, Klong Luang, Pathumthani 12120, Thailand

Haylor, G., Institute of Aquaculture, University of Stirling, Stirling FK9 4LA, UK

Heidhues, F., Department for Agricultural Economics and Social Sciences in the Tropics and Subtropics (490), University of Hohenheim, D-70593 Stuttgart, Germany

Hussain, M.G., Bangladesh Fisheries Research Institute (BFRI), Mymensingh 2201, Bangladesh

Innes-Taylor, N., AIT Aqua Outreach-Lao PDR, Livestock and Fisheries Section, Division of Agriculture and Forestry, PO Box 16, Savannakhet, Lao PDR

Ireland, M.J., CARE Bangladesh, Dhaka 1209, Bangladesh

Islam, M., Extension Officer, Northwest Fisheries Extension Project, Parbatipur, Dinajpur District, Bangladesh

Jeney, Z., Fish Culture Research Institute, PO Box 47, Szarvas, Hungary

Kodithuwakku, S.S., Department of Agricultural Economics, Faculty of Agriculture, University of Peradeniya, Peradeniya, Sri Lanka

Lin, C.K., Aquaculture and Aquatic Resources Management, School of Environment, Resources and Development, Asian Institute of Technology, PO Box 4, Klong Luang, Pathumthani 12120, Thailand

Lithdamlong, D., Regional Development Committee, Livestock and Fisheries Section, Division of Agriculture and Forestry, PO Box 16, Savannakhet, Lao PDR

Little, D.C., Institute of Aquaculture, University of Stirling, Stirling FK9 4LA, UK

Long, D.N., Freshwater Aquaculture Department, College of Agriculture, Cantho University, Cantho, Vietnam

Lopez, T., International Center for Living Aquatic Resources Management (ICLARM), PO Box 500, 10670 Penang, Malaysia

Luu, L.T., Research Institute for Aquaculture No. 1, Dinh Bang, Tu Son, Bac Ninh, Vietnam

Mair, G.C., University of Wales Swansea, Swansea SA2 8PP, UK

Mardall, N., Extension Advisor, Northwest Fisheries Extension Project, FMS, House 42, Road 28, Gulshan, Dhaka, Bangladesh

Mazid, M.A., Bangladesh Fisheries Research Institute (BFRI), Mymensingh 2201, Bangladesh

McAndrew, K.I., CARE Bangladesh, Dhaka 1209, Bangladesh

Meusch, E., AIT Aqua Outreach-Cambodia, Aquaculture Office, Department of Fisheries, PO Box 835, Phnom Penh, Cambodia

Morales, E.J., Freshwater Aquaculture Center, Central Luzon State University, Neuva Ecija 3120, Philippines

Munzir, A., Department for Agricultural Economics and Social Sciences in the Tropics and Subtropics (490), University of Hohenheim, D-70593 Stuttgart, Germany

Murray, F., Institute of Aquaculture, University of Stirling, Stirling FK9 4LA, UK

Oficial, R., International Center for Living Aquatic Resources Management (ICLARM), PO Box 500, 10670 Penang, Malaysia

Olah, J., Fish Culture Research Institute, PO Box 47, Szarvas, Hungary

Pant, J., Aquaculture and Aquatic Resources Management, School of Environment, Resources and Development, Asian Institute of Technology, PO Box 4, Klong Luang, Pathumthani 12120, Thailand

Pekar, F., Fish Culture Research Institute, PO Box 47, H-5541 Szarvas, Hungary

Phong, N.T., Department of Mariculture, College of Agriculture, Cantho University, Cantho, Vietnam

Prein, M., International Center for Living Aquatic Resources Management (ICLARM), PO Box 500, 10670 Penang, Malaysia

Promthong, P., Aquaculture and Aquatic Resources Management, School of Environment, Resources and Development, Asian Institute of Technology, PO Box 4, Klong Luang, Pathumthani 12120, Thailand

Radheyshyam, KVK/TTC, Central Institute of Freshwater Aquaculture, Kausalyaganga, Bhubaneshwar-751002, India

Rahman, A., Bangladesh Fisheries Research Institute (BFRI), Mymensingh 2201, Bangladesh

Roos, N., Research Department of Human Nutrition, The Royal Veterinary and Agricultural University, Rolighedsvej 30, DK-1958 Frederiksberg C, Denmark

Roy, T.K., CARE Bangladesh, Dhaka 1209, Bangladesh

Satapornvanit, A., Aquaculture and Aquatic Resources Management, School of Environment, Resources and Development, Asian Institute of Technology, PO Box 4, Klong Luang, Pathumthani 12120, Thailand

Setboonsarng, S., Aquaculture and Aquatic Resources Management, School of Environment, Resources and Development, Asian Institute of Technology, PO Box 4, Klong Luang, Pathumthani 12120, Thailand

Sevilleja, R.C., Freshwater Aquaculture Center, Central Luzon State University Neuva Ecija 3120, Philippines

Sollows, J.D., International Center for Living Aquatic Resources Management (ICLARM), GPO Box 500, 10670 Penang, Malaysia

Thilsted, S.H., Research Department of Human Nutrition, The Royal Veterinary and Agricultural University, Rolighedsvej 30, DK-1958 Frederiksberg C, Denmark

Trang, P.V., Research Institute for Aquaculture No. 1, Dinh Bang, Tu Son, Bac Ninh, Vietnam

Tu, N.V., Faculty of Fisheries, University of Agriculture and Forestry, Ho Chi Minh City, Vietnam

Turongruang, D., Aquaculture and Aquatic Resources Management, School of Environment, Resources and Development, Asian Institute of Technology, PO Box 4, Klong Luang, Pathumthani 12120, Thailand

Varadi, L., Fish Culture Research Institute, PO Box 47, Szarvas, Hungary

Wahab, Md. A., Department of Fisheries Management, Bangladesh Agricultural University, Mymensingh, Bangladesh

Yesmin, K., CARE-Bangladesh, Dhaka 1209, Bangladesh

Foreword

It gives me great pleasure to have the opportunity to write an introduction to this text, partly as it makes available to a wider readership the very valuable products of the concepts and efforts of much respected friends and colleagues. More importantly perhaps, these contributions, taken together, represent a real advance in understanding of the nature and scope of aquaculture development in Asia, particularly as it affects poorer families and communities. Rather than the crude concepts of aquaculture and its trends and growth, and the broad supposition that Asia's tradition in aquaculture has assured the supply of aquatic products as an integral part of rural activity, we have here a practical and realistic description of what is actually happening, who does it, what development approaches are being engaged, what benefits are being derived, and who obtains the benefits.

As Manager of the DFID Aquaculture and Fish Genetics Research Programme, which has underwritten some of the research described in this text, I am particularly pleased to be able to see the outputs of our programme, and our many colleagues' related work, reach a wider public. As a professional group, we share a view that well directed and clearly targeted development in aquaculture and associated managed fisheries, based on a clear understanding of their natural resource base, and the livelihoods of the people concerned, will deliver real, very positive and potentially large scale benefits, now and for succeeding generations. With better understanding of the sector, and the directions in which it can be developed, producers, development agents, policy specialists and sectoral managers will be able to contribute to this very significant opportunity. Research and development in these themes continues, and I hope we will be able to explain and report further, in the not too distant future, some of the emerging outcomes of this present work. In doing so, I would like to extend a very open invitation to readers of this text and those who might be stimulated by its insights and explanations to join us in this very fascinating, challenging and rewarding endeavour.

James F. Muir
Institute of Aquaculture
University of Stirling, UK
Manager, DFID Aquaculture and Fish Genetics Research Programme

This document includes outputs from a project funded by the UK Department for International Development (DFID) for the benefit of developing countries. The views expressed are not necessarily those of DFID.

Preface

The term 'rural aquaculture' derives from the traditional dichotomy of development between rural/agricultural and urban/industrial areas. In line with the emergence of the term 'rural development' to address widespread poverty and inequity in rural areas, implicit in the term are aquaculture systems that meet the needs of, and fit in with the resources available to, small-scale farming households.

The chapters of the book are based on papers presented on rural aquaculture at the Fifth Asian Fisheries Forum, International Conference on Fisheries and Food Security Beyond the Year 2000, held in November 1998 in Chiang Mai, Thailand. Half of the papers were presented in the Special Session on Rural Aquaculture organised by the Asian Institute of Technology with financial support provided by the Department for International Development (DFID) of the UK. The remainder were solicited by the editors for their relevance from authors of papers from other sessions of the conference. The papers have been reviewed and updated for inclusion in this book. The Special Session concluded with an open discussion on rural aquaculture.

Rural aquaculture is complex and diverse as reflected by the coverage of the chapters, which are arranged under six main topics. The first two chapters in 'environmental context' argue that, for small-scale farmers, the key issue for livelihood is the availability of fish, regardless of its origin, and that aquaculture should be promoted to provide fish for small-scale farming households only after careful assessment of the existing wild fishery. The third chapter provides a framework for assessing the development of aquaculture in irrigation systems. Chapters 4 to 7 cover the role of 'integration with agriculture' in lowland rice-growing areas, both culture of fish in paddies and ponds, and pond culture as a component of watershed management in upland areas. The next four chapters, 8 to 11, deal with the role of 'specialised and intensive systems' in rural aquaculture, specifically a rice-prawn-fish system in Bangladesh, fertilisation of ponds with inorganic fertilisers in Thailand, and cage culture in Bangladesh and Indonesia. In the section on 'seed', Chapters 12 to 15 present experiences on the development of fish seed networks in the Lao PDR and a village-level seed production centre in India, outline fish seed quality in general, and discuss the role of 'improved' quality tilapia seed, respectively. Chapters 16 to 18 on 'social aspects' cover farmers' perceptions of the value of tilapia in Bangladesh, the nutritional role of small indigenous species in Bangladesh, and the role of aquaculture in improving the situation of women in Thailand. In the section on 'development models', Chapters 19 and 20 present findings towards developing extension systems for rural aquaculture in Southern Vietnam. Chapter 21 outlines experience in targeting communities with significant numbers of poor households rather than individual poor households in Bangladesh, while Chapter 22 describes experiments with a 'distance extension' approach in Thailand.

The final chapter, Issues in Rural Aquaculture, is a synthesis by the editors of topical issues and the degree to which the papers have contributed towards their resolution. Included in Chapter 23 is also a reference list of other key papers to facilitate access by readers to the widely scattered literature in rural aquaculture.

Aquaculture has contributed in the past, and continues to contribute today, towards the alleviation of poverty in rural areas where it is traditional practice. It also has significant unfulfilled potential there and elsewhere in areas where aquaculture is a relatively new farming practice. Whilst the papers presented here reveal the complexity and constraints facing rural aquaculture, they also indicate positive attributes that make it an attractive entry point for research and development programmes to improve the livelihoods of the rural poor.

Professor Peter Edwards, Dr David C. Little and Dr Harvey Demaine
Editors

Chapter 1

The Ricefield Catch and Rural Food Security

R. Gregory and H. Guttman
*Aquaculture and Aquatic Resources Management, School of Environment,
Resources and Development, Asian Institute of Technology, PO Box 4, Klong
Luang, Pathumthani 12120, Thailand*

Abstract

Longitudinal studies in South-eastern Cambodia of three different areas with
differing water regimes indicated that the average amount of fish and other
aquatic animals caught in rice fields and surrounding areas was over 380kg per
household year^{-1}. The amount varied significantly ($P < 0.001$) between areas
with better (mean 604kg) and poorer (mean 158kg) water resources. Despite
this trend, the annual consumption of 37kg per person and the amount of fish
purchased annually did not vary significantly between the different areas. The
amount of fish sold, however, increased almost tenfold between water-poor and
water-rich areas, 34kg and 356kg, respectively. This suggests that the desirable
level of *per caput* fish consumption in these areas is around 37kg and most
households will attain this one way or another.

The value of the rice field catch was on average over 80% of that of the rice
harvest, though this varied significantly between areas with differing water
resources. It can be argued that a one third increase in rice production would be
detrimental if it reduced local fish catches by 40% or more.

Aquaculture is often promoted to fill the gap between declining fish
production from natural sources and the increasing demand from the growing
populations. We argue that this is a gross over-simplification, as sustainable
management of the ricefield fishery and associated resources needs to be high
on the national agenda for many countries such as Cambodia. This important
resource should feature in any agricultural and infrastructure development
initiatives aimed at addressing rural food security and helping the rural poor. To
ignore it is to risk achieving food security on paper alone.

Introduction

Much of lowland Southeast Asian agriculture is dominated by the cultivation of rice, providing the bulk of carbohydrates and protein in the diet. For many rural populations, fish and other aquatic animals caught in rice fields and adjacent water bodies represent an important, and often the most important, source of animal protein. So dominant are the two food sources that lowland areas of Southeast Asia are often denoted as being 'rice-fish cultures'.

With rapidly growing populations, rural food security remains a high priority for many governments. However, with limited scope for new rice areas to be brought under cultivation, the opportunity for increased production lies in the intensification of rice cultivation. Herein lies a hidden danger. Through the introduction of more intensive agricultural practices and associated increases in pesticide use, it is possible that gain in increased rice production may be offset by the resultant losses in aquatic animal production, resulting in a net nutritional loss for the community involved.

In lowland Southeast Asia, rice cultivation, starting in Cambodia some 2,000 years ago (Helmers, 1997), has shaped many of the societies in the region. Today, a national policy of food security and food self sufficiency is often equated with 'rice security' and rice production. What has perhaps not been clearly identified is the associated aquatic produce in areas of wet rice cultivation in lowland Southeast Asia, where on a rural community level 'fish security' is second only to 'rice security'. Asking a farming household in lowland floodplains if they do any fishing is almost like asking if they grow any crops, and asking if they eat any fish is almost like asking if they eat.

Fish is, by far, the most important source of animal protein for the lowland rural households in the lower Mekong River basin. The term fish is used here to denote aquatic animals including snails, frogs, shrimp, shells, crabs, etc. that comprise 10-20% of the catch of rural households (Gregory et al., 1996). While the importance of animal proteins in the diet is controversial, certain amino acids (e.g. lysine) contained in fish are very important under certain circumstances. The diet in Southeast Asia primarily consists of rice, fish and leafy vegetables and lacks the cereals and pulses with more complete essential amino acid profiles which, for example, are found in South Asia.

Official estimates often tend to grossly underestimate the actual amounts of fish consumed. Consumption studies suggest that rural households consume about 20-30kg $caput^{-1}$ $year^{-1}$ in Northeast Thailand and 20-50kg $caput^{-1}$ $year^{-1}$ in lowland Cambodia (Table 1.1), more than double that given in official statistics.

Most of the fish consumed by rural households in the region is not purchased but caught (Prapertchop, 1989; Gregory et al., 1996; Edwards et al., 1997). In this article, we identify the importance of ricefield fisheries as an important source of fish for household consumption in contrast to the observations made by Lightfoot et al. (1993).

Table 1.1. *Per caput* fish consumption in Northeast Thailand and lowland Cambodia. Amounts are quoted as weight of aquatic animal products as consumed.

Area	Fish consumption (kg $caput^{-1}$ $year^{-1}$)	Reference
Northeast Thailand	32.7	Prapertchop (1989)
	20.0	AIT (1992)
	22.6	Setboonsarng (1993)
	31.6	Choowaew *et al.* (1994)
	19.6	Saengrut (1998)
Cambodia	13.6-19.0	Touch (1993)
	42.0	Gregory *et al.* (1996)
	40.0	CIAP, unpublished data
	38.0	APHEDA (1997)
	57.0	Ahmed *et al.* (1998)

A Seasonal Fishery

Seasonal flooded rice fields and swamps can be very productive fisheries; yields are believed to very variable but with estimates in the range of 25-303kg ha^{-1} (Table 1.2).

Table 1.2. Estimates of aquatic animal production in lowland Southeast Asia.

Area	Production (kg ha^{-1} $year^{-1}$)	Reference
Malaysia	88-175	Ali (1990)
Cambodia	51	Gregory and Guttman (1999)
Northeast Thailand	56-298	Leelapatra and Sollows (1990)
	214-303	Middendorp (1992)
	25-125	Fujisaka and Vejpas (1990)

The seasonal cycle of inundation is utilised by aquatic animals in a highly opportunistic way. During the rainy season, perennial water bodies become linked to newly inundated fields and swamps. Fish and other aquatic animals migrate from perennial water bodies with monsoon rains to newly inundated areas, which are rich in food, where they also breed. A certain proportion of animals falls prey to fishing traps or natural predators during this migration (Fig. 1.1).

Once in the vast areas of shallow inundated areas, the organisms exploit the nutrients there and an 'explosive' production is the result as stored nutrients combined with high photosynthetic production rates are converted into animal biomass. The fishing pressure is rather low during this period as the area in which the animals thrive is very large.

As the rains finish and the inundated areas begin to dry up, the fish and other organisms begin to move towards perennial water bodies. During this time a very productive fishery using traps and trap ponds removes the bulk of the fish population, allowing only a fraction to return to the perennial water bodies.

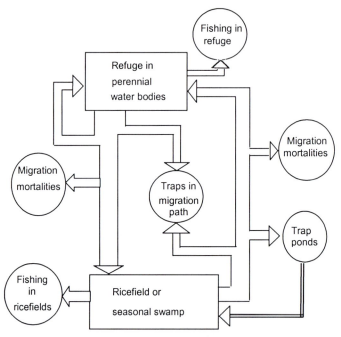

Fig. 1.1. Diagrammatic representation of seasonal migrations by aquatic animals in floodplain systems (Guttman, 1998).

As the dry season progresses, the only areas where water remains are subject to high fishing pressure. Partial or total dewatering can result in most fish being removed from the water body. These dry season refuges provide shelter for the fish that will breed in the coming wet season.

Guttman (1998) provided a generalised description of these seasonal migrations. Furthermore, there is a need to define terminology relating to floodplain and ricefield fisheries (Box 1.1). The characteristics of the habitat favours re-selected species with short life span, rapid growth and early maturity. These populations rely on the ability to colonise new habitats and increase rapidly to exploit unpredictable environments. These populations can sustain high levels of exploitation but are susceptible to sudden collapse (Hoggart and Halls, 1997).

Box 1.1. Floodplain fisheries and ricefield fisheries.

Floodplain fisheries are often defined as fishing activities in a geographically defined area, a floodplain. The term ricefield fisheries means fishing undertaken by rural farming households in lowland areas, in and around rice fields. It is usually a part-time activity of families considering themselves to be rice farmers. The ricefield fishery is not restricted to fishing in ricefields, as households fish in swamps and rivers, particularly in the dry season. It is also characterised by who undertakes the activity, separating it from more commercial/large-scale fisheries and small-scale/artisanal (full-time) fisheries on the floodplains.

Food Security and Ricefield Catch

Heckman (1979) concluded that one important contributing factor to the apparent general well being, in terms of nutritional status, of rural households in Northeast Thailand was the amount of fish and other aquatic animals gathered in and around rice fields. This is also true for lowland Cambodia (Ovesen *et al.,* 1996; Ahmed, 1997). Surprisingly enough, his call for more research into this issue 20 years ago has not been heeded; and the extent and importance of the foods collected in and around rice fields is still poorly understood.

The main results in this chapter are drawn from two longitudinal studies undertaken in Svay Rieng province in South eastern Cambodia in 1995/96 and 1997/98. Earlier studies of the aquatic resources in the province indicated three zones differentiated by the extent of water resources and fish production, as represented by trap pond yields (Gregory and Bunra, 1995). Fig. 1.2 depicts the three study areas, where Zone 1 has the lowest and Zone 3 the highest production of fish.

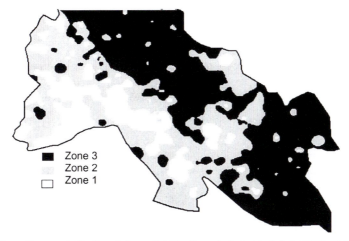

Zone 3
Zone 2
Zone 1

Fig. 1.2. Thematic map indicating fish availability through trap pond fish catch of Svay Rieng province, Cambodia (*n* = 1,267). Legend indicates the fish harvest m^{-2} at the end of the rainy season (authors' unpublished data). Zone 1 = < 0.15kg m$^{-2,}$ Zone 2 = 0.15-0.30kg m^{-2}, Zone 3 = > 0.30kg m^{-2}.

The study followed 66 households in the three zones over a period of 12 months with visits every 1-3 weeks to collect information on catch, fishing sites, fishing effort, consumption and purchase of fish and other aquatic animals. Although of importance to the nutritional status of the households, information on harvest of aquatic plants and other foraging activities, e.g. for birds and other wildlife, was not collected due to time constraints.

Harvesting the Ricefield Catch

The longitudinal studies showed that the amount of aquatic animals caught in rice fields and surrounding areas was over 380kg per household year^{-1}. The amount varied significantly ($P < 0.001$) between areas with better, Zone 3 (mean 604kg), and those with poorer, Zone 1 (mean 158kg), water resources (Fig. 1.3). Households in areas intermediate in terms of water resource availability (Zone 2) had an annual catch of 321kg.

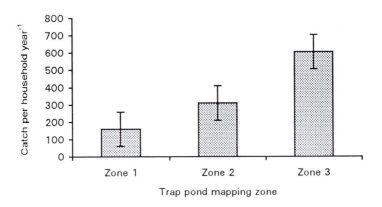

Fig. 1.3. Average household's annual catch of fish and other aquatic animals. Bars represent 95% confidence interval, n = 65.

The availability of fish and other animals varied depending on the surrounding water resources, but *per caput* annual consumption (mean 37.4kg) did not vary significantly between the different areas. What was lacking in the household catch was bought from the local market. The distribution of fish consumption was slightly skewed, but 55% of households had an average *per caput* consumption of between 20 and 50kg per person year^{-1} (Fig. 1.4).

The amount of fish bought did not vary significantly between the different areas, averaging 27kg per person year^{-1}, though there was a trend of households in areas with poorer water resources buying more fish. There was no significant difference between zones with respect to the source of bought fish where around 60-70% were bought from the market and the remainder from within the village. The amount of fish sold, however, increased almost tenfold between water-rich and water-poor areas, 34 compared to 356kg. This illustrates the complexity of the system where often the more valuable part of the catch is sold, providing cash, and if necessary cheaper fish is purchased for consumption.

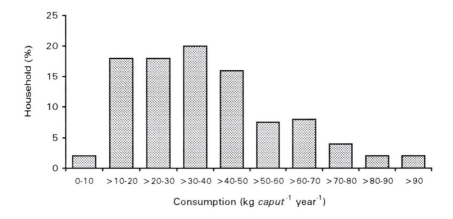

Fig. 1.4. *Per caput* consumption of fish and other aquatic animals, *n* = 65.

Fish and aquatic animals were the most important animal protein sources, representing almost 90% of the total animal protein intake (on a gross wet weight of meat basis). It should, however, be noted that this figure does not include consumption of livestock produced by the household and non-aquatic animals foraged (an omission in the questionnaire). Although likely to be an overestimate, it is consistent with findings of other studies (Ahmed *et al.,* 1998; Shams and Hong, 1998).

On average 97 days were spent fishing by the households. The time spent catching fish differed between the zones. Households in Zone 2 spent significantly (*P* < 0.01) more time fishing compared to Zone 1, at 116 and 64 days per household, respectively. Zone 3 households were intermediate at 104 days fishing (Fig. 1.5).

A gross estimate of catch per unit effort (CPUE) indicates that there was a significant difference between the zones (Fig. 1.6). For Zone 1 and 2 the CPUE was very similar, possibly because the high effort in Zone 2 reduced the CPUE. A simple interpretation of this is that households in this zone tried to increase the household catch through increased effort, while in Zone 1 they used their time doing other things. Zone 3 had over twice the CPUE of the other zones.

The value of the catch was on average equivalent to 85-125% of the value of the rice harvest, though this varied significantly between the different zones, being higher in areas with better water resources.

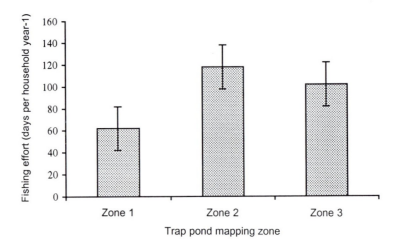

Fig. 1.5. Comparison of average household effort (days) spent on fishing. Bars represent 95% confidence interval, *n* = 65.

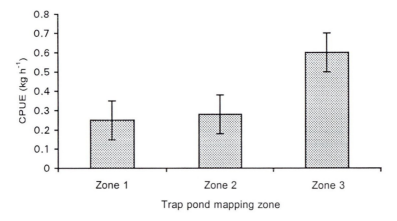

Fig. 1.6. Comparison of average catch per unit effort (CPUE) (kg h⁻¹). Bars represent 95% confidence interval, *n* = 65.

Anecdotal information and other studies indicate a great variation in fish availability between years, primarily depending on the amount and timing of the rains (AIT, 1998). Variation between years is significant but the degree in variation appears to be linked to the perennial water resources. Five households in the study area were followed with respect to fish catches, through a wet year and a drier year. The catch in the drier year, for two households only, was 55-69% of the catch in the wetter year. However, for the remaining three households it was between 79 and 112% of the catch in a wetter year. Although in the same zone, the former two households were in an area with lower household catch than the latter three households. This suggests that in years with poor rains the households in the areas with poorer water resources are more affected than areas with better resources.

Securing the Catch

Assessing food security issues in Cambodia, Tickner (1996) concluded that the nature of food insecurity is more an issue of access than availability. He also suggested that acute undernutrition is not widespread in Cambodia, but that seasonal food shortages and poorly balanced diets contribute to a high prevalence of malnutrition, especially among children. If fish is not readily available, many farming households will spend large amounts of their limited money on purchasing it. Studies in areas away from larger water bodies with limited local production of fish indicate that households spend 50% of their annual income on fish (Cambodia-IRRI-Australia Project, unpublished data). Overall, rural populations are estimated to spend about 20% of their annual income on purchasing animal protein (Tickner, 1996).

Ricefield fisheries clearly must be an important consideration in the national food security equation. However, current government policy in Cambodia is dominated by a need to be self sufficient in, or an exporter of, rice. Furthermore, current official statistics give the wrong signals to planners and aid agencies as they underestimate the importance of fish and aquatic animals; the Government of Cambodia still uses an estimate of 5-7kg $caput^{-1}$ for annual fish consumption (MAFF, 1996). The major efforts undertaken to improve food security are largely focused on increased production of rice. Though this is of great importance, there is a risk of 'killing the patient by applying the cure'.

As the scope for increased paddy rice output through expansion in land under cultivation is limited, any future increase is more likely to be achieved through intensification (Nesbitt, 1997). However, gains in production through intensification in agricultural practices may be offset by reduction in catches of stocks of aquatic animals. Application of more fertilisers may have a neutral, or even positive, impact on the production of aquatic animals but intensification is usually associated with increased use of pesticides and more controlled water management practices which frequently have a negative impact. Developing recommendations and application procedures without taking this into consideration risks nullifying gains in rice production by losses in aquatic animal production.

Another potential threat is infrastructural development as roads and other structures may block fish migration paths and re-direct water. While this sort of development is needed, particularly in a country like Cambodia, it is important that the seasonal flow of water and associated fish migrations are considered and, if necessary, negative impacts are mitigated, otherwise the cost to the surrounding villages may be greater than the benefits.

There is scope to address the issues. Management of key components of the fishery, such as maintaining migration paths, dry season refuge protection, restoration/improvement of key habitats, etc. are feasible. These sorts of activities have been initiated by some fishery and wetlands projects in the region, and will hopefully generate research data and other information to bring the issues up high on the government's agenda.

Is not the fishery overexploited already? Current consumption figures available for Cambodia suggest an annual *per caput* consumption of between 20 and 50kg (Table 1.1). This would imply that 8 million rural households consume between 160,000 and 400,000t of aquatic animals year^{-1}. At an annual household catch of between 100 and 300kg year^{-1}, the total catch by farming households would be over 200,000t. Two million ha of seasonally inundated areas, rice fields and seasonal wetlands, can, at a productivity of about 50-150kg ha^{-1} (Gregory *et al.*, 1996; Gregory and Guttman, 1999; P. Degan, personal communication) potentially support a fish production of 100,000 to 300,000t year^{-1}. In addition the commercial/regulated fishery adds the order of 105,000-155,000t, balancing the picture (van Zalinge *et al.*, 1996; Diep *et al.*, 1998). Although these figures are gross estimates, they are based on independent estimates, indicating that the fishery in Cambodia is indeed heavily exploited. As the fishery is also very productive, it is imperative that issues of how to manage it effectively and sustainably are addressed.

Promotion of small-scale aquaculture has an important role in helping households in areas with poorer water resources to achieve 'fish security' as it complements the catches of wild fish, and can provide additional cash income through fish sales. It is not, however, appropriate to justify aquaculture development as a replacement of a declining fishery as this is not evident. It complements a heavily exploited productive fishery. From the results in this study, it is likely that aquaculture promotion is most appropriate for areas in Zone 1, perhaps changing to some sort of enhanced fisheries in Zone 2 areas while conservation of critical spawning areas, dry season refuge management and protection of migration routes may be more appropriate for Zone 3 areas.

Sustainable management of the rice field fishery needs to be high on the national agenda for Cambodia, which is attempting to improve rice production and achieve national food security. Consideration of the ricefield aquatic production should feature in agricultural and infrastructure development initiatives aimed at improving the food security situation, especially for the rural poor. To ignore it is to risk achieving food security on paper alone.

Acknowledgements

The authors would like to acknowledge the staff of the Royal Cambodia Department of Fisheries and the Svay Rieng Provincial Department of Agriculture involved in collecting data for this study.

Furthermore, the authors would like to extend their gratitude to those persons who have persevered with the constant questioning and unconventional approaches, particularly Peter Degan of the Mekong River Commission and Harvey Demaine of AIT Aqua Outreach.

The AIT Aqua Outreach Programme is supported by the governments of Denmark and Sweden.

References

Ahmed, M. (1997) Fish for the poor under a rising global demand and changing fishery regime. *International Consultation on Policy Research in Fisheries in Developing Countries*. NAGA supplement, ICLARM Quarterly July-December 1997, pp.4-7.

Ahmed, M., Navy, H., Vuthy, L. and Tingco, M. (1998) *Socio-economic Assessment of Freshwater Capture Fisheries of Cambodia: Report on a Household Survey*. Mekong River Commission, Phnom Penh, Cambodia. 186 pp.

AIT (1992) Srisaket baseline study. *Working Paper 15, AIT Aquaculture Outreach*. Asian Institute of Technology, Bangkok. 30 pp.

AIT (1998) Assessment of fish catch by catching teams in Sisaket Province, Thailand (results from 1994-95 survey). *Working Paper No. T-9, AIT Aqua Outreach*. Prepared by H. Guttman, P. Anagporn, P. Mingkano and D. Turongruang. Asian Institute of Technology, Bangkok. 14 pp.

Ali, A.B. (1990) Rice/fish farming in Malaysia: a resource optimization. *Ambio* 19(8), 404-408.

APHEDA (1997) Baseline survey report (Angkor Chey, Bantemay Meas, Chhouk and Kompong Trach District). Report prepared by N.C. Paul, Domestic Fish Farming Program, Australian People For Health Education and Development Abroad (APHEDA)-Department of Agriculture, Forestry and Fishery, Kampot Province, Cambodia. 29 pp.

Choowaew, S., Chandrachai, W. and Petersen, R.C., Jr (1994) The socioeconomic conditions in the vicinity of Huai Nam Un wetland, Songkhram River, Lower Mekong basin, Thailand. *Mitteilungen Internationale Vereinigung für Theoretische und Angewandte Limnologie* 24, 41-46.

Diep, L., Ly, S. and van Zalinge, N.P. (eds) (1998) *Catch Statistics of the Cambodian Freshwater Fisheries*. Mekong River Commission, Phnom Penh. 146 pp.

Edwards, P., Little, D.C. and Yakupitiyage, A. (1997) A comparison of traditional and modified inland artisanal aquaculture systems. *Aquaculture Research* 28, 777-788.

Fujisaka, S. and Vejpas, C. (1990) Capture and cultured paddy fisheries in Khu Khat, Northeast Thailand. *Thai Journal of Agricultural Science* 23, 167-176.

Gregory, R. and Bunra, S. (1995) Trap pond mapping as a survey tool for aquaculture development planning. *Working Paper No. C-4, AIT Aqua Outreach (Cambodia)*. Asian Institute of Technology, Bangkok. 6 pp.

Gregory, R. and Guttman, H. (1999) A diverse monoculture. *Mekong Fisheries Network Newsletter* 5(1), 3-5.

Gregory, R., Guttman, H. and Kekputhearith, T. (1996) Poor in all but fish: a study of the collection of ricefield foods from three villages in Svay Theap District, Svay Rieng. *Working Paper C-5, AIT Aqua Outreach (Cambodia).* Asian Institute of Technology, Bangkok. 29 pp.

Guttman, H. (1998) Rice and fish. *AARM Newsletter* 3(3), 6-7.

Heckman, C.W. (1979) *Ricefield ecology in Northeast Thailand.* Dr Junk Publishing, The Hague. 228 pp.

Helmers, K. (1997) Rice in the Cambodian economy: past and present. In: Nesbitt, H.J. (ed.) *Rice Production in Cambodia.* International Rice Research Institute, Los Baños, pp. 1-14.

Hoggart, D.D. and Halls, A.S. (1997) Fisheries dynamics of modified floodplains in Southern Asia. *Final Technical Report (DRAFT).* ODA Fisheries Management Science Programme, Project R5953. MRAG, University College, London. 275 pp.

Leelapatra, W. and Sollows, J. (1990) Rice fish culture as practised by Northeast Thai farmers: results of the Northeast Fishery Project fish in rice fields survey. *NFP/Technical Report No. 33.* Canadian International Development Agency and Department of Fishery, Ministry of Agriculture and Cooperatives, Bangkok. 40 pp.

Lightfoot, C., Roger, P.A., Cagauan, A.G. and dela Cruz, C.R. (1993) Preliminary steady-state nitrogen models of a wetland ricefield ecosystem with and without fish. In: *Trophic Models of Aquatic Ecosystems.* ICLARM Conference Proceedings 26, 56-64.

MAFF (1996) Food security in Cambodia. A country position paper prepared for World Food Summit, Rome, November 1996. Ministry of Agriculture, Forestry and Fisheries, Phnom Penh.

Middendorp, H.A.J. (1992) Contribution of stocked and wild fish in ricefields to fish production and farmer nutrition in Northeast Thailand. *Asian Fisheries Science* 5, 145-161.

Nesbitt, H.J. (1997) Constraints to rice production and strategies for improvement. In: Nesbitt, H.J. (ed.) *Rice Production in Cambodia.* International Rice Research Institute, Los Baños, pp. 107-112.

Ovesen, J., Trankell, I. and Öjendal, J. (1996) When every household is an island: social organisation and power structures in rural Cambodia. *Uppsala Research Reports in Cultural Anthropology, No. 15.* Uppsala University and Sida, Stockholm. 93 pp.

Prapertchop, P. (1989) Analysis of freshwater fish consumption and marine product marketing in NE Thailand. *Report No. 9, Part II/III.* Khon Kaen University. 35 pp.

Saengrut, T. (1998) Role of wild fish in aquatic resources development in the lower Chi valley of Thailand. MSc Thesis AE-98-1, Asian Institute of Technology, Bangkok. 100 pp.

Setboonsarng, S. (1993) Farmer's perception toward wild fish: 'product not predator' an experience in rice-fish development in Northeast Thailand. Paper presented at the Regional Workshop on Integrated Rice-Fish research and Development, Sukarmandi Institute for Food Crops, 6-11 June.

Shams, N. and Hong, T. (1998) Cambodia's rice field ecosystem biodiversity - resources and benefits. Technical report submitted to Deutsche Gesellshaft fur Technische Zusammenarbeit (GTZ), Kompong Thom Provincial Development Program, Phnom Penh. 60 pp.

Touch, S.T. (1993) Fish supply and demand in rural Svay Rieng, Cambodia. MSc Thesis AE-93-34. Asian Institute of Technology, Bangkok. 114 pp.

Tickner, V. (1996) Food security in Cambodia - a preliminary assessment. *UNRISD Discussion Paper NRISD/DP/80/96/9, ISSN 1012-6511.* United Nations Research Institute for Social Development, Geneva. 88 pp.

van Zalinge, N.P., Diep, L. and Ly, S. (1996) Catch assessment and fisheries management in the Tonle Sap Great Lake and River. Contribution to the Workshop on Fishery Statistic, 18-19 September 1996. Department of Fisheries, Phnom Penh. 24 pp.

Chapter 2

Developing Appropriate Interventions for Rice-Fish Cultures

R. Gregory and H. Guttman

Aquaculture and Aquatic Resources Management, School of Environment, Resources and Development, Asian Institute of Technology, PO Box 4, Klong Luang, Pathumthani 12120, Thailand

Abstract

The AIT Aqua Outreach Programme and the Department of Fisheries in Cambodia are collaborating on fisheries and aquaculture development for small-scale rice farmers in the southern provinces of Cambodia. Examples are provided on how situation appraisal is being carried out; how promising interventions are field-tested and evaluated; how proven recommendations are disseminated; and how the impact of these interventions is monitored. Through this approach, small-scale aquaculture has become accepted by more than 1,300 households in areas where wild fish supplies are normally inadequate to meet total household requirements. In contrast, in areas with more productive ricefield fisheries, interventions are being tested to improve the communal management of wetland areas considered crucial as dry-season refuges for broodstock and fry from which the reseeding of the local wet-season ricefield fishery takes place each year. The appropriateness of using aquaculture as an entry point to working with rice farming communities is being questioned. A broader fisheries development approach involving whole communities, which includes aquaculture as one component, would seem to be more appropriate for rural Cambodia.

One constraint to this approach is that extension staff generally lack a broad enough background to work in this way. The compartmentalisation of fisheries and aquaculture by educational and development institutions has tended to produce specialists who are technically competent in aquaculture or fisheries but who are uncomfortable when faced with issues relating to the other subject matter area. This paper argues that to work effectively on improving fish security in rural Cambodia, the ideal field worker must be equally at home giving advice on small-scale fish culture as on small-scale fisheries; and as comfortable working with individual farm households as with whole communities.

Whether fish originate from culture or capture systems is largely irrelevant for most lowland rice farmers. What is important is that enough fish can be harvested for household needs with perhaps a small surplus, thus preserving the integrity of the rice and fish culture in which these farmers live. Exactly how options are considered and interventions developed to assist rice farmers in meeting this requirement need to be carefully reassessed.

Introduction

In many Southeast Asian countries, the harvest of fish ('black fish', including *Channa striata*, *Clarias* spp. and *Anabas testudineus*, as well as small 'white fish' including *Rasbora* spp., *Puntius* spp. and *Trichogaster* spp., together with other aquatic animals such as crabs, frogs, prawns and a range of edible insects) from rice fields is as much a part of the seasonal calendar as transplanting and harvesting rice. In fact, for many rural communities the harvest of the wet season rice crop harvest is preceded by the peak ricefield fish catch and followed by the harvesting of fish caught in trap ponds in and around rice fields. Many rural households in lowland rice-growing areas remain dependent on these foraged foods as the major source of animal protein in their diet (Ovensen *et al.*, 1996). Fish consumption *per caput* from lowland rice-growing areas has been recorded to be as high as 40kg (Gregory *et al.*, 1996; APHEDA, 1997; Gregory and Guttman, this volume), virtually all originating from rice fields, adjacent canals, ponds and swamps. So dominant is the nutritional staple of rice and fish that lowland communities in much of Southeast Asia are often denoted as 'rice-fish cultures'.

In areas where the aquatic environment has been degraded through swamp reclamation, infrastructure development or regular pesticide application, or in areas where ricefield fish may never have been abundant, small-scale aquaculture has been adopted by many lowland rice farmers. Even in such areas, however, farmers tend to use aquaculture to supplement the natural fish catch or as an off-season source of fish for consumption or sale (Gregory *et al.*, 1996). Most rice farmers seem quite willing to switch between capture and culture practices as it best fulfils the household's nutritional requirement (Gregory and Guttman, 1996, 1997; Gregory, 1997; Guttman, 1999). The target group for small-scale aquaculture and ricefield fisheries lies in the centre of the broad spectrum of the aquaculture and fisheries continuum (Fig. 2.1) which perhaps explains why small-scale farmers are able to switch readily between culture and capture based systems.

The AIT Aquaculture Outreach project in Cambodia began work in the southern districts of Svay Rieng province in 1994 (Fig. 2.2) with aquaculture considered to be the technology to improve the fish supply for the rural poor.

Initially it was planned to carry out on-farm research to identify and then remove constraints to the development of aquaculture and thereby encourage its establishment as a viable activity throughout the province. At that time, apart from a few farmers experimenting with growing *Pangasius* spp. and *Clarias* spp., no farmers cultured carp or tilapia in the areas where the Project began to work. In fact, farmers were very suspicious of this type of aquaculture because of unsuccessful past attempts by a number of non-governmental organisations to promote fish culture. Nor was fish seed available in the villages for those farmers who wanted to try to farm fish.

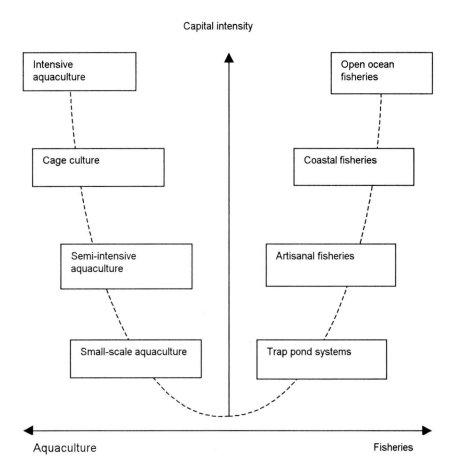

Fig. 2.1. Aquaculture-fisheries continuum related to capital intensity of the activity (Guttman, 1996).

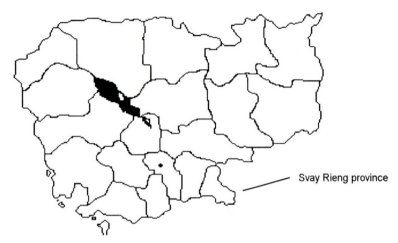

Svay Rieng province

Fig. 2.2. Map of the study area, Svay Rieng province.

Svay Rieng has a population of close to 500,000 and a density of 161 persons km^{-2} (UNPF, 1998). It is varied in terms of water and soil resources. Some parts of the province have few perennial water bodies where farmers struggle to collect enough fish for their household needs (Touch, 1993), but in other areas (sometimes only a few kilometres away), farmers are able to harvest annually up to a tonne of fish from rice fields, canals and trap ponds, making sales of fish the most important source of on-farm income (Gregory *et al.*, 1996). This latter study suggested to the authors that some of these communities comprised not so much rice farmers who caught a few fish but rather fishing communities that planted some rice.

The assumption that Svay Rieng was, overall, a fish deficit area was later challenged (Guttman and Khunty, 1997). This study suggested that the province may be a net exporter of fish, tending to export high value fish (mainly 'black fish') and import lower grade fish (small cyprinids). In such circumstances it was deduced that a blanket approach to aquaculture development was unlikely to be consistently successful as farmer interest in aquaculture would depend on the relative contribution that cultured fish could make to the overall fish supply. What was required were tools to determine which interventions would stand the highest chance of success in any given area, participatory approaches to testing these interventions and finally, ways of widely disseminating knowledge to a potentially large number of farmers in similar situations.

Resource Assessment

Areas with different aquatic resources were initially identified through mapping trap pond productivity (Gregory and Bunra, 1995). Longitudinal studies in which households were monitored throughout a year for catch and use of aquatic products were also carried out (Gregory *et al.*, 1996; Gregory and Guttman, this volume). Thus, the Project has been able to approximately zone areas according to ricefield fish supply, thereby better identifying fish deficit areas, fish surplus areas and intermediate areas. The zoning of areas was useful in selecting fish deficit areas which might have greater potential for small-scale aquaculture development where extension staff could focus on promoting fish culture. Fish deficit areas usually have: few natural water resources; few ponds; small ponds, typically less than $100m^2$; silty soils with poor water retention properties with a tendency to cause acute pond turbidity; and few nutrient-rich, on-farm by-products which could be used as significant pond inputs. Although such areas did not appear, at first sight, to have much potential for aquaculture, these were the areas where farmers had the greatest need for fish. Therefore the technical staff had to devise techniques to overcome these constraints, and which the farmers could be expected to readily adopt.

Areas where aquaculture could be expected to be a better option than the natural fishery were few and patchy in distribution. This posed the question as to what the Project should do in other areas where the fisheries were more productive? It was not considered appropriate from a poverty perspective to just wait until the ricefield fisheries had declined further to the point at which it would be feasible to promote aquaculture. Experience has shown in many countries in Asia that it is the poorer households that are slow to adopt aquaculture due to uncertainty about the new technology and the tendency to avoid risk (Lewis *et al.*, 1996). On the other hand, it is usually the poorer households that are most active in ricefield fisheries due to both their more open access and reliable returns to labour. To have ignored the natural fisheries would have meant ignoring the problem of declining fish catch facing the poor in favour of a fish production technology better suited to the better-off villagers. The Project decided to search for an alternative approach to work with communities on preventing the degradation of the more productive fishing areas.

Aquaculture Development Process

The process of successfully promoting aquaculture in areas identified as fish deficit areas proved to be fairly straightforward. Models developed in Bangladesh (Lewis *et al.*, 1996) and Thailand (AIT, 1997) were copied (Fig. 2.3). Development of appropriate techniques to extend to farmers were developed in a participatory way involving farmers who had been successful in fish culture (Gregory and Turongruang, 1996).

There has been a dramatic (30-fold) increase in the number of households culturing carp (*Cyprinus carpio* and *Barbodes gonionotus*) and Nile tilapia

(*Oreochromis niloticus*) in the province (Fig. 2.4). The areas where farmers are adopting fish culture in numbers correspond well to the fish deficit areas identified on the trap pond map (Fig. 2.5).

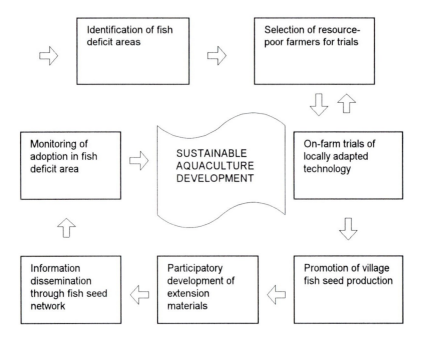

Fig. 2.3. The process of sustainable aquaculture development.

Fish seed production in the province has also increased dramatically (Fig. 2.6) although this remains a major limiting factor to new entrant farmers attempting fish culture. Fish seed produced by villagers makes the largest contribution to the overall fish seed supply situation in the area (Fig. 2.7).

Ricefield Fisheries Development Process

Guttman (1998) drew up a conceptual framework for the fishery dynamics which facilitates identification of possible management options. This led to the hypothesis that a potentially manageable component of the fishery is the dry-season refuge and the out-migration of fry and broodstock from it onto the floodplain with the first good rains. This is in effect a 'bottleneck' in the ricefield fishery system. The logic of this approach is that improved management of fish refuges in the dry season should allow for faster recruitment onto the floodplain each year, leading to increased fish production. Inadequate or absent management of local refuges with a consequent delay in the links to the surrounding fishery mean that the ricefield fish must come from more distant refuges with a consequent delay in the colonisation of the ricefields and subsequent reduced production.

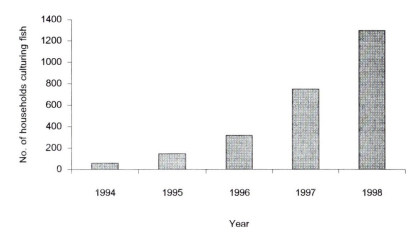

Fig. 2.4. Total number of farming households culturing fish in southern Svay Rieng province from 1994 to 1998.

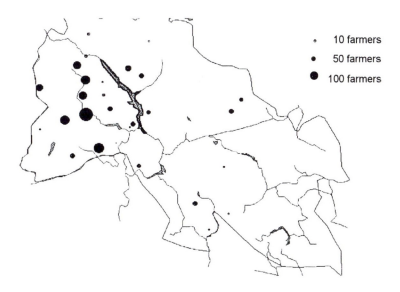

Fig. 2.5. Number of fish farming households by commune in 1997.

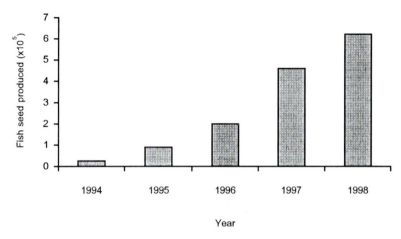

Fig. 2.6. Total number of fish seed produced in Svay Rieng from 1994 to 1998.

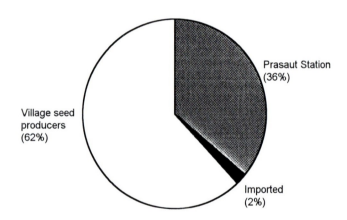

Fig. 2.7. Origin of fish seed produced in Svay Rieng province in 1998.

To test this hypothesis the Project has begun to work with communities to improve the management of dry-season refuges and attempt to monitor impact. Selection of refuges is important, as a viable refuge should be large enough in size to have an impact on the surrounding fishery, should have water in even the driest year, must be community owned or controlled and should be distant from other perennial water bodies to facilitate impact monitoring. The Project short-listed three communities for the initial trials during 1998, but disagreement in one village over management options and the complete drying out of the refuge in a second village, left only one village, Phum Kandal, in which to test this idea.

The dry-season refuge at Phum Kandal is a 1ha water body about 1m in depth at the end of the dry season, increasing to a depth of around 3m in the wet season. It is surrounded by tall screw pine (*Pandanus* sp.) and has an abundance of aquatic macrophytes such as lotus and reeds. It is actually an old pond with embankments that are 20-40m wide and it is linked to the surrounding ricefields by a number of canals through the pond embankment. There are a number of small ponds within the pond embankment which are periodically pumped dry to catch fish during the dry season. Moreover the canals linking the pond to the ricefields are used to trap fish on both the out-migration and the back-migration. It remained the only source of surface water in the village or within a radius of several kilometres during the dry season of 1998.

Initial discussions with the village headman and a number of elders suggested that the community might be interested in working with the Project to attempt to improve fish supplies in the village through improved management of the refuge pond. An initial community meeting was held and the benefits of conserving the resource at certain times of year were discussed. The villagers seemed to follow the logic of broodstock management easily enough but the classic commons problem was raised whereby it would not pay any one individual to act more responsibly in a more forward looking manner, if others in the community would not act accordingly. Whilst this is a common problem globally, it is particularly acute in Cambodia where very little trust between rural people has survived the recent traumatic past (Meas, 1995).

A simple mapping exercise was carried out with the community in which the key fishing areas and other important landmarks were identified. This helped Project staff as outsiders to understand better the current management strategies used by the village. The villagers also explained what they understood of fish migrations to and from their fishery and identified a nearby seasonal swamp as an important cool-season fishery, and that it was often dewatered during the dry season.

At a second meeting the Project staff were able to facilitate discussion in the community enabling it to agree on two major issues: firstly that all fishing in the refuge would stop from April 1998 onwards; and secondly that the canals linking the refuge to the fields would be left open to allow free movement of the fish during the out-migration. As the stock had been depleted through overfishing, the Project agreed to stock 200kg of snakehead (*Channa striata*) and walking catfish (*Clarias macrocephalus*) broodstock into the refuge to mimic the effects of leaving 200kg of such fish in the refuge through conservation. The decision to stock fish might be frowned upon by some but in retrospect it proved important in additional ways apart from providing a reasonable number of broodfish. Stocking the refuge clearly showed the villagers that the Project was serious about working with them and made conserving the resource worthwhile. It also raised the profile of the project as monks and other local dignitaries attended the fish release ceremony.

Under no pressure from the Project, the most active villagers formed a committee with responsibility for organising meetings and ensuring that agreements were respected. The villagers also requested the Project to install a pipe through the main road in the area to improve drainage of their rice-growing areas and improve access for refuge fish to the swamp areas. The Project agreed to this on condition that no one would set permanent nets across this important migration route.

Within a few weeks of stocking the refuge, snakehead fry began to appear in large numbers. Over the next 2 months, the villagers counted 57 'nests' of fry which could result in excess of 100,000 fingerlings (assuming a fecundity of 2,500 eggs per female and high survival to fry of 75% due to parental care exhibited by the males), although it was obvious that cannibalism of the fry by other snakehead was occurring. The refuge canals were opened with the first good rains, first with bamboo screens to allow only the passage of fry/fingerlings but later these screens were removed and the broodstock were also allowed free access to the floodplain. Following this, 'nests' of snakehead began to appear in ricefields and other ponds in the village, which by now had filled with water. Monthly meetings with the community allowed a consensus on fishing practices in and around the refuge. Areas that could be fished and gears that could be used were discussed and common policy for the following month was agreed upon. The Project staff at this stage acted mainly as facilitators, offering the occasional suggestion but allowing the community to freely discuss options open to them.

Four monitoring systems have been put in place by the Project to try and pick up the impact of improved management of the refuge pond. These involve regular sampling of fields at distances from the refuge to assess ricefield aquatic biodiversity and biomass; regular visits to 40 households in eight villages to collect data on fish catches throughout the season; a survey of fishing in the seasonal swamp from where is assumed that many of the recruits from the refuge have migrated; and a study of trap pond catches in the area to determine whether more fish attempt to return to the refuge area than in other areas with no management controls.

Although the refuge pond may have improved the local ricefield fishery through being a valuable source of recruits, it may prove impossible to measure this impact through the monitoring systems in place. At present the villagers are convinced that they are catching more fish than they did during the same period in the previous year, but unless we can clearly show that improved refuge management benefits the villagers of Phum Kandal more than others in the area or more than in years with no management, then the sustainability of the approach will remain uncertain. However, even if as a result of this first intervention we cannot show benefit, there have been many indications that the Project is on the right lines through this type of approach. Enough to certainly warrant involving many more villages in similar refuge management projects in the years to come.

Discussion

Involvement in both aquaculture and wild fisheries in Svay Rieng province has been stimulating for the field staff, requiring them to work with both typical aquaculture issues such as small-scale hatcheries, but also more fisheries oriented issues such as dry-season refuge management as well. This led to posing the question, 'are these approaches really so different?'. Both aim to improve the management of aquatic resources to increase fish seed supplies from which increased fish for food are derived. Although one uses hormones and needles with private farmers on the one hand, and a fish conservation approach through community development on the other, both have the same long-term goal. This then prompted the question and the main stimulus for this chapter 'are we right to separate these two subject matter areas when working with lowland rice farming communities?'. In answer to this question, we have recently seen a number of farmers in Phum Kandal village carrying out more conventional aquaculture in their small ponds in addition to being involved in the dry-season refuge management exercise.

Whilst these development trends are encouraging and suggest the Project has been successful in its original objective, one should not lose sight of the fact that the 50t of fish produced by small-scale aquaculture farmers in Svay Rieng during 1998 amounts to less than 1% of the ricefield fisheries catch in 1997, estimated at between 5,000 and 10,000t (Gregory and Guttman, this volume).

Models to promote aquaculture are well tested and documented. However, the same cannot be said about ricefield fisheries. Most authors either choose to ignore it, dismiss it as unimportant (Lightfoot *et al.*, 1993) or assume that nothing can be done to exert influence over it. The latter attitude is understandable given the open nature of the floodplain and the logic that the greatest factor affecting fish production is always likely to be the duration and amount of rainfall in any year. Heckman (1979) in his excellent study of ricefield ecology provides a rich source of qualitative information but does not begin to address the issue of improved ricefield fisheries management.

Although it has proved to be fairly straightforward to stimulate aquaculture development in many areas of Asia through a fairly conventional approach, this study has shown through innovative village-based trials that there is a need to better manage small-scale fisheries to provide fish for rural food security. Furthermore, this could greatly reduce the need for small-scale aquaculture in the lowland rice-growing areas of Cambodia, if it can be shown that management of ricefield fisheries works. Involvement in this innovative project has prompted those associated with it to assess how we, as fisheries scientists, approach the issue of fish security in these rice farming systems. We now believe that the distinction between small-scale aquaculture and floodplain and ricefield fisheries is meaningless when considering communities such as these in Cambodia.

It is clear that the rice farmers in this area are quite comfortable moving freely between fisheries and aquaculture as they work to harvest enough fish to meet their household needs. This chapter argues that as fisheries scientists we should try and be as equally flexible, exploring broader integrated options than those which our specialism dictates.

Acknowledgements

The authors would like to thank all the staff of the Royal Cambodian Department of Fisheries for their hard work and dedication. Further thanks are extended to the staff of the provincial staff of the Department of Agriculture in Svay Rieng and Takeo provinces.

Finally, the authors are grateful for the inputs provided by Dr Harvey Demaine for comments and stimulating ideas.

The AIT Aqua Outreach Programme is supported by the governments of Sweden and Denmark.

References

AIT (1997) *Policy Paper No. 1: Issues in Developing Fish Seed Supply*. AIT Aqua Outreach, Asian Institute of Technology, Bangkok. 4 pp.

APHEDA (1997) Baseline survey report (Angkor Chey, Bantemay Meas, Chhouk and Kompong Trach District). Report prepared by N.C. Paul, Domestic Fish Farming Program, Australian People For Health Education and Development Abroad (APHEDA)-Department of Agriculture, Forestry and Fishery, Kampot Province. 29 pp.

Gregory, R. (1997) *Ricefield Fisheries Handbook*. Cambodia-IRRI-Australia Project, Phnom Penh. 38 pp.

Gregory, R. and Bunra, S. (1995) Trap pond mapping as a survey tool for aquaculture development planning. *Working Paper No. C-4, AIT Aqua Outreach (Cambodia)*. Asian Institute of Technology, Bangkok. 6 pp.

Gregory, R. and Guttman, H. (1996) Capture or culture? Management of ricefield fisheries in Southeast Asia. *ILEIA Newsletter* 12(2), 20-21.

Gregory, R. and Guttman, H. (1997) Capture and culture ricefield fisheries in Cambodia. In: Nesbitt, H.J. (ed.) *Rice Production in Cambodia*. International Rice Research Institute, Los Baños, pp. 99-106.

Gregory, R., Guttman, H. and Kekputhearith, T. (1996) Poor in all but fish: a study of the collection of ricefield foods from three villages in Svay Theap

District, Svay Rieng. *Working Paper C-5, AIT Aqua Outreach (Cambodia)*. Asian Institute of Technology, Bangkok. 29 pp.

Gregory, R. and Turongruang, D. (1996) Participatory development of an information leaflet for fish farmers in Cambodia. *Working Paper C-3, AIT Aqua Outreach (Cambodia)*. Asian Institute of Technology, Bangkok. 10 pp.

Guttman, H. (1996) Wild fisheries resources. *AASP Newsletter* 1(3), 3.

Guttman, H. (1998) Rice and fish. *AARM Newsletter* 3(3), 6-7.

Guttman, H. (1999) Rice field fisheries - a resource for Cambodia. *NAGA* 22 (2), 11-15.

Guttman, H. and Khunty, R. (1997) Survey of fish marketing in Svay Rieng Province, Cambodia. *Working Paper C-6, AIT Aqua Outreach (Cambodia)*. Asian Institute of Technology, Bangkok. 19 pp.

Heckman, C.W. (1979) *Ricefield Ecology in Northeastern Thailand*. Dr Junk Publishers, The Hague. 228 pp.

Lewis, D.J., Wood, D.G. and Gregory, R. (1996) *Trading the Silver Seed; Local Knowledge and Market Moralities in Aquacultural Development*. Intermediate Technology Publications, London. 106 pp.

Lightfoot, C., Roger, P.A., Cagauan, A.G. and dela Cruz, C.R. (1993) Preliminary steady-state nitrogen models of a wetland rice field ecosystem with and without fish. *ICLARM Conference Proceedings* No. 26, 56-64.

Meas, N. (1995) Towards restoring life. JSRC, Phnom Penh.

Ovesen, J., Trankell, I. and Öjendal, J. (1996) When every household is an island: social organisation and power structures in rural Cambodia. *Uppsala Research Reports in Cultural Anthropology, No. 15*. Uppsala University and Sida, Stockholm. 93 pp.

Touch, S.T. (1993) Fish supply and demand in rural Svay Rieng, Cambodia. MSc Thesis AE-93-34, Asian Institute of Technology, Bangkok. 114 pp.

UNPF (1998) *General Population Census of Cambodia 1998: Provisional Population Tables*. United Nations Population Fund and National Institute of Statistics, Ministry of Planning, Phnom Penh. 36 pp.

Chapter 3

A Framework for Research into the Potential for Integration of Fish Production in Irrigation Systems

F. Murray[1], D.C. Little[1], G. Haylor[1], M. Felsing[1], J. Gowing[2] and S.S. Kodithuwakku[3]

[1]*Institute of Aquaculture, University of Stirling, Stirling FK9 4LA, UK*
[2]*Centre for Land Use and Water Resources Research, Porter Building, University of Newcastle, Newcastle upon Tyne NE1 7RU, UK*
[3]*Department of Agricultural Economics, Faculty of Agriculture, University of Peradeniya, Peradeniya, Sri Lanka*

Abstract

Water is a critically important but increasingly scarce resource in much of the developing world. Increases in overall productivity in relation to water use are desirable in the context of rising pressure to utilise water more efficiently. This highlights the need for improved integration amongst water use systems. This paper presents a framework for research towards integrating fish production and irrigation systems, focusing on opportunities for poor people as beneficiaries. Any assessment of farming systems in Asia quickly recognises that 'irrigated' and 'rainfed' zones are far more complex with respect to water availability than such simple terms suggest. The constraints and opportunities for the poor to benefit from integration of fish production into large, institutionally-managed systems are likely to be very different for farmers managing various forms of micro-irrigation. We categorise irrigation as being under the control of: (i) the household or immediate community; or (ii) an outside institution, typically an irrigation authority or department, as the basis of a framework leading to the development of improved systems integrated with fish production for both. We outline approaches that use technical and social methodologies both to understand current systems and to develop innovative approaches in participation with stakeholders. Engineering and management options are examined in an interdisciplinary mode in which action research with local communities follows situation analysis. Projects are being undertaken in India,

Pakistan and Sri Lanka but technical and management guidelines, and the wider policy issues developed will have broader relevance in developing countries.

Background

It is estimated that almost half the world's poorest people, nearly 500 million, live in drought prone areas and depend on irrigated agriculture to provide them with much of their food supply (HDR, 1997). This will rise to 20% of the world's population by 2050 (Engleman and LeRoy, 1993). The area irrigated by major systems has increased by an order of magnitude during the 20th century; however, recent estimates (Yudelman, 1994) suggest that scope for further increases of the area irrigated in Asia may be exhausted in the next 20-30 years. Population pressure, competition with urban and industrial users, and increased frequency of drought will combine to make water a dwindling *per caput* resource in these areas.

Agriculture is responsible for some 70% of global water use (FAO, 1995) and Postel (1993) estimates average irrigation efficiency (defined as the ratio of the quantity of water delivered to that actually consumed by crops during their growth cycle) to be as low as 37% world-wide, indicating enormous potential for water savings in these systems. Improved design and management will continue to be a major response to these problems. However, a new development priority has also emerged, following a recent paradigm shift: the impacts of irrigation systems are being reassessed at the wider watershed level in contrast to the 'on-farm soil and water management' priorities of earlier strategies. This approach recognises the importance of collective action in sustainable management of natural resources and thus offers greater options for individual on-farm management of water. Molden (1997) observed that increases in irrigation efficiency do not always coincide with increases in overall basin level productivity of water, which may be more readily achieved through the simultaneous multiple use and high concurrent re-use of water (Wolters and Bos, 1990).

Although the development of water resources for irrigation has often had a negative effect on natural fisheries, there has been little research into the production potential of extensive culture-based fisheries, and intensive cage culture in irrigation storage systems and little development focus on appropriate systems to benefit the resource-poor. Two thirds of the predicted shortfall in world fish production (20-30 million t by the year 2000 according to FAO, 1995) will occur in the semi-arid tropics, an extensive bio-climatic zone, home to large numbers of the world's poor, further underlining the need for such research effort.

Project Purpose

The purpose of this Project is to investigate the potential for integrated production of all aquatic species, plant and animal, of nutritional or other economic value within large and small-scale irrigation systems; and the potential for this production to benefit the poor. Principal collaborators in India are the NGO Samuha and Tamil Nadu Agricultural University; in Pakistan the Department of Fisheries and the International Irrigation Management Institute (IIMI); and in Sri Lanka the Agribusiness Centre, Peradeniya University, and the Mahaweli Development Authority. A number of categories of irrigation systems representative of a wide range of 'water stressed' areas will be studied in each of these countries.

Direct beneficiaries could include the poorest marginal, landless and women's groups, all of whom have traditionally derived least benefit from irrigation developments. A key objective is to identify options to diversify the existing livelihood strategies of such groups through low-risk aquatic food production. Indirect benefits could accrue to fish processors and the wider population in the irrigated area through potential control of aquatic weeds, disease vectors, generation of subsidiary income generating activities and increased provision of a cheap, high quality protein source.

The Project seeks to promote a participatory approach to the design and implementation of research with individual households and communities. Stakeholder workshops and situation analyses will be used to guide initial selection of research areas, institutional collaboration and prioritising of key researchable constraints. Situation analyses are currently under way in India (Karnataka, Tamil Nadu), Pakistan (the Punjab) and Sri Lanka (North West Province). Research outputs will include technical guidance to engineers, policy guidance to planners and donors, research guidance to scientists, and extension guidance to producers.

The Relevance of Research Areas to Wider Development Needs

Research outputs from case studies in India, Pakistan and Sri Lanka will be relevant to extensive areas of 'rainfed drylands' located in broad sub-equatorial belts north and south of the equator. Large sections of the population in these regions have less than the WHO recommended daily intake of 2,800 calories due mainly to the erratic and short duration of rainfed growing seasons and nutrient-poor soils which reduce the productivity of local agriculture (Myrada and IIRR, 1997). Yet because of their wide extent, rainfed drylands support large populations and produce a large proportion of national food requirements in many developing countries. Rainfed drylands are home to one sixth of the world's population and represent one of the largest bio-climatic zones within the semi-arid tropics (ICRISAT, 1997). Rainfed drylands in India provide almost 44% of the country's food and support nearly 40% of the population (Myrada and IIRR, 1997). Irrigation systems are widely used to increase the cropping potential or simply to reduce the risk of crop losses in these impoverished areas.

Other Project Objectives

Gender Focus

Our approach will consider gender in targeting beneficiaries, in addition to marginal and landless groups. Although rural women contribute over half the labour used to produce the food required in Asia (SPPRGA, 1997), they come last in the distribution of productive resources and social services and often form the poorest and most vulnerable sub-sector of marginal communities (Engle, 1987; Mehendale, 1991). Only 3% of extension time and resources are allocated to women in Asia compared to a global average of 5% (UN, 1997). The tendency to ignore women (who may have different inputs, ideas and needs than men) in development planning, renders many development projects unsustainable (Fatima, 1991; Agarwal, 1997). Many developing country development projects have actually marginalised women further by depriving them of their control over productive resources and authority within the household, whilst failing to lighten their traditional workloads (Afshar, 1991; Momsen, 1991). Since the UN Decade of Women 1975-1985, women have been increasingly singled out for special attention in development. Yet despite the increasing provision of favourable policy and legal reforms, awareness of gender issues and wider adoption of participatory techniques within development programmes, women continue to have little opportunity to participate in the planning of development projects. Moser (1989) suggested that there remains a widespread reluctance by development planning authorities to consider gender as an important planning issue. In addition, the preoccupation of feminist writers with describing the complexity of gender relations makes it difficult to translate gender awareness into a practical gender-planning framework. Amongst research priorities, the adoption of analytical frameworks that facilitate better understanding of gender relations in rural areas is required.

Although recommendations on how to incorporate gender concerns into aquaculture development planning are virtually non-existent, two main approaches do exist with respect to women's involvement within the wider development arena. The first, known as the 'women in development' approach, focuses on special, women-only projects considering 'traditional' women's roles related to domestic activities. The more recent 'gender in development' approach concentrates on efforts to promote women's empowerment by involving men and women alike (Moser, 1989; Humble, 1998). King (1989) identified areas for the training of women in aquaculture, which included pond management (preparation, stocking, feeding, fertilisation and harvesting), and handling, transport and marketing of fish. Our approach involves a gender impact assessment to evaluate existing gender roles, workloads, access to resources and the potential negative and positive impacts of aquaculture interventions on both sexes. Although many barriers exist, there are also reasons to expect good potential for women's participation in aquaculture initiatives. Firstly, where fish production is adopted as a new activity, men will not already dominate it. Secondly, there is good evidence to suggest that poorer women do take on non-traditional roles, even in conservative cultures. Woman may also utilise other aquatic products for income generating household activities such as reeds harvested for basket making.

Sustainability

The fragile ecosystems characteristic of rainfed dry lands typically suffer accelerated degradation where poor resource management accompanies increasing population pressure (HDR, 1997). Improved understanding of whole-farm systems gained in participation with farmers may aid in the development of sustainable strategies involving lower-input, integrated crop and aquatic production rather than high-input, monocropping. Although it is recognised that resource-poor farmers' short-term needs may not readily correspond with those which promote sustainability, greater efficiency of water management gained through closer integration could reduce the costs of producing both terrestrial and aquatic crops. Aquaculture extension in many developing countries currently focuses on high-yielding technologies dependent on off-farm inputs (Haylor *et al.,* 1997). Such approaches exclude the majority of poorer farmers who possess only limited resources which must be managed with minimal risk. We expect that suitable technologies would include semi-intensive production of herbivorous/omnivorous species low in the food chain.

Biodiversity

Both south-west India and Sri Lanka have been recognised as bio-diversity 'hotspots': areas where high concentrations of endemic fish species are experiencing unusually high rates of habitat modification or loss (Kottelat and Whitten, 1996). This has specific ecological impacts as well as negative long-term impacts on wild capture fisheries. The potential of indigenous species and indigenous farming practices to enhance sustainable development will be evaluated.

Methods

A participatory research approach with the following components will be adopted:

• A wide-ranging situation analysis (Box 3.1) to determine relevant initiatives which impact the poor. Situation analysis progresses from regional to local level, using secondary information and key informant interviews as well as other participatory rural appraisal (PRA) approaches in villages. Validity of PRA results may be enhanced by use of triangulation techniques, and in some cases non-parametric statistical tests, to determine whether different stakeholders hold significantly different viewpoints.

•   Beneficiaries, institutional collaborators, opportunities for, and researchable constraints to, fish production can be identified using methods such as stakeholder workshops. Mechanisms including stakeholder analysis (Lawrence *et al.,* 1999) can be used in conflict management and resolution leading to a multi-perspective, participatory approach involving irrigation, aquaculture, socio-economic and development specialists.

•   Farmers who wish to research these opportunities on their farms can be supported to monitor and evaluate such research. Such village level activities can be established through community based organisations.

Methodologies need to be constantly reassessed and adapted to both understand current systems and develop innovative systems in participation with stakeholders, which are relevant to a wider range of situations.

Box 3.1. Major components of a situation analysis for aquaculture-related development (adapted from Haylor *et al.,* 1997).

Regional situation analysis:
•   institutional support (governmental, non-governmental organisation, research and training, credit)
•   fisheries production, including history, production by sector, seed production
•   market analysis, including consumer preferences, infrastructure, wholesale and retail systems
•   relevant political and economic situation, i.e. demography, social disintegration

Local situation analysis (includes village PRA):
An aquaculture situation analysis contains information about:
•   the economy, landholdings, social structure and priorities of the local community
•   farming systems and the role of women in these systems (seasonal patterns, workloads)
•   physical nature of the area (temperature range, rainfall, soil types, water resources)
•   relevant indigenous knowledge and perceptions and demand for fish and aquaculture

---

Aquaculture Potential and Poverty Focus in Large and Small-scale Irrigation Systems

Irrigation systems comprise the following functional sub-sections: water collection, water delivery, on-farm application and wastewater removal. Potential exists to incorporate aquatic production in each of these components (Haylor, 1994). Important differences exist in water delivery policy and control structures. The impact of these differences on aquaculture potential has not been considered. This is in large part due to past failures to value the multiple uses of water (Gowing, 1998).

The majority of the world's poorest farmers in the most marginal regions continue to employ traditional minor and micro-irrigation systems (Table 3.1) to increase the productivity of their land such as village tanks, farm ponds (Fig. 3.1), open wells (Fig. 3.2), check dams (Fig. 3.3) and irrigation ponds (Fig. 3.4). Dryland development in India is increasingly being undertaken on a watershed basis on which the Indian government spends some US$300 million year^{-1} (Barr, 1998). Integrated components include the construction of farmer-managed, soil and water-harvesting structures. Much of this substantial and expanding resource remains unquantified and there has been little exploration of the potential for integrated fish production into these water bodies.

Table 3.1. Administrative classification of irrigation systems in project nations. 1 = household or community managed, 2 = formally managed by outside institution.

Criteria	Major (2)	Medium (1 or 2)	Minor (1)	Micro (1)
Description	Large dams and canals constructed on perennial rivers	Reservoirs fed by run-off (or cross basin diversions)	Reservoirs fed by ephemeral surface or ground water	Rain and silt harvesting devices or ground wells
Seasonality	Perennial	Perennial or seasonal	Mostly seasonal	Perennial only with ground supply
Water spread (command area)	> 200ha (> 600ha)	50-200ha (80-600ha)	1-50ha (< 80ha)	< 0.1-1ha (< 1ha)
Planning and management	State	State or Community	Community	Community or individual farmers
Construction	Outside contractors	State	Community using local materials	Community or individual farmers
Limnology	Natural productivity usually low. CPUE[1] for stocked species[2] is low	Higher natural productivity due to vast draw-down	Increasing natural productivity. Highest CPUE for stocked species	Manageable by farmer interventions

[1] Catch per unit effort.
[2] Where species are not naturally recruiting.

Contrary to popular perceptions of 'small-scale simplicity', because of their inherently less predictable nature, some aspects of small-scale irrigation systems (SSIs) are sometimes considered more difficult to manage than larger systems

(Carter, 1992). Smaller storage systems, which harvest surface run-off waters, are most likely to be seasonal in nature (often holding water for less than 6 months). Such seasonality is a critical factor when considering aquaculture potential, bringing with it a range of constraints and benefits (Box 3.2). Small-scale water bodies which exploit ground water resources, are more likely to be perennial, but are often used as a source of drinking water.

Box 3.2. Benefits and constraints imposed on aquaculture by seasonality.

Constraints:
- requirement for annual inputs of seed (often advanced fingerlings to take advantage of the shortened growing season)
- seasonal availability of large volumes of fish reduces prices and producers' returns
- increased potential for conflict with irrigation and other primary uses of water bodies

Benefits:
- improved manageability, including ease of predator control and high catch per unit effort (fishing need not be a full-time occupation)
- draw-down areas used for pasture increasing natural productivity

---

Large or small, the economic and environmental valuation of different water uses might be used to quantify the relative costs and benefits of aquaculture, particularly with respect to the primary objective of each system, i.e. water conveyancing in irrigation supply canals, and analysis conducted to determine who benefits. Other potentially competing uses of water include human and livestock consumption, domestic, industrial needs and bathing. The challenge is to find complementarities between integrated uses which may improve both equity and productivity over single production outputs. Some of the principal factors that should be considered during site-specific evaluation are:

- Cost and benefits of integration
- The scale of investment required
- Equity of benefits accruing from aquaculture production
- The degree of synergy or antagonism of multiple uses
- Production factors, i.e. access, water quality, quantity and reliability, availability of suitable species, of inputs and of markets.

The constraints to, and opportunities for, the poor to benefit from integration of fish production into large and small-scale irrigation systems are likely to be different as a consequence of both technical and socio-economic factors. With increasing system size, the potential of individual households to participate in water management is progressively replaced by larger groups, whole communities and ultimately outside institutions (Table 3.1). In many instances reduced autonomy and participation of individual farmers accompanies this progression.

Figs. 3.1-3.4. Small-scale irrigation devices in Karnataka State, India with potential for aquaculture. Farm ponds are micro-irrigation and ground water recharge devices, open wells are traditional ground water harvesting devices. **Fig. 3.1.** Farm pond (highly seasonal); **Fig. 3.2.** Open well (perennial); **Fig. 3.3.** Check dam (seasonal); **Fig. 3.4.** Irrigation pond (perennial).

Where aquaculture is integrated into major systems and SSIs in close proximity, synergies can be expected to bring indirect benefits to poor people:

- Enclosed fingerling culture in perennial systems used to stock seasonal water bodies, or production of seed in seasonal water bodies to stock perennial systems
- Reduction in fish prices through increased supply
- Creation and strengthening of production, processing and trading networks.

The poorest farmers are often located at the 'tail-end' of large engineered systems and suffer particularly from unequitable and unreliable water allocation, and land degradation due to salinisation, seepage and liability to flooding, all of which compromise their ability to grow crops successfully. Production of fish in borrow-pits (resulting from excavation of soil for other purposes) of small on-farm reservoirs, seepage zones or emergency irrigation ponds may offer such farmers a means of using their limited resources more productively.

Common property resources (CPRs) within both large and small-scale systems may offer opportunities for the landless to participate, for instance through stocking village tanks or small-scale cage culture in reservoirs and canals. However, many issues surround the capacity of the poor to make productive use of CPRs without these being appropriated by more powerful individuals or groups. Legal issues will need to be investigated. Levies placed on production could contribute to the maintenance of communally owned water resources, potentially improving social cohesion. Women are often constrained by a need to stay close to their homes to attend to domestic tasks traditionally undertaken by women. Identification of suitably located CPRs or smaller on-farm water bodies could facilitate their participation.

However, traditional power structures may undermine attempts to utilise irrigation systems for novel uses. Changes may be required at a range of levels, including water and power authorities, local administration and communities themselves. Accommodating alternative users may therefore require new institutional as well as technical solutions.

References

Afshar, H. (1991) Women and development: myths and realities; some introductory notes. In: Afshar, H. (ed.) *Women Development and Survival in the Third World.* Longman, New York, pp. 1-10.

Agarwal, B. (1997) Gender, environment, and poverty inter-links: regional variations and temporal shifts in rural India, 1971-91. *World Development* 25 (1), 23-52.

Barr, J. (1998) Current issues in water management on-farm, a review of options. DFID Research Project R6759 (integrated aquaculture in Eastern India). Unpublished.

Carter, R. (1992) Small-scale irrigation a balanced view. In: Priorities for water resources allocation and management. Natural Resources Advisers Conference, July 1992, Southampton. Unpublished.

Engle, C.R. (1987) Women in training and extension services in aquaculture. In: Nash, C.E., Engle, C.R. and Crosetti, D. (eds) *Women in Aquaculture.* Proceedings of the ADCP/NORAD Workshop on Women in Aquaculture Rome, FAO, 13-16 April 1987. ADCP Report 28. FAO, Rome, pp. 67-82.

Engleman, R. and LeRoy, P. (1993*) Sustaining Water: Population and the Future of Renewable Water Supplies.* Population Action International, Washington DC. 10 pp.

ESCAP (1997) Women in Sri Lanka. *UN Statistical Profiles* 13, 3-35.

FAO (1993) *The State of Food and Agriculture.* FAO, Rome. 46 pp.

FAO (1995) *The State of World Fisheries and Aquaculture.* FAO, Rome. 58 pp.

Fatima, N. (1991) The plight of rural women. In: Sebastion, L.R. (ed.) *Quest for Gender Justice: a Critique of the Status of Women in India.* T.R. Publications for Satya Nilayam Publications, New Delhi, pp. 12-26.

Gowing, J. (1998) Integration of aquaculture within irrigation systems (a proposal). DFID Engineering Division, Technology Development and Research, Unpublished. DFID, London.

Haylor, G. (1994) Fish production from engineered water systems in developing countries. In: Muir, J. and Roberts, J. (eds) *Recent Advances in Aquaculture V.* Blackwell Science, London, pp. 1-103.

Haylor, G., Lawrence, A. and Meusch, E. (1997) *Identification of Technical, Socio-economic Constraints to the Rearing of Fish in Ricefields of Lao PDR.* ODA Natural Systems Program R6380CB, ODA, London. 85 pp.

HDR (1997) *Human Development Report.* Oxford University Press, New York. 359 pp.

Humble, M. (1998) Assessing PRA for implementing gender and development. In: Guijt, I. and Shah, M.K. (eds) *The Myth of Community.* Intermediate Technology Publications, London, pp.35-45.

ICRISAT (1997) *This is ICRISAT.* International Crop Research Institute for the Semi-Arid Tropics, Patancheru, Andra Pradesh. 63 pp.

IIMI (1997) *Irrigation, Health and the Environment: a Literature Review with Examples from Sri Lanka.* Discussion paper No.42. International Irrigation Management Institute (IIMI), Colombo, Sri Lanka. 25 pp.

King, H. (1989) Fisheries development programs and women. *Naga* 12(2), 6-7.

Kottelat, M. and Whitten, T. (1996) *Freshwater Biodiversity in Asia with Special Reference to Fish.* World Bank Technical Paper 343. World Bank Washington, DC. 34 pp.

Lawrence, A., Barr, J. and Haylor, G. (1999) *Stakeholder Approaches to Planning Participatory Research by Multi-institution Groups.* ODI Agricultural Research and Extension Network, AGREN. Network Paper 91. 9 pp.

Mehendale, L. (1991) The integrated rural development program for women in developing countries: what more can be done? A case study from India. In:

Afshar, H. (ed.) *Women Development and Survival in the Third World.* Longman, New York, pp.223-238.

Molden, D. (1997) *Accounting for Water Use and Productivity.* SWIM Paper 1. International Irrigation Management Institute, Colombo. 16 pp.

Momsen, J.H. (1991) *Women and Development in the Third World.* Routledge, London. 115 pp.

Moser, C.O.N. (1989) Gender planning in the third world: meeting practical and strategic gender needs. *World Development* 17(11), 1799-1825.

Myrada and IIRR (1997) *Resource Management in Rainfed Drylands – an Information Kit.* Myrada, Bangalore and International Institute of Rural Reconstruction, Silang, Cavite. Bangalore. 244 pp.

Postel, S. (1993) Water and agriculture. In: Gleick, P. (ed.) *Water in Crisis.* Oxford University Press, Oxford, pp.56-66.

SPPRGA (1997) *A Global Program on Participatory Research and Gender Analysis for Technology Development and Organisational Innovation.* Network Paper no. 72, January 1997. System-wide Program on Participatory Research and Gender Analysis (SPPRGA), ODI Agricultural research and Extension Network. OD1, London.11 pp.

UN (1997) Improvement of the situation of women in rural areas. Report of the Secretary General. www.un.org/womenwatch.

Wolters, W. and Bos, M.G. (1990) Interrelationships between irrigation efficiency and the reuse of drainage water. Symposium on Land Drainage for Salinity Control in Arid and Semi-Arid regions, Cairo, 25 February–2 March 1990. Vol. 3: 237-245. Presented papers. Ministry of Public Works and Water resources, Cairo.

Yudelman, M. (1994) Demand and supply for foodstuffs up to 2050 with special reference to irrigation. *IIMI Review* 8(1), 4-14.

Chapter 4

Economics and Adoption Patterns of Integrated Rice-Fish Farming in Bangladesh

M.V. Gupta[1], J.D. Sollows[1], M.A. Mazid[2], A. Rahman[2], M.G. Hussain[2] and M.M. Dey[1]

[1]*International Center for Living Aquatic Resources Management (ICLARM), GPO Box 500, Penang 10670, Malaysia*
[2]*Bangladesh Fisheries Research Institute (BFRI), Mymensingh 2201, Bangladesh*

Abstract

The chapter presents the costs and benefits of integrating aquaculture with rice farming in Bangladesh during the rainfed and irrigated seasons. The analysis is based on data collected from 256 farmer implemented on-farm trials and a survey of households that have independently adopted integrated rice-fish farming. On average, farmers obtained a fish production of 233 and 184kg ha^{-1} during the rainfed and irrigated seasons, respectively, with a net benefit of Tk4,948 (US$1 = Taka40.28) during the former season and Tk4,799ha^{-1} during the latter season. Rice yields were higher in 82.4% of the integrated farms compared to farms with only rice. The survey indicated that adoption was by the relatively well off farmers with larger land holdings, higher cropping intensity and higher literacy, as was the case in the early years of the green revolution.

Introduction

Rice farming is the main occupation for the majority of rural households in most Asian countries. As the potential to expand the area under rice farming is limited, farmers are shifting to high yielding varieties of rice to meet the increasing demand. There are concerns that intensification of rice cropping is adversely affecting the ecology of rice fields, and yields have reached a plateau or may even be declining (Pingali, 1991). In recent years, rice production has become less profitable for farmers due to stagnant yields and high input costs.

Hence, there is a move towards diversification out of rice monoculture (Pingali, 1992). This is leading to renewed interest in research and development on alternatives to rice monoculture. One of these is the age-old practice of integrating fish culture with rice farming.

Fish culture in rice farms can be traced back to the Eastern Han Dynasty (AD 25-222) in China (Li, 1992). With the introduction of high yielding varieties of rice which require a high input of pesticides and fertilisers, less time for maturity, and controlled water management regimes, the practice declined to a large extent in the 1960s and 1970s. In recent years, with the introduction of integrated pest management and identification of aquaculture species which can grow in shallow waters and reach marketable size within the rice-growing period, the practice is again gaining popularity in many countries. It shows a promising potential for increasing fish production and improving the nutrition of farming households. The potential for integration can be gauged by the fact that 615,000ha in Bangladesh (Dewan, 1992), 5 million ha in China (Li, 1992; Wang, 1992; Xu and Guo, 1992), 2 million ha in India (Ghosh, 1992), 1.57 million ha in Indonesia (Koesoemadinata and Costa-Pierce, 1992), 326,000ha in Vietnam (Mai *et al.,* 1992), and 254,000ha in Thailand (Fedoruk and Leelapatra, 1992), were estimated to be suitable for integrating fish culture with rice farming. Currently, almost 1 million ha in China and 94,000ha in Indonesia are reported to be under integrated rice-fish farming (Lightfoot *et al.*, 1992).

Rice and fish are the staple food of the 114 million people of Bangladesh. While rice production increased with the introduction of high yielding varieties of rice and increased area under irrigation and flood control during the 1980s and 1990s, fish production has not increased to the same extent due to the overexploitation of natural aquatic resources and environmental degradation. In the past, poor rural households were catching fish from the common property, inland open water resources, to meet their needs. With the decline in catches from open waters as a result of overexploitation and a decrease in floodplain area due to flood control, the availability of fish to the rural poor has declined. Before the intensification of agriculture, farmers were capturing wild fish that entered the rice fields through flooding by excavation of a sump in the low-lying area of their farms. This source also has dried up with intensification of agriculture and flood control.

The decline in wild stocks, coupled with increasing demand for fish, has led researchers in Bangladesh in the 1970s and 1980s to study the feasibility of integrating fish culture with rice farming. A number of on-station and researcher implemented on-farm studies were undertaken and these studies concentrated on biological aspects: species, species combinations and stocking densities without emphasis on the economic evaluation of the integration (Das, 1982; Haroon *et al.*, 1989, 1992; Haroon and Alam, 1992; Ali *et al.*, 1993; Kohinoor *et al.*, 1994). In spite of all these studies, there was hardly any adoption of integrated rice-fish farming. CARE (Bangladesh) has been active in advocating the benefits of adopting integrated rice-fish farming to farmers as a part of its integrated pest management programme (CARE, 1993; Kamp and Gregory, 1993). The Third Asian Regional Rice-fish Farming Research and Development Workshop concluded that on-station and researcher managed, on-farm trials do not reflect the real situation and stressed the need for assessing the economic viability of the technology through farmer implemented trials for identifying the economic benefits to the farmers and constraints to adoption if any, before policy makers are asked to invest in dissemination of the technology (dela Cruz, 1994).

The International Centre for Living Aquatic Resources Management (ICLARM) in collaboration with the Bangladesh Fisheries Research Institute (BFRI) and the assistance of the Department of Agriculture Extension (DAE), undertook farmer implemented trials during 1992-1995 in Mymensingh to assess the costs and benefits of integrating fish culture with rice farming and the constraints to adoption, if any. Data collected during these trials are used in this paper to assess the costs and benefits of integrated farming and the impact on the households that have taken advantage of the integrated rice-fish farming technology.

Methodology

Data were collected from a total of 256 farms where rice-fish farming trials were undertaken by farmers during five rice farming seasons: three rainfed autumn (*aman*) seasons and two irrigated winter (*boro*) seasons during 1992-1994 in 11 *thanas* (sub-districts) of Mymensingh district and three *thanas* of Jamalpur district in Bangladesh. Farmers for trials were selected based on their interest in participating in the trials and were given one day training in various aspects of integration, which included plot preparation (excavation of sump and strengthening of dikes), species and densities, feeding, fertilisation and integrated pest management practices. No inputs were provided to them. The data collected were used for quantifying the costs and benefits of integrating aquaculture with rice farming.

Subsequently, an intensive random survey of 47 farmers who were practising rice-fish farming during 1994/95 was undertaken in three *thanas* of Mymensingh district. The survey covered profiles of farmers, management practices, production obtained and farmers' perceptions on integration. The farmers were classified according to: (i) type of farmer — research farmers were those who participated in on-farm trials conducted earlier and adopters were those who adopted the technology after seeing neighbours; and (ii) season of farming — integrated fish farming during *aman* season only, *boro* season only or both seasons.

Of the 47 farmers surveyed, 20 participated in on-farm research earlier and the remaining 27 adopted the technology after seeing the operations of the research farmers. The data collected were used to study the profiles of farmers and assess the socio-economic impact on the households that adopted the rice-fish farming technology, either after seeing neighbouring farmers who were involved in the research or from information provided by the personnel of the Department of Agriculture Extension (Gupta *et al.*, 1998). Information from the Bangladesh Bureau of Statistics (BBS, 1996) was used to compare the profiles of the households that adopted the technology with the rest of the population in the area. The data were analysed using FoxPro, SPSC PC+ and SAS/ETS software.

Results and Discussion

To integrate fish into rice farming, the farmers had to strengthen the dikes of the rice plots and raise them above flood level, install screens in embankments to prevent the escape of fish during heavy rains, and excavate a sump as a refuge for fish during low water level in the fields. The sump occupied an average of 1-5% of the rice plot area, though some farmers either had bigger natural depressions or excavated bigger sumps.

The data indicated that common carp (*Cyprinus carpio*) was the main species stocked during the *boro* season and the silver barb (*Barbodes gonionotus*) during the *aman* season. The choice of the species was mainly based on the availability of these species during the two seasons. The other species stocked were; rohu (*Labeo rohita*), mrigal (*Cirrhinus mrigala*), catla (*Catla catla*), silver carp (*Hypophthalmichthys molitrix*) and Nile tilapia (*Oreochromis niloticus*). Average stocking density during the *boro* season was 3,825 fingerlings ha^{-1}, while it was 4,082 during the *aman* season. The average size of fingerlings stocked varied between 4.8 and 10.1cm for different species. The few farmers who had larger sumps stocked at much higher densities.

The fish rearing period ranged from 34 to 113 days with an average of 75 days during the *boro* season, and 45 to 138 days with an average of 85 days during the *aman* season. At the time of harvest, the fish attained an average weight of 121g during the *boro* season and 90g during the *aman* season. Farmers obtained an average fish production of 233kg ha^{-1} during the *boro* season and 184kg ha^{-1} during the *aman* season (Table 4.1), with a net benefit of Tk4,948 ha^{-1} during the former and Tk4,799 ha^{-1} during the latter season.

Table 4.1. Stocking densities, recovery and fish production in integrated rice-fish farming. Standard deviations are in parentheses.

Season	Number of cases	Stocking density (number ha^{-1})	Average weight at harvest (g)	Recovery (%)	Fish production (kg ha^{-1})
Boro	145	3,825 (2,814)	121 (96)	55.6 (23.4)	233 (197)
Aman	98	4,082 (2,198)	90 (97)	59.0 (22.3)	184 (179)

Taking both cash and non-cash costs into consideration, the cost-benefit ratios were 1.71 and 1.80, respectively (Table 4.2). The non-cash costs included on-farm inputs such as rice bran, cattle manure and household labour, while the cash costs included fingerlings, feed, fertilisers, irrigation and hired labour. The cost of plot preparation was very low during the *aman* season as the farmers already had a ditch/sump in their rice fields for collecting wild fish. No irrigation was undertaken, hence there was no cost for irrigation.

Table 4.2. Cash and non-cash costs (Taka) of fish production and benefits ha^{-1} from integrated rice-fish farming. Standard deviations are in parentheses.

Season	Number of cases	Feed and fertiliser	Finger-lings	Plot prepa-ration	Irriga-tion	Total	Gross benefit	Net benefit
Boro	144	296	1,693	745	152	2,886	7,834	4,948
		(311)	(1,575)	(976)	(255)	(2,121)	(6,077)	(5,094)
Aman	104	574	1,789	-	10	2,661	7,460	4,799
		(603)	(1,307)			(2,270)	(7,328)	(6,147)

The comparative costs of rice monoculture and integrated rice-fish farming are indicated in Table 4.3. In each case the data were collected from two adjacent plots that were agroecologically similar, one with only rice farming and the other integrated with aquaculture. Both belonged to the same farmer and were subject to the same management practice for rice cultivation.

Table 4.3. Production costs of plots stocked with fish and comparable unstocked plots. Standard deviations are in parentheses.

Season	Number of cases	Cost of production (Tk ha^{-1})				
		Control plot : rice	Stocked plot : rice	Difference from control (%)	Stocked plot : rice and fish	Difference from control (%)
Boro	22	13,135	11,983	-9.4	14,925	+15.4
		(2,538)	(2,909)	(12.6)	(3,834)	(23.1)
Aman	10	5,088	4,543	-10.1	5,981	+17.5
		(934)	(809)	(9.6)	(953)	(15.5)

The cost of production for rice in integrated farming was lower than for monoculture of rice by 9.4 and 10.1% during the *boro* and *aman* seasons, respectively. This was due to the lower use of fertilisers and pesticides and the lower cost of weeding. Integration resulted in reduced weed infestation (Fig. 4.1). Infestation of pests such as stem borer (*Scirpophaga incertulas, Chilo suppressalis, Rupea albinella*), rice bug (*Leptocorisa oratorius*), green leafhopper (*Nephotettex* sp.), white leafhopper (*Cofana spectra*), short-horned grasshopper (*Oxya* sp.), golden cricket (*Euscyrtus concinnus*), gall midge (*Orseolia oryzae*), rice earhead bug (*Leptocorisa acuta*), rice skipper (*Pelopidas mathias*) and black bug (*Scotinophara* sp.) was much less on integrated farms (Fig. 4.2). The cost of inorganic fertilisers in integrated plots was an average of

15 and 46% lower than in rice monoculture, while the cost of weeding was lower by 29 and 23% during the *boro* and *aman* seasons, respectively. Due to the lower cost for rice cultivation in integrated farming, the overall costs as a result of integration of aquaculture were higher only by 15.4 and 17.5% compared to rice monoculture during the *boro* and *aman* seasons, respectively.

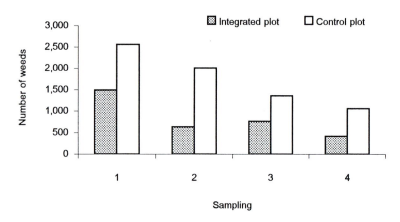

Fig. 4.1. Number of weeds collected from 3m^2 area each from seven plots during *boro* season (source: Gupta *et al.*, 1998).

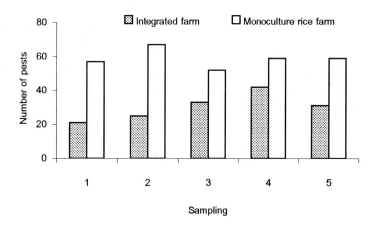

Fig. 4.2. Number of pests collected from 20 hills each from four plots during *boro* season (source: Gupta *et al.*, 1998).

The analysis also indicated that rice yields (calculated on the basis of total area including sump) from integrated farms were higher in 82.4% of the farms during the *boro* season and in 56.2% of the farms during the *aman* season, compared to monoculture of rice in adjacent plots. The yields were higher by an average of 10.25 and 11.6% during the *boro* and *aman* seasons, respectively (Table 4.4). The lower average cost for rice cultivation combined with the higher rice yields on integrated farms resulted in an average increase in net benefits from rice production of 22.6% during the *boro* season and 11.9% during the *aman* season compared to rice monoculture, but the differences were not statistically significant.

Table 4.4. Rice yields from integrated rice-fish and monocropped rice farms. Ranges are in parentheses.

Season	Number of cases	Rice yield (kg ha^{-1})		Integrated farms with higher rice yields (%)	Mean difference in yield of rice (%)
		Control	Integrated		
Boro	34	4,555 (3,046-6,000)	4,980 (3,264-6,571)	82.4	+10.2
Aman	25	3,498 (1,976-6,250)	3,811 (2,058-4,940)	56.2	+11.6

Data from the subsequent survey undertaken to assess the adoption of the technology provided interesting insights into the socio-economics of the segment of the population that has adopted integrated rice-fish farming. The average household size of the farmers that adopted the technology was much higher at 9.4 (consisting of 3.7 men, 2.7 women and 3.0 children) compared to the average household size of 5.3 in the study area and the national average of 5.5 (BBS, 1996).

Compared to the literacy rate of 17.6% in the study area and the national average of 23.8%, 89.4% of farmers who adopted the technology were literate, with 53.2% having completed secondary and higher secondary levels of schooling. The mean land holding of these farmers was 2.0ha of which 1.7ha was cultivated compared to average land holdings of 0.9ha both in the study area and the national average.

An analysis of management practices followed by the farmers who adopted the technology and those involved in the farmer implemented trials earlier indicated considerable differences. The farmers who adopted the technology had a higher proportion of their rice-cultivated area under integrated farming and had bigger ditches/sumps in their plots, to achieve higher fish yields (the data has indicated a positive linear relationship between proportion of the sump area in the rice field and fish production) (Table 4.5). The researchers and extension agents had suggested a stocking density of 3,000 fingerlings ha^{-1} but the farmers adopting the technology stocked their fields at very high densities, as much as 400-500% higher than the suggested stocking density (Table 4.6). When the use of feed and fertilisers was compared, farmers who adopted the technology used an appreciably higher quantity of inputs compared to the research farmers: 759kg versus 561kg ha^{-1} of rice bran; 1,725kg versus 764kg ha^{-1} of cattle manure; and 50.6kg ha^{-1} of inorganic fertilisers versus nil (Table 4.7). Concomitant with the higher sump size, higher stocking densities and higher input use, the adopting farmers obtained higher fish production (Table 4.8).

Table 4.5. Area under rice-fish farming and sump area for two categories of farmers.

		Boro				Aman		
Category of farmers	Number of cases	Plot area (ha)	Sump area (m²)	Integrated area in total rice cultivated area (%)	Number of cases	Plot area (ha)	Sump area (m²)	Integrated area in total rice cultivated area (%)
Research farmers	11	0.2 (0.13)	129 (59)	33.4 (42.35)	12	0.2 (0.11)	165 (89)	22.0 (25.65)
Adopters[1]	11	0.2 (0.16)	229 (175)	36.3 (33.02)	24	0.3 (0.24)	566 (824)	45.4 (44.12)

[1]Those who did not participate in on-farm research, but adopted technology subsequently.

Table 4.6. Stocking densities by season. Standard deviations are in parentheses.

	Season			
	Boro		Aman	
Class of farmer	Mean density (number ha^{-1})	Number of cases	Mean density (number ha^{-1})	Number of cases
Research farmers	8,709 (6,916)	8	5,202 (4,161)	12
Adopters	14,281(10,401)	8	20,996 (17,306)	19
Both	11,495 (9,005)	16	14,882 (15,723)	31

The farmers who adopted rice-fish farming had bigger land holdings, larger household size and higher literacy. Their cropping intensity (percentage of total cropped area against land area) was higher at 190% compared with the national average of 179%. These findings are similar to some of the studies undertaken in Bangladesh on the adoption of modern rice farming technology. Hossain *et al.* (1994) observed a higher adoption of rice farming technology, especially the capital-intensive *boro* farming, among larger farms. From the early years of the green revolution until about 1974, the evidence that more larger farmers were adopting the technology sooner seemed overwhelming (Lipton and Longhurst, 1989). Of 12 quantitative studies undertaken in Bangladesh during the 1970s to assess the adoption of modern rice varieties, seven showed a positive correlation between adoption and farm size and only one a negative link (Herdt and Garcia, 1982). A positive correlation between literacy and adoption of modern rice technology was observed by Hossain *et al.* (1994).

Table 4.7. Pattern of feed and fertiliser application by research farmers and adopters during *boro* (B) and *aman* (A) seasons. Standard deviations are in parentheses.

Feed/ fertiliser	Farmer category	Total number of farmers B	A	Farmers applying (%) B	A	Quantity (kg ha^{-1}) B	A
Rice bran	Research farmers	12	14	92	79	561 (450)	562 (457)
	Adopters	12	24	92	83	759 (1,277)	700 (725)
Cattle	Research farmers	12	14	92	86	764 (743)	774 (695)
Manure	Adopters	12	24	92	67	1,725 (2,712)	1,373 (2,124)
Duck	Research farmers	12	14	42	21	2.9 (8.4)	137 (335)
Weed	Adopters	12	24	42	21	33.5 (85.8)	10.2 (34.8)
Mustard	Research farmers	12	14	58	50	101 (160)	68.7 (129)
oil cake	Adopters	12	24	83	76	260 (538)	278 (331)
Inorganic	Research farmers	12	14	25	21	0	6.7 (19.7)
fertiliser	Adopters	12	24	50	46	50.6 (84.0)	55.6 (118)
Others[1]	Research farmers	12	14	33	29	11.1 (33.4)	27.5 (79.5)
	Adopters	12	24	50	29	125 (288)	423 (1,250)

[1]Termites, wheat, bran, kitchen waste, etc.

Table 4.8. Cultured fish production from rice-fish plots as reported by farmers. Standard deviations and sample size are in parentheses. *n* = number of observations.

	Fish production (kg ha^{-1})	
Category of farmer	*Boro* season	*Aman* season
Research farmers	442 (471) (*n* = 9)	303 (238) (*n* = 13)
Adopters	1,856 (2,393) (*n* = 8)	1,534 (2,417) (*n* = 20)
Both classes	1,107 (1,774) (*n* = 17)	1,049 (1,966) (*n* = 33)

Integrated rice-fish farming requires a higher labour input to strengthen embankments and excavate the sump. One of the reasons for adoption of the technology by households with larger family size could be the availability of family labour. Most of the adult male members of marginal/poor farming households in Bangladesh also work as labourers on other farms or have off-farm employment in addition to attending to their small farms. This would limit the time available to provide additional labour to attend to the needs of integrating aquaculture with rice cultivation. Hossain (1988) observed a positive relationship between the availability of family labour and the adoption of modern rice farming technology in the *aman* season in Bangladesh.

The analysis showed that the adopters of the technology used higher inputs than the research farmers. Usually adopting farmers use less input and have lower yields (yield gap) than research farmers who are guided by researchers. In this case, it has been the opposite, indicating that the relatively better-off farmers not only adopted the technology but also intensified the operations.

There are many reasons for marginal farmers not adopting integrated farming in the initial stages of technology development: lack of knowledge, as there is a tendency among government extension workers to target literate and financially better-off farmers as their understanding and adoption of technologies is expected to be better; lack of additional resources needed for integrating aquaculture; lack of credit availability; lack of time, as extra labour is needed for excavation of the sump and maintenance of dikes and most marginal farmers in Bangladesh also work outside their farms; difficulties in getting fingerlings of the required species of appropriate size at the time of stocking; and unwillingness to take risks as there is possibility of loss of fish due to the breach of dikes during heavy rainfall.

A number of non-governmental organisations (NGOs) are active in Bangladesh in working with farmers at the grassroots level to try to understand their problems; to provide appropriate technologies, training, credit; to ensure availability of needed inputs for adoption/implementation of the technology; and to monitor them periodically. Their target group is landless farmers, defined as households with less than 0.5acre (0.2ha) land holding. The adoption of new technologies among marginal farmers appears to be better through the influence of their work. Earlier surveys undertaken to study the impact of adoption of new technologies disseminated by the NGOs for the culture of Nile tilapia (*O. niloticus*) and silver barb (*B. gonionotus*) in seasonal ponds/ditches indicated that the poor farmers were able to benefit from the technologies (Gupta *et al.*, 1992; Gupta and Rab, 1994). A comparison (albeit with different technologies), of two dissemination strategies — rice-fish farming by observation and knowledge dissemination by the government extension agency, and culture of Nile tilapia and silver barb by NGOs — indicated that the former was adopted by relatively better-off farmers using more inputs and obtaining higher production, while the latter was adopted by marginal farmers using much lower inputs than suggested and obtaining lower production, although with good net benefits. With regard to input use, farmers used mostly on-farm inputs (cattle manure, rice/wheat bran) in the case of rice-fish farming, while for the other technology, even though use of inputs was low, the farmers had to purchase part of the input used. This indicated that the adopters in the latter case were resource-poor farmers who lacked on-farm resources such as rice bran and cattle manure. This clearly indicates that in the early stages of technology adoption, small farmers require support in terms of training, input supply and credit, which is normally ensured by NGOs for their target group, while government extension agencies concentrate on transfer of knowledge but not input supply and credit.

Conclusions

Integration of aquaculture with rice farming during the rainfed and irrigated seasons in Bangladesh is a viable, environment friendly activity with multiple benefits and an entry point to integrated pest management. The sustainability of the operation is indicated by the fact that, in spite of the small size of fish at harvest (due to the short rearing period and low water depth), many farmers adopt integrated farming. The study also indicates that the benefits of new technology will accrue to relatively prosperous farmers unless institutional support is provided for marginal farmers to take advantage of new technologies and research outputs. However, as observed in the case of the green revolution, small farmers will take to new technologies once the economic viability is established and extension services, easy availability of seed of required species and credit are ensured (Hayami and Ruttan, 1985; Lipton and Longhurst, 1994).

References

Ali, M.H., Miah, N.I. and Ahmed, N.U. (1993) *Experience in Deep-water Rice-Fish Culture*. Bangladesh Rice Research Institute, Gazipur, Bangladesh, Mimeo. 28 pp.

BBS (1996) *Bangladesh Bureau of Statistics 1995 Statistical Yearbook of Bangladesh*. Ministry of Planning, Dhaka. 646 pp.

CARE (1993) NOPEST pilot project results; rice-fish cultivation. 1992 Boro season. CARE Bangladesh. 46 pp.

Costa-Pierce, B.A. (1992) Rice-fish systems as intensive nurseries. In: dela Cruz, C.R., Lightfoot, C., Costa-Pierce, B.A., Carangal, V.R. and Bimbao, M.P. (eds) *Rice-Fish Research and Development in Asia*. ICLARM Conference Proceedings 24, 117-130.

Das, N.C. (1982) Paddy-cum-fish culture. MSc Thesis, Department of Aquaculture and Management, Bangladesh Agriculture University, Mimensingh. 64 pp.

dela Cruz, C.R. (ed.) (1994) Role of fish in enhancing ricefield ecology and in integrated pest management. *ICLARM Conference Proceedings 43*. 50 pp.

Dewan, S. (1992) Rice-fish systems in Bangladesh: past, present and future. In: dela Cruz, C.R., Lightfoot, C., Costa-Pierce, B.A., Carangal, V.R. and Bimbao, M.P. (eds) *Rice-Fish Research and Development in Asia*. ICLARM Conference Proceedings 24, 11-16.

Fedoruk, A. and Leelapatra, W. (1992) Ricefield fisheries in Thailand. In: dela Cruz, C.R., Lightfoot, C., Costa-Pierce, B.A., Carangal, V.R. and Bimbao, M.P. (eds) *Rice-Fish Research and Development in Asia*. ICLARM Conference Proceedings 24, 91-104.

Ghosh, A. (1992). Rice-fish farming development in India: past, present and future. In: dela Cruz, C.R., Lightfoot, C., Costa-Pierce, B.A., Carangal, V.R. and Bimbao, M.P. (eds) *Rice-Fish Research and Development in Asia*. ICLARM Conference Proceedings 24, 27-44.

Gupta, M.V., Ahmed, M., Bimbao, M.P. and Lightfoot, C. (1992) Socioeconomic impact and farmers' assessment of Nile tilapia (*Oreochromis niloticus*) culture in Bangladesh. *ICLARM Technical Report 35*. 50 pp.

Gupta, M.V. and Rab, M.A. (1994) Adoption and economics of silver barb (*Puntius gonionotus*) culture in seasonal waters of Bangladesh. *ICLARM Technical Report 41*. 39 pp.

Gupta, M.V., Sollows, J.D., Mazid, M.A., Rahman, A., Hussain, M.G. and Dey, M.M. (1998) Integrating aquaculture with rice farming in Bangladesh: feasibility and economic viability, its adoption and impact. *ICLARM Technical Report 55*. 90 pp.

Haroon, Y. and Alam, M. (1992) Integrated paddy-cum-fish/shrimp farming. Final Report. Fisheries Research institute, Mymensingh, Bangladesh. 41 pp.

Haroon, A.K.Y., Alam, M. and Mazid, M.A. (1989) An experimental study on integrated paddy-cum-fish/shrimp farming practices in Bangladesh. 1. Effects of stocking densities on growth and yield. *Bangladesh Journal of Fisheries* 12(1), 100-108.

Haroon, A.K.Y., Dewan, S. and Karim, S.M.R. (1992) Rice-fish production systems in Bangladesh. In: dela Cruz, C.R., Lightfoot, C., Costa-Pierce, B.A., Carangal, V.R. and Bimbao, M.P. (eds) *Rice-Fish Research and Development in Asia*. ICLARM Conference Proceedings 24, 165-171.

Hayami, Y. and Ruttan, V.W. (1985) *Agricultural Development: International Perspective*. John Hopkins University Press, Baltimore. 506 pp.

Herdt, R.W. and Garcia, L. (1982) Adoption of modern rice technology: the impact of size and tenure in Bangladesh. Mimeo, International Rice Research Institute, Los Baños. Philippines.

Hossain, M. (1988) *Nature and Impact of Green Revolution in Bangladesh*. IFPRI and BIDS Research Report No. 67. International Food Policy Research Institute, Washington D.C. 156pp.

Hossain, M., Quasem, M.A., Jabbar, M.A. and Akash, M.A. (1994) Production environments, modern variety adoption and income distribution in Bangladesh. In: David, C.C. and Otsuka, K. (eds) *Modern Rice Technology and Income Distribution in Asia*. Lynne Rienner Publisher, Boulder. pp. 221-279.

Kamp, K. and Gregory, R. (1993) Rice-fish cultivation as a means to increase profitability from rice fields: implications for integrated pest management. In: dela Cruz, C.R. (ed.) *Role of Fish in Enhancing Ricefield Ecology and in Integrated Pest Management*. ICLARM Conference Proceedings 43, 29-30.

Koesoemadinata, S. and Costa-Pierce, B.A. (1992) Development of rice-fish farming in Indonesia: past, present and future. In: dela Cruz, C.R., Lightfoot, C., Costa-Pierce, B.A., Carangal, V.R. and Bimbao, M.P. (eds) *Rice-Fish Research and Development in Asia*. ICLARM Conference Proceedings 24, 45-62.

Kohinoor, A.H.M., Saha, S.B., Akhteruzzaman, Md. and Gupta, M.V. (1994) Suitability of short-cycle species *Puntius gonionotus* for culture in rice fields. Role of fish in ricefield ecology and integrated pest management. ICLARM Conference Proceedings 43. 50pp.

Li, K. (1992) Rice-fish farming systems in China: past, present and future. In: dela Cruz, C.R., Lightfoot, C., Costa-Pierce, B.A., Carangal, V.R. and

Bimbao, M.P. (eds) *Rice-Fish Research and Development in Asia*. ICLARM Conference Proceedings 24, 17-26.

Lightfoot, C., Costa-Pierce, B.A., Bimbao, M.P. and dela Cruz, C.R. (1992) Introduction to rice-fish research and development in Asia. In: dela Cruz, C.R., Lightfoot, C., Costa-Pierce, B.A., Carangal, V.R. and Bimbao, M.P. (eds) *Rice-Fish Research and Development in Asia*. ICLARM Conference Proceedings 24, 1-10.

Lipton, M. and Longhurst, R. (1989) *New Seeds and Poor People*. John Hopkins University Press, Baltimore. 473 pp.

Mai, V.Q., Duong, L.T., Son, D.K., Minh, P.N. and Nghia, N.G. (1992) Ricefield aquaculture systems in the Mekong delta, Vietnam: potential and reality. In: dela Cruz, C.R., Lightfoot, C., Costa-Pierce, B.A., Carangal, V.R. and Bimbao, M.P. (eds) *Rice-Fish Research and Development in Asia*. ICLARM Conference Proceedings 24, 105-116.

Pingali, P. (1991) Agricultural growth and environment: conditions for their compatibility in Asia's humid tropics. In: Vosti, S.A., Reardon, T. and von Unff, W. (eds) *Agricultural Sustainability, Growth and Poverty Alleviation: Issues and Policies*. Zentralstelle fur Ernaehrung und Landwirtschaft, Feldafing, Germany, pp. 295-310.

Pingali, P. (1992) Trends in agricultural diversification: regional perspectives. *World Bank Technical Paper* 180, viii + 214 pp.

Wang, J. (1992) Methodology in extending rice-fish farming systems in China. In: dela Cruz, C.R., Lightfoot, C., Costa-Pierce, B.A., Carangal, V.R. and Bimbao, M.P. (eds) *Rice-Fish Research and Development in Asia*. ICLARM Conference Proceedings 24, 325-332.

Xu, Y. and Guo, Y. (1992) Rice-fish systems research in China. In: dela Cruz, C.R., Lightfoot, C., Costa-Pierce, B.A., Carangal, V.R. and Bimbao, M.P. (eds) *Rice-Fish Research and Development in Asia*. ICLARM Conference Proceedings 24, 315-324.

Chapter 5

Promotion of Small-scale Pond Aquaculture in the Red River Delta, Vietnam

L.T. Luu[1], P.V. Trang[1], N.X. Cuong[1], H. Demaine[2], P. Edwards[2] and J. Pant[2]

[1]*Research Institute for Aquaculture No.1, Dinh Bang, Tu Son, Bac Ninh, Vietnam*
[2]*Aquaculture and Aquatic Resources Management, School of Environment, Resources and Development, Asian Institute of Technology, PO Box 4, Klong Luang, Pathumthani 12120, Thailand*

Abstract

A survey of 191 VAC farming households (integration of crops, fish and livestock quarters from Vietnamese words vuon, ao, chuong for garden, pond, livestock quarters, respectively) in 1994 across lowland, intensive and suburban agro-ecological zones (AEZs) revealed considerable variations in the performance of aquaculture. Major pond inputs of grass, pig manure, and rice bran varied within and across AEZs in polyculture of carps (common carp, grass carp, mrigal and rohu). Extrapolated yields ranged from about 50kg ha^{-1} to over 6t ha^{-1}, with an average of around 2.5t ha^{-1}, indicating tremendous potential for the improvement of fish production. A set of recommendations involving simple changes in existing resource allocation patterns and stocking and harvesting strategies was introduced to 40 households in 1996 and expanded to 80 households in 1997 and 1998. Major changes were noted based upon introduction of recommendations: stocking large-sized fingerlings at adequate density, multiple stocking and harvesting, and increase in the amount of pond inputs. Impressive increases in fish production occurred as average yields recorded after project intervention almost doubled from 1994. Cost:benefit (C/B) ratios of fish production were influenced by year-to-year variation in types and amount of inputs applied and their effects on fish yields. The C/B ratio over the 3-year period ranged from around 1:1.5 to 1:3.0, higher than that of 1994. As the trials also suggested that VAC farm households had successfully increased fish production close to the maximum attainable yield with the current culture systems, a monoculture of Nile tilapia in ponds fertilised with pig manure and inorganic fertilisers was introduced to farmers who wished to further intensify fish production in their VAC system.

Introduction

The Red River delta in northern Vietnam with an average population density of over 1,000 persons km^{-2} is one of the most densely populated areas of the world. Although the delta comprises only about 5% of Vietnam's total landmass, its population is over 20% of the country's total (Cuc and Vien, 1993). Thus, the delta has minute agricultural holdings of only 0.3-0.5ha per household. Although 80% of the population of the Red River delta makes a living from agricultural activities, limitation of arable land is a major constraint to improving the livelihoods of its rapidly growing population. Farming systems in the Red River delta are rice-based, often with double cropping of rice with a combined average yield for two crops exceeding 11t ha^{-1} — close to the maximum potential yields (Rambo and Cuc, 1993). Therefore, opportunities to increase productivity of rice through adoption of modern technologies remain rather scanty (Uehara and Patanothai, 1996) as there is virtually no 'yield gap' (Rambo and Cuc, 1993). Nevertheless, the potential for agricultural diversification is still recognized because of rich soils, year-round access to irrigation water and ever growing demand for a wide range of horticultural, livestock and aquatic products.

Amongst approaches suggested to improve the economic situation, promotion of integrated farming centred around the development of a fish pond has long been considered to be among the most promising. Fish culture is also viewed by the government of Vietnam as having the potential to significantly improve rural livelihood (Ministry of Fisheries, 1999). Fish culture in the Red River delta has a long tradition, probably in response to high population density and overexploited wild fisheries (Chevy and Lemasson, 1937). Aquaculture in the Red River delta is dominated by small-scale, household-level integrated garden systems commonly known as VAC. This acronym derives from the Vietnamese for garden (*vuon*), pond (*ao*), and livestock quarters (*chuong*). These systems were expanded during the period of the recent war as part of a food security strategy. In the earlier organisation of agriculture into communes, VAC systems were the core of private plots and made a major contribution to rural livelihoods. In recent years, VAC has also been promoted as a strategy to improve food supply and nutritional standards in small-scale farming communities. The fish culture component of the VAC systems in the Red River delta has considerable potential for wider dissemination and intensification (Luu, 1992; Luu *et al.*, 1992; Pham, 1994), as fish culture in the delta remains underdeveloped (Uehara and Patanothai, 1996). Whilst a number of development programmes have promoted VAC activities, the essence of the integrated system which is recycling of nutrients between the sub-systems was not adequately considered. On the contrary, so called VAC promotional programmes promoted the culture of high value aquatic species with virtual reliance on off-farm formulated feed. Consequently, traditional integrated VAC systems were breaking down. Hence, there was a need to study the prospect for their improved performance.

Fig. 5.1. Fish pond integrated with crops grown on the dike. Cassava and banana leaves are fed to grass carp.

Fig. 5.2. Fish pond integrated with pigs. The manure from the pigsty with the thatched roof to the left is used to fertilise crops but urine and washings drain to the pond, creating plankton-rich, green water.

The Research Institute for Aquaculture No.1 (RIA No. 1) and the Asian Institute of Technology (AIT) began a research project on integrated agriculture-aquaculture systems in 1994 under the AIT Aqua Outreach programme to identify constraints to, and opportunities for, increased productivity of VAC systems through appropriate approaches to research and development of small-scale aquaculture. After gathering baseline information in the first phase in 1994, a second phase was launched in 1996 as farmer managed on-farm trials with the introduction of a low-cost technological package to selected VAC farming households. In subsequent years (1996-98), the performance of VAC systems was monitored in relation to the change in resource allocation pattern after project intervention. In this paper, we present the results of the baseline survey conducted in 1994 and the changes in inputs-outputs in VAC systems as a result of project intervention over three subsequent years.

Methodology

Eight districts representing three different agro-ecological settings of the Red River delta were identified through rapid rural appraisal (RRA). Following selection of study areas, a baseline survey was carried out in 1994 through a structured questionnaire applied to 241 households across the three agro-ecological zones (AEZs), namely lowland, intensive and suburban zones (Table 5.1). Data were collected on the profiles of respondents, their resource base, and inputs-outputs in different sub-systems. Data from 191 households possessing pond(s) were used for analyses.

Table 5.1. Agro-ecological zones (AEZ) of the VAC systems study area.

AEZ	Profile of the area	Districts	Provinces
Lowland	Low lying area Double cropping of rice but second rice often subject to flood Low rice yields (2-3t ha^{-1} crop)	Gia Luong, Binh Luc	Bac Ninh, Ha Nam
Intensive	Double cropping of rice followed by a dry season crop Rice yields up to 5-6t ha^{-1} crop	Kien Xuong Thuan Thanh, CamGiang, My Van	Thai Binh Bac Ninh, Hai Duong, Hung Yen
Suburban	Adjacent to Hanoi market Intensive production systems dominated by horticultural enterprises Access to urban industrial by-products (brewery waste, night soil) as agricultural inputs	Gia Lam Tu Liem	Hanoi Hanoi

Based upon the findings of the baseline survey, recommendations aimed at improvement of performance that required only simple changes/modifications in the existing resource allocation pattern of the VAC farming households were devised. In general, VAC farm households were already using relatively high inputs to ponds. Therefore, emphasis was given to improvement of fish stocking and harvesting strategies. Stocking larger-sized fingerlings, and subsequent multiple harvesting and supplementary stocking were the two major recommendations considered to be useful.

A total of 40 farmers representing different socio-economic groups from the three AEZs were selected for trials in 1996, provided with recommendation(s) based on their resource base, and asked to monitor the inputs and outputs in their VAC systems. The trials continued in 1997 and 1998 with some expansion in the number of trialist farm households. Trialists were provided with a record book to record inputs, outputs and major activities performed in their VAC systems on a regular basis. The researchers from RIA No. 1 made frequent visits to each farm household to ensure that data were recorded in an appropriate way. AIT Aqua Outreach staff visited the area on an occasional basis to provide advice.

Data analysis was performed using SPSS 7.5, a Windows-based statistical software package. Inputs and productivity data were calculated on a per hectare basis and returns expressed in US dollars. Costs and benefits of fish production were assessed using data on gross costs and gross benefits. Descriptive statistics such as mean, standard deviation, median, range, percentage and frequency distribution were used for analyses. Analysis of data was compromised by changing numbers of farmers and trial households between years and incomplete data collection.

Results and Discussion

VAC Systems Prior to Intervention

General description

The average size of VAC farm households was five persons with a land holding of nearly 0.5ha. No apparent variation was evident in size of household or size of land holdings across AEZ, but variation was noted in the land allocation pattern for major VAC sub-systems. The area of paddy was largest in the lowland zone, but smallest in the suburban area. The converse held for the size of garden. The average size of garden in the suburban zone was nearly three times larger than that in lowland and intensive zones. The total area covered by pond(s) was largest in the intensive zone followed by lowland and suburban zones (Table 5.2). A larger sized garden in suburban areas was a response to demand for a wide range of fresh vegetables and herbs in the Hanoi market. By contrast, risk of floods and relatively difficult access to the market might be reasons for a smaller garden area in lowland VAC farms. Larger pond areas in the intensive than in other zones indicated the greater significance of fish culture over other enterprises in this area.

Table 5.2. Size of VAC farm sub-systems (m^2) by agro-ecological zone (AEZ). Figures in parentheses denote standard deviations. n = number of farming households.

AEZ	Pond		Rice		Garden		Total	
	Mean	Median	Mean	Median	Mean	Median	Mean	Median
Lowland	1,286	1,080	4,087	3,600	473	360	5,837	5,400
(n = 51)	(1,061)		(1,718)		(400)		(2,387)	
Intensive	1,829	828	2,828	2,484	558	360	5,188	4,302
(n = 105)	(2,279)		(1,910)		(1,252)		(3,688)	
Suburban	861	720	2,012	1,800	1,555	720	4,429	3,719
(n = 35)	(693)		(1,135)		(4,591)		(4,720)	
Total	1,506	770	3,015	2,525	719	360	5,222	4,428
(n = 191)	(1,837)		(1,875)		(2,203)		(3,626)	

Distribution of households by major farming occupations was in line with the variation noted in land allocation pattern. Rice cultivation accounted for over 60% of the total area of individual households and was dominant in all zones. However, over a third of respondents in the suburban zone regarded gardening as their main occupation. Likewise, a similar proportion of respondents in the lowland zone considered aquaculture to be their main occupation. In general, livestock raising was widely regarded as a secondary occupation among VAC farm households although a few respondents, particularly in the intensive and lowland zones, gave livestock raising as their primary occupation (Table 5.3). Nearly 15% of respondents considered aquaculture as their major occupation with a much larger proportion in the lowland zone. Aquaculture was the second major occupation of another 15% of VAC farm households. This may also be interpreted to mean that two thirds of households possessing a pond did not pay sufficient attention to it for aquaculture to be considered even a major secondary occupation.

Hence, VAC systems in suburban areas were much more oriented towards horticulture, in the intensive zone towards livestock rearing, and in the lowland zone towards fish culture. These findings were basic to the different packages of recommendation made for the on-farm trials carried out in 1996-1998.

Aquaculture sub-system in VAC

The VAC farm households used various stocking, feeding and harvesting strategies in the aquaculture sub-system which generally reflected variations in resource base, both in different agro-ecological settings and in specific circumstances — farmers' goals, preferences and resource constraints of the farm.

Table 5.3. Primary and secondary occupations of VAC farm households by agro-ecological zone in 1994. *n* = number of farming households.

Activity	Occupations											
	Lowland				Intensive				Suburban			
	Main		Subsidiary		Main		Subsidiary		Main		Subsidiary	
	(*n*)	(%)	(*n*)	(%)	(*n*)	(%)	(*n*)	(%)	(*n*)	(%)	(*n*)	(%)
Rice crop	31	61	12	24	75	71	20	19	23	66	7	20
Gardening	1	2	21	41	3	3	24	23	11	31	7	20
Livestock	4	8	12	24	11	11	31	30	-	-	19	54
Fish culture	15	29	6	12	15	14	28	27	1	3	2	6

Stocking strategy

Stocking density, ratio of stocked species, and size of the stocked fingerlings were the major points considered. Overall, stocking density ranged between 0.04 and 14 fish m^{-2} with an average of 2.2 fish m^{-2} (Table 5.4). However, variation in average stocking density was noted between AEZs with the highest, 2.7 fish m^{-2} in the lowland zone and the lowest, 1.6 fish m^{-2} in the suburban zone.

Table 5.4. Average stocking density (fish m^{-2}) in VAC systems by agro-ecological zone (AEZ) in 1994. *n* = number of farming households.

AEZ	Mean (fish m^{-2})	Median
Lowland (*n* = 40)	2.7	1.8
Intensive (*n* = 88)	2.1	1.3
Suburban (*n* = 21)	1.6	0.8
Total (*n* = 149)	2.2	1.3

Virtually all farmers practiced polyculture with common carp, Chinese carp (grass and silver) and Indian major carp (mrigal and rohu) the major species stocked in all zones. Considering the overall proportion of species stocked, rohu ranked first (38%) followed by silver carp (33%), mrigal (25%), grass carp (16%) and common carp (13%). Over two-thirds of respondents indicated that they primarily stocked three major species, namely grass carp, rohu and silver carp. The number of farmers stocking common carp and mrigal was relatively low. Some farmers also stocked bighead carp, mud carp and other fish species but these accounted for an insignificant proportion of their stock.

There was wide variation in size of stocked fingerlings, largely associated with species. Average weight of stocked fingerlings was largest for grass carp (128g per fingerling) but was similar for the other species within a narrow range of 43-48g per fingerling (Table 5.5). Some variation in size of fingerlings at stocking was also noted across AEZ. Farmers in the lowland zone stocked relatively larger-sized fingerlings regardless of species, while farmers in the other two zones stocked relatively smaller-sized fingerlings. This may have been because several farmers in the lowland zone were engaged in fry production and nursing.

Table 5.5. Average size (g) of stocked fingerlings by species in VAC system ponds in 1994 by agro-ecological zone. *n* = number of farming households. SD=standard deviation.

Species	Mean (g per fish)	SD	Median	Range Minimum	Maximum
Common carp					
Lowland (*n* = 18)	70	49	50	10	200
Intensive (*n* = 50)	44	56	20	1	252
Suburban (*n* = 15)	35	15	40	10	50
Total (*n* = 83)	48	51	30	1	252
Grass carp					
Lowland (*n* = 38)	172	324	100	5	2,000
Intensive (*n* = 78)	110	109	100	2	600
Suburban (*n* = 17)	114	92	100	16	300
Total (*n* = 133)	128	196	100	2	2,000
Silver carp					
Lowland (*n* = 37)	57	46	50	1	200
Intensive (*n* = 77)	46	55	20	1	250
Suburban (*n* = 17)	40	21	40	12	100
Total (*n* = 131)	48	49	30	1	250
Mrigal					
Lowland (*n* = 21)	71	44	80	5	150
Intensive (*n* = 41)	34	24	25	1	100
Suburban (*n* = 10)	24	8	23	12	40
Total (*n* = 72)	43	34	30	1	150
Rohu					
Lowland (*n* = 36)	71	74	55	1	400
Intensive (*n* = 84)	34	31	20	0.5	130
Suburban (*n* = 6)	32	8	30	20	40
Total (*n* = 126)	44	50	28	0.5	400

Pond inputs

Three main types of pond inputs, namely agricultural by-products (virtually all rice bran), grass and pig manure and were recorded as major inputs. However, the frequency of use and the amount of these inputs used varied across AEZ. Farmers in the lowlands used all three inputs to a greater extent with average amounts of rice bran, pig manure, and grass used in these areas estimated at around 7, 16 and 91t ha^{-1}, respectively (Table 5.6). Farmers in the intensive zone applied an almost similar amount of manure on average, but grass and rice bran were applied at only about half the average amount applied in the lowland zone. Total amount and frequency of application of all these pond inputs were lower on average in the suburban zone.

Table 5.6. Major pond input (t ha^{-1}) in VAC system ponds by agro-ecological zone (AEZ) in 1994. *n* = number of farming households.

AEZ	Grass			Pig manure			Rice bran		
	Mean	Median	*n*	Mean	Median	*n*	Mean	Median	*n*
Lowland	91.0	27.8	29	15.6	15.6	35	6.8	1.4	33
Intensive	53.0	17.0	60	15.2	9.3	76	3.1	0.9	55
Suburban	33.1	13.9	15	6.5	2.8	19	0.6	0.4	11
Total	60.7	18.5	104	14.0	6.6	130	4.0	0.9	99

Fish production

Fish yields in VAC system ponds ranged between < 0.1 and 6.7 with an average yield of 2.6t ha^{-1} per season which extended for about 9 months, from April to December. The highest average fish yield was recorded in the lowland zone (3.4t ha^{-1}) followed by intensive (2.5t ha^{-1}) and suburban areas (1.9t ha^{-1}) (Table 5.7). Variation in fish yield across AEZ was apparently related to variation in stocking strategy and pattern of input use (Tables 5.4, 5.5 and 5.6). Compared to intensive and suburban zones, farmers in the lowland zone stocked relatively larger sized fingerlings at higher densities and applied an adequate amount of pond inputs, which was reflected in higher fish yields in the lowland zone. There was a significant relationship between stocking density and the fish yield ($P < 0.001$) in VAC farm ponds (Fig. 5.3).

Table 5.7. Fish production (kg ha^{-1} year^{-1}) in VAC system ponds by agro-ecological zone (AEZ) in 1994. *n* = number of farming households.

AEZ	Mean (fish m^{-2})	Median	Range	
			Minimum	Maximum
Lowland (*n* = 40)	3,395	3,783	100	6,722
Intensive (*n* = 90)	2,477	1,876	33	6,349
Suburban (*n* = 21)	1,904	1,196	147	4,944
Total (*n* = 151)	2,641	1,980	33	6,722

Costs and benefits of fish production

Average gross returns from fish production were estimated at around US$1,700ha^{-1} per season (Table 5.8). However, there was variation across AEZs in line with those of stocking and feeding strategies (Tables 5.4, 5.5 and 5.6). Thus, gross returns from fish production were higher in the lowland zone followed by the intensive and the suburban zones. The gross returns from fish production in the lowland zone were about 50 and 100% higher than those of intensive and suburban zones, respectively.

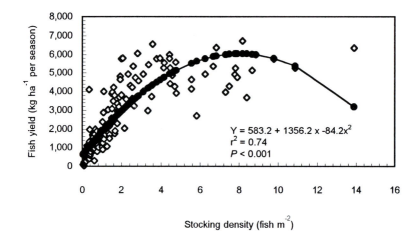

Fig. 5.3. Relationship between fish stocking density and fish production in VAC systems in 1994. ◇ = observed; ● = predicted values.

Economic analysis also showed that cost benefit ratios (C/B) of fish production in the VAC system also varied with AEZ. However, higher gross benefit from fish production was not necessarily associated with higher net returns. Although gross benefit of fish production was highest in the lowland zone, the C/B ratio was only 1 : 1.2. In contrast, although the lowest gross return from fish production was in the suburban zone, the C/B ratio of 1 : 2.1 was the highest net benefit among all the AEZs. Cost of production apparently increased in the lowland zone due to application of an excessive amount of rice bran (Table 5.6), relatively costly off-farm input used also as livestock feed. Buouhahom *et al.* (1993) found that despite a high fish yield of 9t ha^{-1} in the Red River delta, the C/B ratio of fish production was rather low (1.0 : 1.3) due to a high input cost. Hence, the baseline survey suggested the need for more judicious use of pond inputs to achieve adequate fish yield in an economically viable manner.

Table 5.8. Costs and benefits (US$ ha^{-1} year^{-1}) of fish production in VAC systems by agro-ecological zone (AEZ) in 1994. n = number of farming households.

AEZ	Gross benefit		Net benefit		Cost : benefit
	Mean	Median	Mean	Median	
Lowland (n = 40)	2,062	1,685	311	261	1.0 : 1.2
Intensive (n = 90)	1,655	790	759	290	1.0 : 1.8
Suburban (n = 21)	1,005	609	518	308	1.0 : 2.1
Total (n = 151)	1,673	854	607	297	1.0 : 1.6

VAC Systems after Project Intervention (1996-1998)

The recommendations provided to selected VAC farm households following the baseline survey involved simple changes in the existing resource allocation pattern in their aquaculture system to improve performance without a substantial increase in production costs. It was suggested in particular that rice bran, a relatively costly input, be substituted to a considerable extent by low-cost alternative resources such as grass and manure. In addition, a number of farmers were recommended to apply a modest amount of inorganic fertilisers, e.g. urea and triple superphosphate to ponds. Another change recommended was to introduce multiple stocking. Generally, the first stocking took place between February and March, with the supplementary stocking between July and August. The farmers also stocked larger-sized fingerlings and reduced fish stocking density. Subsequent sections present stocking strategy, input-output, and economic performance of aquaculture systems of VAC systems across AEZs for three consecutive years of on-farm trials (1996-1998).

Species stocked and size of fingerlings at stocking

The five fish species (common carp, grass carp, silver carp, mrigal and rohu) remained dominant as in the baseline survey as 50-60% of project farm households in all zones stocked these species. Virtually all project farmers practiced polyculture over the 3 years. In terms of particular species, grass carp ranked the first regardless of AEZ or year of trial, followed by silver carp. Rohu, mrigal and common carp ranked third, fourth and fifth, respectively (Table 5.9). Tilapia appeared only in 1997 in some household ponds when promoted by the project, while hybrid catfish, which had been promoted earlier, virtually disappeared from the system.

Average stocking density of around 1.5 fish m^{-2} was lower than in the baseline survey (Tables 5.6 and 5.10) and in general did not vary over time across AEZ, indicating that project farmers followed technical recommendations. A slightly higher stocking density was noted in the suburban zone in 1997 but it was reduced to close to the average in 1998.

Table 5.9. Average size (g) of stocked fingerlings of the major fish species in VAC system ponds by agro-ecological zone in 1996-1998. *n* = number of farming households.

Species	1996			1997			1998		
	Mean	Median	*n*	Mean	Median	*n*	Mean	Median	*n*
Common carp									
Lowland	-	-	-	24	14	11	43	20	3
Intensive	-	-	-	50	50	25	50	50	10
Suburban	19	20	4	44	25	3	38	38	4
Total	19	20	4	42	25	39	46	50	17
Grass carp									
Lowland	-	-	-	53	50	18	85	100	17
Intensive	-	-	-	54	47	34	71	66	23
Suburban	40	20	4	59	50	5	71	75	5
Total	40	20	4	55	50	57	74	67	45
Silver carp									
Lowland	-	-	-	40	25	17	35	15	4
Intensive	-	-	-	42	40	33	40	50	24
Suburban	32	25	3	31	30	5	66	50	5
Total	32	25	3	40	40	55	43	50	33
Mrigal									
Lowland	-	-	-	26	14	14	20	-	1
Intensive	-	-	-	37	28	32	29	23	16
Suburban	17	20	3	30	-	1	-	-	-
Total	17	20	3	34	25	47	29	20	17
Rohu									
Lowland	25	-	1	33	27	12	15	15	4
Intensive	-	-	-	37	30	31	43	32	22
Suburban	10	-	1	30	25	5	40	50	5
Total	17	-	2	36	28	48	39	30	31

Size of stocked fingerlings showed a slight variation with species but not with AEZ. A comparison of fingerlings between 1997 and 1998 showed some variation as most project farmers stocked slightly larger fingerlings in 1998 than in 1997. One notable change between baseline and trials was a decrease in the size of grass carp fingerlings although for other species fingerlings size remained similar to that of the baseline survey (Table 5.9).

Table 5.10. Average stocking density of fish (number m^{-2}) in VAC project farms by agro-ecological zone (AEZ) in 1997/98. *n* = number of farming households.

AEZ	1997			1998		
	Number m^{-2}	Median	*n*	Number m^{-2}	Median	*n*
Lowland	1.5	1.5	16	1.7	1.5	6
Intensive	1.5	1.2	32	1.4	1.2	23
Suburban	2.3	2.0	5	1.9	1.5	5
Total	1.6	1.4	53	1.5	1.5	34

Major pond inputs

Grass, pig manure and rice bran remained major pond inputs of trialist farm households as in the baseline survey in all AEZs (Table 5.11). A number of farm households also supplemented pond inputs with urea. In general, all types of pond inputs were applied in relatively larger amounts than those recorded in the baseline survey. Variation in input use pattern across AEZ was also noted but in contrast to lowland and intensive zones, data on pond inputs in the suburban zone were available from only a few project farm households. Hence, interpretation of input use pattern for this zone could not be made.

In general, an increasing number of households applied grass inputs to their ponds and a steady increase in the total amount of grass applied by trialists was noted between 1996 and 1998. Variation in the amount of grass applied was also evident between AEZs with lowland farm households applying a higher amount of grass than intensive zone households (Table 5.10). The average amount of grass input in the lowland zone in 1996 was estimated at 65t ha^{-1} per season; it declined slightly in 1997 but increased remarkably in 1998, reaching nearly 100t ha^{-1} per season. Whilst farm households in the intensive zone had applied a fairly low quantity of grass (average 20t ha^{-1}) as a pond input in 1996, it increased markedly in subsequent years with an average amount of grass applied of over 50t ha^{-1}.

The most apparent change was that farm households began to use more organic manure and reduced the use of rice bran (Table 5.11). On average, farmers applied around 13t ha^{-1} of manure in 1996, which increased to 20t ha^{-1} in 1997 and nearly double the 1996 level by 1998. Such changes were most notable in the lowland zone. The amount of pig manure applied in this zone was estimated at around 10t ha^{-1} in 1996, it increased slightly in 1997 but increased over fivefold in 1998. Unlike the lowland zone, the amount of pig manure applied in the intensive zone did not follow a clear trend. The trialist farm households in this zone applied pig manure at an average rate of 9t ha^{-1} in

1996, which doubled in 1997 but was reduced to 14t ha^{-1} in 1998. As there was abundant manure in the intensive zone, an area rich in livestock, this sharp decline in the use of manure as a pond input from 1997 to 1998 might have been associated with the farmers' attempts to make appropriate adjustments in input use.

Table 5.11. Major pond inputs applied in VAC system ponds by agro-ecological zone (AEZ) in 1996-1998. Quantities are fresh weights. *n* = number of farming households.

AEZ		Grass (t ha^{-1})			Pig manure(t ha^{-1})			Rice bran (t ha^{-1})		
		1996	1997	1998	1996	1997	1998	1996	1997	1998
Lowland	Mean	65.8	59.8	99.7	10.3	14.2	56.3	4.7	3.2	12.0
	Median	55.6	66.7	64.0	4.3	13.1	41.9	4.3	1.5	5.5
	n	5	15	8	7	8	8	5	8	8
Intensive										
	Mean	19.5	52.1	51.3	8.9	21.4	14.3	1.4	2.6	3.3
	Median	15.9	40.6	37.7	8.7	17.5	10.9	1.3	2.1	2.4
	n	4	33	21	3	30	20	7	28	20
Suburban										
	Mean	-	94.1	8.4	18.6	25.7	3.2	11.8	5.9	0.2
	Median	-	129.9	8.4	7.4	23.2	3.2	1.9	5.3	0.2
	n	-	5	2	5	6	2	3	6	2
Total										
	Mean	45.2	58.2	61.0	12.8	20.7	24.7	4.6	3.2	5.4
	Median	40.7	47.2	38.5	6.9	17.1	11.8	1.5	2.4	2.4
	n	9	53	31	15	44	30	15	42	30

The pattern of application of rice bran over the period of 1996-1998 also varied with AEZ. In general, VAC farm households in the lowland zone applied rice bran at the highest rate. In contrast, the amount of rice bran applied in the intensive zone was relatively low. Furthermore, year-to-year variation in the amount of rice bran applied was also evident within a particular AEZ. Lowland farmers reduced the amount of rice bran to 3t ha^{-1} in 1997 from around 5t ha^{-1} in 1996 but in 1998 it increased markedly to 12t ha^{-1}. In the intensive zone, there was a steady increase in the amount of rice bran applied between 1996 and 1998. The amount of rice bran applied in 1996 was a mere 1.4t ha^{-1} but it nearly doubled in 1997 to 2.6t ha^{-1} and further increased to 3.3t ha^{-1} in 1998.

Although urea was recommended pond input, only a few farmers applied it in ponds, with year-to-year variation in the number of households using it. In 1996 and 1998, only a few farm households supplemented their pond inputs with urea, irrespective of AEZ. In contrast, a much larger number of households in both lowland and intensive zones applied a modest amount of urea in their ponds in 1997. The average amount of urea applied was estimated to be over 200kg ha^{-1}. Only a single farm household in the suburban zone applied urea in a pond.

Fish production

In line with the large increase in pond inputs in successive years of project intervention, fish yield also showed a steady upward trend in all AEZs. From around 2.5t ha^{-1} in 1994, yields increased to 3.9t ha^{-1} in 1996, to 4.5t ha^{-1} in 1997, and reached 5.0t ha^{-1} 1998 (Table 5.12). However, fish yields were relatively low in the lowland zone and much higher in the suburban zone in all years.

Table 5.12. Average fish yield (kg ha^{-1} per season) and C/B of fish production (US$ ha^{-1}) in VAC ponds by agro-ecological zone in 1996-1998. *n* = number of farming households.

	1996			1997			1998		
	Mean	Median	*n*	Mean	Median	*n*	Mean	Median	*n*
Fish yields									
Lowland	3,050	2,778	7	4,205	3,713	16	5,083	5,123	10
Intensive	3,793	4,058	13	4,166	4,167	32	4,235	3,984	24
Suburban	4,941	4,895	8	7,023	6,215	6	7,854	6,981	6
Total	3,935	3,199	28	4,506	4,167	54	4,990	4,639	40
Gross benefit									
Lowland	2,668	1,956	9	2,574	2,373	18	3,978	3,720	10
Intensive	4,024	2,745	14	2,932	2,778	33	2,802	2,656	24
Suburban	3,529	3,068	8	4,736	5,231	6	4,489	4,490	6
Total	3,503	2,706	31	3,008	2,744	57	3,349	3,016	40
Gross cost									
Lowland	1,662	1,215	9	922	902	18	1,531	1,253	10
Intensive	1,382	805	14	1,011	1,012	33	1,041	723	24
Suburban	1,599	1,035	8	2,790	2,388	6	1,567	1,406	6
Total	1,519	1,032	31	1,170	1,051	57	1,243	939	40
Cost : benefit ratio									
Lowland		1.0 : 1.6			1.0 : 2.8			1.0 : 2.6	
Intensive		1.0 : 2.9			1.0 : 2.9			1.0 : 2.7	
Suburban		1.0 : 2.2			1.0 : 1.7			1.0 : 2.9	
Total		1.0 : 2.3			1.0 : 2.6			1.0 : 2.7	

There was also agro-ecological zonal variation in fish yield. In the lowland zone, a linear pattern of increase in fish yields was noted where yields were estimated to be 3.1, 4.2, and 5.1t ha^{-1} in 1996, 1997, and 1998, respectively. In the intensive zone, there was limited variation with mean yields of 4.1 to 4.2t ha^{-1}. Although the suburban zone had the lowest yields in 1994, after project intervention there was a remarkable increase in mean fish yields, which reached the highest levels among the three zones. There was a remarkable increase from 1996 to 1997 from 4.9 to 7.0t ha^{-1}, with a further modest increase in 1998 to 7.9t ha^{-1} (Table 5.12).

The impressive yields in the suburban zone realised after project intervention were probably associated with a combination of factors: stocking larger fingerlings, relatively high stocking densities and an adequate amount of pond inputs (Tables 5.9, 5.10 and 5.11). Although the average amounts of pond inputs applied in lowland and intensive zones were comparable to those in the suburban zone, stocking densities were relatively low in the former two zones resulting in lower yields. An appropriate balance between stocking density and amount of pond inputs is important for the improved performance of VAC systems. The baseline survey results also showed a strong correlation between stocking densities and fish yields (Fig. 5.3).

Gross return and benefits from fish production in VAC systems

Costs and benefits of fish production in VAC systems assessed through a partial economic analysis using data for the 3-year period (1996-1998) are presented in Table 5.13. Gross benefits ranged between US$3,000 and 3,500ha^{-1}, at least twice the gross returns in 1994 (Tables 5.8 and 5.12) due mainly to a substantial increase in fish yields after project intervention.

Year-to-year variation in gross benefits was also evident in all AEZs. In the lowland zone, there was a slight decline in gross benefits in 1997 but a remarkable increase in 1998. Similar trends of gross benefits were noted in the suburban zone. In contrast, a continuous decline in gross return, a rather surprising trend as there was a steady increase in fish yields over a period of three years (Table 5.12), occurred in the intensive zone. The reason for this inverse relationship between fish yield and gross benefit might have been due to the decline in fish prices in the local market.

Higher gross benefits were often associated with smaller C/B ratios, and *vice versa*. Apparently, higher fish yields were achieved at the cost of increased amounts of pond inputs, which inevitably increased the gross costs. Although types and amounts of inputs applied were the major factors influencing the C/B ratio of fish production (Table 5.12), the relationship between input use and C/B ratios was rather uncertain.

Considering agro-ecological zonal variation, the C/B ratio in the intensive zone remained fairly stable. In the lowland zone, the C/B ratio sharply increased in 1997 but declined slightly in 1998. In contrast, a slight decline in

C/B ratio was noted in 1997 in the suburban zone although it increased in 1998. Clearly, the VAC farm households, following project intervention, were attempting to make appropriate adjustments in types and amounts of input use, which has important consequences in year-to-year variation in C/B ratios.

Table 5.13. Fish yields (kg ha^{-1} per season) and C/B (US\$ ha^{-1}) of tilapia-based culture systems in VAC ponds by agro-ecological zones in 1997/98. n = number of farming households.

	1997			1998		
	Mean	Median	n	Mean	Median	n
Fish yields						
Lowland	320	-	1	7,273	-	1
Intensive	5,929	5,952	9	4,150	3,984	6
Suburban	7,129	5,934	5	6,964	5,060	4
Total	5,955	5,934	15	5,457	4,409	11
Gross benefit						
Lowland	141	-	1	4,091	-	1
Intensive	4,408	3,968	9	2,708	2,656	6
Suburban	4,599	5,201	5	3,605	3,510	4
Total	4,187	3,968	15	3,160	3,038	11
Gross Cost						
Lowland	53	-	1	2,250	-	1
Intensive	1,399	1,145	9	918	816	6
Suburban	2,735	1,999	5	1,428	1,406	4
Total	1,755	1,573	15	1,225	1,276	11
Cost : benefit ratio						
Lowland		1 : 2.7			1 : 1.8	
Intensive		1 : 3.2			1 : 2.9	
Suburban		1 : 1.7			1 : 2.5	
Total		1 : 2.4			1 : 2.6	

The type of culture system also influenced fish yields and C/B ratios of fish production. Although aquaculture in VAC was predominately a carp-based polyculture system, tilapia was introduced from 1997 in selected trialist farm households, particularly in the intensive and suburban zones. Therefore, a separate analyses of households stocking tilapia (monoculture/polyculture with carp) (Table 5.13) and carp (Table 5.14) was performed using 1997 and 1998 data. Fish yields and C/B ratios in the tilapia-based systems were higher than the carp-based systems (Tables 5.13 and 5.14). Furthermore, agro-ecological variations in yield and C/B ratios were also evident. In spite of relatively low fish yields in the tilapia-based system, C/B ratios were apparently high in the intensive zone (Tables 5.13 and 5.14). However, in the suburban zone, fish yields were estimated to be higher but the C/B ratios were relatively low. The C/B ratio of the tilapia-based system in the lowland zone was smaller compared to the overall average C/B ratios for that zone; however, only one farmer stocked tilapia in 1997 and 1998 in the lowland zone and yields were low possibly due to inadequate experience of the farmer in fertilising ponds to produce plankton-rich, green water.

Table 5.14. Fish yields (kg ha^{-1} per season) and C/B (US\$ ha^{-1}) of carp-based culture systems in by agro-ecological zone in VAC ponds in 1997/98. n = number of farming households.

	1997			1998		
	Mean	Median	n	Mean	Median	n
Fish yields						
Lowland	4,464	3,953	15	4,839	5,079	9
Intensive	3,476	3,333	23	4,265	3,819	18
Suburban	6,495	-	1	9,634	9,633	2
Total	3,934	3,659	39	4,813	5,000	29
Gross benefit						
Lowland	2,717	2,430	17	3,965	3,472	9
Intensive	2,378	2,179	24	2,834	2,625	18
Suburban	5,420	-	1	6,257	6,257	2
Total	2,588	2,373	42	3,421	2,995	29
Gross Cost						
Lowland	973	984	17	1,451	1,157	9
Intensive	865	755	24	1,082	718	18
Suburban	3,063	-	1	1,845	1,845	2
Total	961	893	42	1,249	833	29
Cost : benefit ratio						
Lowland		1 : 2.8			1 : 2.7	
Intensive		1 : 2.7			1 : 2.6	
Suburban		1 : 1.8			1 : 3.4	
Total		1 : 2.7			1 : 2.7	

Hence, a switch to monoculture of tilapia in ponds fertilized with manure and inorganic fertilisers can be foreseen as a viable alternative to carp polyculture in intensive zones. For the other two zones, a further assessment of the economic viability of tilapia-based over carp-based systems is required.

Trends and Constraints in Aquaculture Development

The results of the on-farm trials carried out on VAC systems of small-scale farmers in the Red River delta indicated the significant positive contribution of aquaculture to rural livelihoods. The on-farm trials clearly demonstrated the improved performance of aquaculture following introduction of Project recommendations. With adoption of improved technologies, an increase of between 50 and 80% in investment costs, mostly in the form of available on-farm inputs, produced two or three times the previous returns in two of the three zones covered in the on-farm trials. This obvious benefit provoked a good deal of interest among the neighbours of the trialists in communes in which trials took place, and many followed their lead spontaneously without direct Project assistance.

Most of the conditions for the successful promotion of small-scale aquaculture exist in the Red River delta of northern Vietnam. The basic technology package is available, although it must be adapted to the specific conditions of farmers. Basic infrastructure in the form of seed supplies, road transportation, and market are adequately developed, and the Government of Vietnam has shown significant commitment to the promotion of aquaculture both for family consumption and for market.

However, a number of constraints still need to be solved. As farmers seek to intensify, it is necessary to monitor carefully the process. Comparison of the results of the on-farm trials conducted in 1996, 1997 and 1998 indicate that there can be diminishing returns in investing steadily increasing levels of inputs into current systems. A farmer review workshop held early in 1998 featured considerable discussion among trialist farm households about the levels of inputs of grass and other vegetable by-products in the grass carp dominated system. There was at least one case of mass mortality of grass carp due to poor water quality caused by excessively high organic matter loading rate of grass in the pond. Several farmers were concerned about the possible spread of red-spot disease to their systems, the incidence of which was reported to have increased recently, possibly due to intensification of production. Because most farmers had followed recommendations and adequately prepared their ponds, they themselves had not suffered, but others in their community had. Work needs to be done on the limits to intensification and to promote good husbandry measures for disease prevention. It appears that the productivity of the grass carp system itself has reached its upper limit in small ponds in the Red River delta, and a switch to a tilapia-based system in fertilised ponds perhaps would be a better option for further intensification. As reported in the workshop, quite a number of farm households have already started stocking tilapia since the 1997 trials. Results of the trials over 2 years (1997/98) clearly suggested that a tilapia-based system can be promoted as an alternative to existing carp-based culture systems in the intensive zone. However, for the other two zones, further assessment is required before making any recommendations on switching to tilapia-based systems.

Other constraints to aquaculture development lie in the socio-economic sphere. Intensification of aquaculture systems may result in a surplus beyond local needs. Therefore, knowledge of the market becomes important, including the development of strategies to release production at times when prices are favourable, the relative prices of different species and cost-return trade-offs between early supply of relatively expensive seed and later stocking of cheaper fingerlings. This is a particular issue in the expansion of tilapia production since North Vietnam's cold winters make it more difficult to produce early fingerlings than is the case with carp. The market issue also has an infrastructural and organisational dimension in the ability of local fish trading networks to deal with transport of fresh fish to more distant urban markets.

Finally, there is the question of the ability of the extension system in Vietnam to disseminate the successful results of on-farm trials such as those carried out by the Project to a wider spectrum of farmers. Vietnam's national agricultural extension system was only established in 1993 and is based within the Ministry of Agriculture and Rural Development's provincial and district offices. Development of aquaculture is based within a separate ministry, the Ministry of Fisheries, which does not have its own provincial offices in inland provinces. The relationship between the two needs to be clarified and then training/retraining of provincial and district staff undertaken, not only in technical issues in aquaculture, but also in development approaches. The RIA-1/AIT Aqua Outreach programme has now begun this effort with provincial staff in Red River delta provinces, seeking to pass over responsibilities for the whole process of on-farm trials and extension to them. In some cases, this effort will need to be linked to the provision of credit if farmers currently unable to invest in ponds are to be given the opportunity to benefit from aquaculture.

References

Buouhahom, B., Phat, B.H., Ha, C.V., Thu, L.V., Lanh, L.V. and Thao, P.X. (1993) The role of livestock in the agroecosystem. In: Cuc, L.T. and Rambo, A.T. (eds) *Too Many Peoples, Too Little Land: the Human Ecology of a Wet Rice-growing Village in Red River Delta of Vietnam.* East-West Center, Honolulu, pp. 117-129.

Chevy, P. and Lemasson, J. (1937) *Contribution à L'étude des Poissons des Eaux Douces Tonkinoises.* Institute Océanographique de L' Indochine. 33e Note, Gouvernement Général de L'Indochine, Hanoi. 182 pp. In French.

Cuc, L.T. and Vien, T.D. (1993) An overview of the Red River delta environment. In: Cuc, L.T. and Rambo, A.T. (eds) *Too Many Peoples, Too Little Land: the Human Ecology of a Wet Rice-growing Village in Red River Delta of Vietnam*. East-West Center, Honolulu, pp. 1-10.

Luu, L.T. (1992) The VAC System in Northern Vietnam. In: *Farmer-proven Integrated Agriculture-Aquaculture: a Technology Information Kit*. IIRR, Manila. 4 pp.

Luu, L.T., Khan, N.G. and Hai, N.V. (1992) *Vietnam Household Food Security Project*. Report on the evaluation mission, UNICEF, Hanoi.

Pham, X.N. (1994) The VAC ecosystem: a model for rural sustainable development and environmental protection in Vietnam. *Enfo News* 6-8 December.

Ministry of Fisheries (1999) Aquaculture development program in the period 1999–2010 (non-authorized translated version). Ministry of Fisheries, Hanoi.

Rambo, A.T. and Cuc, L.T. (1993) Prospects for sustainable development. In: Cuc, L.T. and Rambo, A.T. (eds) *Too Many Peoples, Too Little Land: the Human Ecology of a Wet Rice-growing Village in Red River Delta of Vietnam*. East-West Center, Honolulu, pp. 165-186.

Uehara, G. and Patanothai, A. (1996) Potential for further increases in agricultural production. In: Patanothai, A. (ed.) *Soils Under Stress: Nutrient Recycling and Agricultural Sustainability in the Red River Delta of Northern Vietnam*. East-West Center, Honolulu, pp. 91-96.

Chapter 6

Eco-technological Analysis of Fish Farming Households in the Mekong Delta of Vietnam

F. Pekar[1], N.V. Be[2], D.N. Long[3], N.V. Cong[2], D.T. Dung[2] and J. Olah[1]

[1]*Fish Culture Research Institute, Szarvas, Hungary*
[2]*Department of Environment and Natural Resources Management, College of Agriculture, Cantho University, Cantho, Vietnam*
[3]*Department of Freshwater Aquaculture, College of Agriculture, Cantho University, Cantho, Vietnam*

Abstract

An eco-technological survey was conducted in the central freshwater region of the Mekong Delta involving around 260 fish farming households with different types of farming systems. The survey was carried out during the dry season of 1996 to describe and analyse the present state, structure and functioning, as well as constraints to, and management options of, integrated fish farming systems. Systems for the survey were selected to represent as wide a spectrum of farming conditions as possible among the high variety and combination of integrated fish farming systems existing in the central region of the Mekong Delta. Two basic types of integration were selected for the survey: AC/VAC systems of fishpond-livestock with/without a vegetable and/or fruit-tree garden, AR system of rice-fish, and a non-integrated system, namely fish culture in ponds and/or garden canals (A) was also investigated. Farmer-managed trials were conducted on 25 selected farms during the culture period of 1997-1998 to test improved technologies based on the results of the survey by partial modification of the limiting technological parameters and management methods.

Introduction

To support sustainable aquaculture development in the Mekong Delta of

Vietnam, the West-East-South Project (WES Project) started in September 1994, financially supported by the Netherlands' Government. The project was implemented by the Fish Culture Research Institute (HAKI), Szarvas, Hungary, mainly through the Fisheries Departments of the College of Agriculture at Cantho University, and in close collaboration with AIT and MRC, Bangkok, Thailand and ICLARM, Manila, Philippines. The ultimate objective of the project was to support rural households in the southern part of Vietnam to improve and diversify production, nutrition and income sources through integration of aquaculture practices in their farming systems. The immediate objectives were to strengthen and upgrade the educational, adaptive research and extension capacity and capability in aquaculture in the southern part of Vietnam; to improve access to current knowledge and experience in aquaculture development; and to increase institutional cooperation between national partners and international development organisations involved in aquaculture. In view of the complexity of factors influencing the sustainability of farming systems, the project adopted a multidisciplinary, problem-oriented approach towards aquaculture development.

One of the key elements of the success of aquaculture development is sound knowledge of the structure, functioning and management of the target agro-ecosystems. A comprehensive survey on the eco-technological and socio-economic situation of fish farming systems was considered as one the most important activities of the WES Project since such data were not readily available for the Mekong Delta. The survey provided information on the constraints for aquaculture development and for the planning of research and extension activities.

Fish are cultured in a wide range of farming systems in the Mekong Delta, and aquaculture is often integrated with other farm activities. In many cases fish culture is only a secondary or complementary activity to the production of rice, field crops, fruit, poultry and livestock farming. Since it is practically impossible to analyse all of these diverse farming systems in detail, typical aquaculture systems with a significant role and future importance in the aquaculture development of the Mekong Delta were selected for baseline survey and analysis. In parallel with the eco-technological survey, a socio-economic survey of the same fish farming households was also undertaken to increase understanding of the rural agricultural sector, and identify constraints to, and opportunities for, the development of integrated fish farming systems (WES Project Report, 1996-1997). After the completion of the baseline surveys, farmer-managed trials were conducted on selected farms to test and evaluate improved technologies by applying the results of the surveys.

Eco-technology may be described as an applied discipline emerging from rapidly developing quantitative ecology and stimulated by technological misuse and abuse in resource consumption. Its main function is to develop and spread more ecologically sound practice in resource consumption, keeping in mind the key concept that human beings are a part of nature rather than separate from nature (Mitsch and Jorgensen, 1989). This definition dates back to the 1960s. In our perception, eco-technology is an interdisciplinary science, which integrates the environment and human production activities. While ecological quantification measures the structural components and the functional processes of ecosystems, eco-technological quantification additionally quantifies technological parameters and processes in the resource consumption process. Thus, eco-technology is a merger of ecology and technology; it is the science of ecological resource consumption (WES Project Report, 1996-1997). Similar to the integrative nature of ecology, eco-technology applies the principles of all those structural and operational disciplines, which constitute its basic science, ecology, and it also integrates all of the achievements of the technologies. Therefore, eco-technology is a promising approach to quantify the environmental sustainability of aquaculture technologies. This quantification is based on the following principles:

- It has to be carried out, as much as possible on-farm by personal or group interviews, and by *in situ* field measurements.
- Only the most indispensable parameters could be included from those that need sampling and further laboratory work.
- The simplest field methods and procedures, as well as approaches of quantification, have to be adopted.
- In spite of the principles of on-farm survey and simplest methodology, all those parameters necessary to quantify the metabolism of the integrated systems have to be included.
- It should be accepted also that some of the metabolic parameters could only be estimated approximately.

This paper describes the structure, management and functioning of the most important types of aquaculture production systems of the Mekong Delta, combining the results of the eco-technological analysis of the fish farming households surveyed and farmer-managed trials conducted.

Methods

The fish farming systems surveyed were selected from four districts of two provinces of the target region of the WES Project, namely Omon and Phunghiep districts in Cantho province, and Tambinh and Longho districts in Vinhlong province. These systems represented as a wide spectrum of farming conditions as possible among the large variety and combinations of integrated fish farming systems existing in the central region of the Mekong Delta.

Two basic types of integration were selected for the survey: the AC/VAC system of fishpond-livestock with/without a vegetable and/or fruit-tree garden; the AR system of rice-fish; and a non-integrated system, namely fish culture in ponds and/or garden canals (A) was also investigated. The name 'AC' originates from the Vietnamese words '*ao*' (fish pond) and '*chuong*' (animal house). There is an additional production component in the 'VAC' systems indicated by the initial letter of a Vietnamese word '*vuong*' which means garden. The name of the integrated system locally called 'AR' originated from the Vietnamese words '*ao*' (fish pond) and '*ruong*' (rice field).

Pre-tested questionnaires and field data sheets were used during the implementation of the detailed survey. The survey was organised at the end of the dry season of 1996 from mid-March to the beginning of May. At least 20 questionnaires and field data sheets for each farming system were completed from each of the four studied districts of the two provinces. Therefore, data were collected on a total of 262 farms randomly selected with the help of agricultural extension workers at district level (Table 6.1). The survey parameters were as follows: farm size and total water area; pond age and fish culture experience; pond morphometry; water supply system; pond preparation; fish stocking, growth, production and survival rates; nutrient applications; fish feeding; water quality; and natural fish food resources.

Table 6.1. Numbers and types of farming systems surveyed.

District/province	Farming system			
	A[1]	AC/VAC[2]	AR[3]	Total (1 + 2 + 3)
Omon/Cantho	23	23	18	64
Phunghiep/Cantho	20	19	23	62
Longho/Vinhlong	20	23	24	67
Tambinh/Vinhlong	23	24	22	69
Total	86	89	87	262

[1] Fish culture in ponds and/or garden canals - non-integrated fish farming systems.
[2] Fishpond-livestock with/without vegetable and/or fruit-tree garden - integrated fish farming systems.
[3] Fish-cum-rice culture - integrated fish farming systems.

After the completion and evaluation of the eco-technological survey, 25 fish farms were selected in the above provinces for farmer-managed trials to test improved production technologies. Each type of aquaculture production system of the survey was represented in the farmer-managed trials as follows: 5 A, 11 AC, 3 VAC and 6 AR systems. During the whole culture period of 1997-1998, besides the respective technological parameters, the following pond ecological parameters were measured in the production systems at least once a month: chemical environment, water nutrient status, suspended solids, sediment nutrient status, chlorophyll-a and primary production.

Results and Discussion

Structure of the Production Systems

Fishpond age and culture experience

The components of the basic physical structure of the fish pond agro-ecosystems, like pond morphometry, dike system, water and sediment depth, depend on the local natural and infrastructure resources, as well as on the tradition and history of culture practice. The Mekong Delta had been well known as a region rich in aquatic resources and with a high potential for aquaculture development. For a long time fish abundance in its natural waters limited interest in farming, although fish culture dates back centuries. Initially, fishery activities were trapping and holding of wild seeds brought into ponds by tidal water, but a yield decrease in capture fisheries motivated a more rapid development of aquaculture over the last decade. Intensification of aquaculture started during the first half of the 1980s.

The age of the surveyed ponds clearly showed the development trend of fish farming practices in the Delta. A total of 262 ponds were surveyed for age in the four districts and 117 ponds were older than 7 years but younger than 12 years, which meant that the majority had been constructed and culture started in the 1980s. Seventy-six ponds had an age of between 4 and 6 years and 69 ponds had an age of less than 3 years. This indicated growing recent interest in fish farming activity. The age structure was different in the various farming systems. Fish farms without integration (system A) had the oldest ponds and almost 60% were built more than 7 years ago. The percentage was only 40 in integration of fish ponds with animals (AC, VAC) and 35 in integration of fish culture in rice fields (AR).

Altogether, 262 households of integrated fish farming were surveyed for farmers' experience in fish culture. Farmers of fish pond and garden integration had a culture experience of 6.1 ± 5.2 years, with integration of fish ponds and animals 4.6 ± 4.4 years, and fish culture in rice fields 4.2 ± 3.4 years.

Pond morphometry

Among the surveyed ponds, more than 90% had rectangular and polygonal shapes and the remaining were square, oval or irregular. Almost the same shape distribution was reported among rural, undrainable fishponds in Orissa State, India (Olah *et al.*, 1990). However, irregular shaped ponds were 16.7% in Orissa due to the natural origin of many of these rural ponds. In the Mekong Delta the ponds were constructed and the rectangular shape was the most practical for management.

The size of the fish rearing units varied in a very wide range from 30 to 10,000m^2 but the average was 200-300m^2 in the A and AC farming systems. The pond size in the same farming systems was higher in Cantho province. The garden canal size in the VAC system was higher in Vinhlong province. The average size of this farming system was smaller than that of the A and AC systems. The actual fish rearing unit, i.e. the ditch area in the AR rice-fish farming system was rather significant, reaching 20-25% in 28% of the surveyed 87 farming systems. These farming systems also had a larger total size with an average of some thousand m^2. The small size of fish farming systems in the Delta restricted the possibilities of small-scale fish culture activity. This may also be explained by the overall small area of land owned by farmers, even in the case of joint properties. Shortage in capital investment may also have contributed to the pond size distribution.

The larger ditch area in ricefields gave more space for fish but less area for rice production. It seemed that the ratio between ditch and total ricefield area depended on the quality of soil. Normally, this ratio was 15-20% but it increased where soil quality was poor because rice yields were usually low in this area. Farmers wished to expand the water area in their ricefield for fish culture to produce more fish for family consumption and to earn cash to compensate for low rice production. On the other hand, where the soil quality was poor due to acid sulphate soil, larger ditches were built to wash out the low pH and high iron content of the water, and to supply good quality water to improve soil quality.

The water depth in the fish farming systems depended primarily on the amount of rainfall in the dry and wet seasons and on the tidal regime, as well as on the geographical characteristics of the area and the water management practice of the farmers. It was generally accepted that the most suitable water depth for fish culture in the Mekong Delta was 0.8 to 1.5m. If water depth were below 50cm, the cultured fish may have suffered from high water temperature during summer, and yellow coloured turbid water would persist, maintaining a decreased rate of primary and oxygen production. The overall result of shallow water was a very serious growth-retarding and stressful effect. On the other hand, if water were too deep, the sediment received less oxygen, reduced chemical compounds accumulated and anaerobic conditions may have developed at the sediment-water interface. These conditions would also result in a growth-retarding effect. There was no depth-induced stress effect in the surveyed farms as average water depths ranged from 0.75 to 1.25m in farming system A and from 0.95 to 1.20m in farming systems AC/VAC. However, the water depth of the ditches in the ricefield was rather low at 0.65-0.70m.

The quality and quantity of sediment depended on several factors. The maternal soil played a decisive role, determining the texture and chemistry of the sediments. The method of pond construction, nature and architecture of the embankment, the macrophyte cover, the bottom relief, the pond productivity and the percentage of bottom feeding fish stocked which stirred up sediments, all contributed to the nature of the sediment and its thickness. In older ponds, beside the soft sediment, a solid sediment layer was also present as a result of diagenesis and fossilisation of the nutrients locked in this layer. The total thickness of the soft plus solid sediment layers had a direct relationship with the age of the fish ponds. A thick sediment rich in nutrients, anaerobic conditions with very slow bacterial decomposition and mineral cycling needs to be properly managed for fish culture. The average of the soft and hard sediments in A, AC/VAC and AR farming systems ranged from 27 to 43 and from 34 to 87cm, respectively. The soft sediment layer was thickest in the AC/VAC system because organic matter from animal husbandry was continuously supplied to the ponds. Less organic matter was supplied to the ponds in other farming systems so their soft sediment layer was not thick.

Water supply

Fish ponds in the Mekong Delta are located near to public canals due to the dense river and canal network in the region. The advantage is that water can be supplied to the ponds by gravity with the help of tidal fluctuations; but the disadvantage is that effluent water of fish ponds is directly discharged to the public canals without treatment. However, except for some intensively fed *Pangasius* culture, the nutrient supply to the ponds was low; furthermore, frequent water exchange diluted the nutrient content of the effluent.

The close proximity of the water supply increased the hazard level of annual floods. In the Mekong Delta the flood usually occurs in September to October every year. However, sometimes it occurs in August and finishes later in November. Flooding was very common in almost all pond types of farming systems, reaching a high value of more than 30% in September and October. The floods were frequently very severe, damaging dikes and flushing away all the fish stock from the ponds.

The source of water was almost exclusively the surface water of the dense river and canal network. Ditch water was used only in a few farms at the end of the dry season. Pond filling and draining were realised by gravity; pumping was very rare. The water exchange rate was high and the water retention time short (Table 6.2). Every fish rearing pond, paddy and garden canal changed its water every 2-3 days. This was one of the most important features of aquaculture in the Delta. This practice was generally applied in the surveyed farms although such short water retention would have been required only in the case of the highest intensity *Pangasius* culture with an annual production of around 50t ha^{-1}.

Except for the dense and organically loaded *Pangasius* culture, short water retention had a negative impact on culture. It kept the water permanently turbid, preventing light penetration which in turn decreased photosynthetic activity of phytoplankton. The final consequences were depressed primary and secondary production, as well as decreased biological oxygenation of the fish production system. At the same time, however, the turbid river water brought nutrients to the fish ponds and supplied the production processes with nitrogen (N) and phosphorus (P).

Table 6.2. Water exchange rate and water retention time.

Farming system	Province	Water exchange rate (% day[-1])	Water retention time (days)
A[1]	Cantho	36 ± 16	2.78
	Vinhlong	27 ± 16	3.70
AC/VAC[2]	Cantho	37 ± 14	2.70
	Vinhlong	34 ± 16	2.94
AR[3]	Cantho	36 ± 16	2.78
	Vinhlong	31 ± 16	3.23

[1] Fish culture in ponds and/or garden canals – non-integrated fish farming systems.
[2] Fishpond-livestock with/without vegetable and/or fruit-tree garden – integrated fish farming systems.
[3] Fish-cum-rice culture – integrated fish farming systems. ± = standard deviation.

Chemical environment

Three basic types of water colour have been distinguished in the surveyed fish ponds in the Mekong Delta (Table 6.3). The yellow colour was the flooding river water of the Mekong. In the fish ponds, if the retention time were not too short, this silt content of the river water would start to sediment and a green colour would develop. This was the reason why the very low transparency of around 13-14cm of the yellow water did not increase following sedimentation. The inorganic, silt-containing yellow water had been converted into organic, alga-green coloured water during the process of primary production. The transparency of these green ponds was in the range of 20-25cm. The grey coloured fishpond water with a range of transparency of 21-26cm was the result of an organic load or heavy fish stocking, particularly in catfish ponds.

The pH varied in a rather narrow range. It was consistently higher in traditional fish ponds, lower in fish ponds with animal integration and lowest in rice-fish systems. However, it was never high enough to endanger fish health with high pH toxifying the ammonia compartment, or low enough to solubilise the trace metals or toxic gases. The measured total ammonia was higher at the beginning of the rainy season but well below toxic levels especially at the measured pH values. The reduced and oxidised inorganic N compartments, represented by ammonia and nitrate, were well balanced with around the same quantities. Neither inorganic phosphate-P or total-P were ever found in limiting concentrations.

The total suspended solids (TSS), inorganic suspended solids (ISS), suspended organic carbon (SOC) and dissolved organic carbon (DOC) indicated the riverine origin of the water and organic matter input to the fishpond systems. The high ISS were the result of the regular flooding of the Mekong River with its high-adsorbed inorganic nutrient compartments. At the same time the particulate organic compartment (POC) of the SOC was also high, surpassing the values of the DOC compartment.

Table 6.3. Water transparency in yellow, grey and green coloured ponds.

Farming system	Yellow	Grey	Green
A[1]	14.5 ± 3.73 (n = 39)	23.5 ± 6.03 (n = 17)	22.6 ± 5.41 (n = 24)
AC/VAC[2]	13.4 ± 3.21 (n = 24)	19.0 ± 7.81 (n = 24)	20.6 ± 7.55 (n = 21)
AR[3]	15.6 ± 3.86 (n = 34)	26.2 ± 9.72 (n = 18)	25.3 ± 5.68 (n = 20)
All systems	14.6 ± 3.72 (n = 97)	22.5 ± 8.30 (n = 59)	22.8 ± 6.47 (n = 65)

[1] Fish culture in ponds and/or garden canals - non-integrated fish farming systems.
[2] Fishpond-livestock with/without vegetable and/or fruit-tree garden - integrated fish farming systems.
[3] Fish-cum-rice culture - integrated fish farming systems. ± = standard deviation.

The 795,000km^2 watershed of the Mekong River basin is still a relatively well forested landscape. Although Cambodia and Laos retain one of the highest proportions of intact forests in Asia, the situation is not very encouraging as the rate of deforestation is accelerating rapidly. The current forest cover is estimated to be between 49-63% in Cambodia and 36-58% in Laos. It was equally difficult to estimate the annual river silt load to the Delta area. The total run-off of 475,000 million m^3 annually produced a sediment load, which was relatively low compared to other major Himalayan rivers such as the Ganges, Irrawady, Yellow and Yangtze. The Mekong carried a suspended sediment load of 132 million t year^{-1} above the Khone Falls. Thereafter, the load decreased: at Phnom Penh the load quantity was 97.55 million t year^{-1} (Pantulu, 1986). The annual average concentration of the total suspended sediment was 294mg dm^{-3} in the Mekong, 607 in the Irrawady and 1,130 in the Ganges River. However, during the flood period in June to October the total suspended sediment could be around 1g l^{-1}.

The organic carbon (C) content of the silt in the Mekong River was high, around 30-40mg g^{-1} dry silt (Pantulu, 1986), which explained the considerable secondary production of fish food organisms in the river. The total N content of the suspended solid was about ten times less, 3-4mg g^{-1} dry matter. The P content of the Mekong silt fluctuated to a great extent according to the published results. The average value of samples between October 1972 to October 1973 from Vientiane, Phnom Penh, Long Xyen, Can Tho and My Tho was 440μg g^{-1} dry silt (Uhera *et al.*, 1974). However, the average value of samples between September 1968 and June 1969 from Phnom Penh only was as low as $2\mu g\ g^{-1}$ dry silt (Martin and Meybeck, 1979).

Calculating with a 100 million t year^{-1} silt load to the Mekong Delta and applying the above values there would be 3 million t year^{-1} of POC, 0.3 million t year^{-1} total particulate-N and 60 kilotonnes of total particulate-P transported yearly to the Delta. This is a huge natural resource which explains why more than half of the total Vietnamese aquatic resources is concentrated here. Three hundred thousand tonnes of river-transported N fertilises the landscape of the Delta every year. Almost half of the total fishery product, including both capture fisheries and aquaculture, around 240,000t, is produced here in or under the influence of the riverine delta. Wetland forestry, rice and other agricultural products rely also upon the natural resource of the transported river nutrients.

The nutrient compartments in the sediments of the fish farming systems were more stable at the beginning of the monsoon season but fluctuated to a great extent thereafter, reflecting the various management practices and farming systems. The excessively high total organic content in the sediment during the second half of the fish rearing season was the result of feeding trash fish to catfish ponds. Besides this special culture system, there was an overall accumulation of sediment nutrients in most of the systems.

Natural fish food resources

The highest phytoplankton densities were found in A and AC/VAC systems. Euglenophyta was the dominant group, which indicated that water in these systems was very rich in nutrients, especially in organic carbon. In AR systems the density of phytoplankton was not so high because of the high turbidity, low nutrient input and most importantly because of the competitive effect of the rice plant for light and nutrients.

The structure of zooplankton in the aquaculture systems depended on pond management. In all investigated systems, rotatoria had the highest number of species, especially in AC/VAC systems. Because rotatoria are filter feeding animals, they develop well in water bodies which have organic suspended solids. The density of zooplankton depended on the nutrient input and the combination of stocked fish species. In the AC/VAC systems, organic matter was supplied continuously into the ponds and zooplankton received more feed for proliferation and growth, thus having higher density. The density of rotatoria in AR systems was lower than those of AC/VAC and A systems because the organic matter concentration in this system was lower.

The density of zoobenthic organisms living on the bottom of the fishponds and their distribution depended on the structure of soils and on the organic nutrients in the sediments. Bivalvia and polychaeta had very low biomass and gastropods had the highest biomass in all systems. On the other hand, oligochaeta and insect larvae were present in all systems but their biomass was not high because they provide food both for carnivorous and omnivorous fish as well as benthic predators. Therefore, it was difficult for them to maintain a high biomass or dominate these water bodies.

Functioning and Management of the Production Systems

Autochthonous organic carbon production

Besides the allochthonous organic C introduced into the fish farming systems either by food or flooding river water, there was an important renewable natural resource in the fish-producing agro-ecosystem, i.e. the photosynthetic algal primary production. The photosynthesising biomass was quantified by laboratory measurements of the photosynthetic green pigment in the fishpond water. In the first half of the fish growing season, both chlorophyll-a and its first degradation product, phaeophytin-a, were higher. The autochthonous primary production, supported by the significant concentration of photosynthesising pigments, was rather high in all investigated systems.

This process seemed to be an important natural agro-ecosystem function also from an economic point of view. In the investigated A and AC/VAC systems, 3.0-3.5g of organic C m^{-2} day^{-1} primary production was detected. This means that 60-70kg ha^{-1} day^{-1} organic matter was produced naturally on a dry matter basis to substitute the artificial food and save money and resources. Only the system of rice-fish farming produced less organic C (1.6g C m^{-2} day^{-1} on average) because of the shallow water and the competitive effect of the rice plant. Rice is effective at taking up nutrients from sediment soil and reducing the level of these compartments in the pond sediment, which is so important for releasing the nutrients into the overlaying water column and supporting the primary production of the phytoplankton.

Lime application for pond preparation

Liming is common practice in fish culture to increase the rate of bacterial decomposition. The organically rich and reduced nature of the fish pond sediment maintains a much lower pH environment compared to the overlaying water column. Liming increases the pH, accelerates bacterial decomposition, kills pathogens and eradicates predators. However, the frequency and rate of lime application were rather low in the Mekong Delta. Among the surveyed 262 fish farms, only 20 farms were recorded to use lime. The range of lime application was 20-2,500kg ha^{-1} with a low average value of 510kg ha^{-1}. In temperate regions, at an average pH and mineral environment, the suggested range of application is 1,000-2,000kg ha^{-1}. However, the application rate in a low mineral and pH environment, especially in the tropics with a high leaching rate, may reach a value of 4,000kg ha^{-1}.

Fish stock management

About 60-70% of the interviewed farmers stocked fish from March to June because it was the spawning season of fish in the Mekong Delta and seed was available. On the other hand, farmers would like to have stocked fish earlier in the year so that they could harvest fish partially as the floods approached.

The average stocking densities in A, AC/VAC and AR systems were 10.3, 15.4 and 1.8 fish m^{-2}, respectively. Since a stocking density of 3 fish m^{-2} is generally appropriate for most tropical polyculture fish farming systems, the stocking densities applied by the farmers in the Mekong Delta were excessively high. The stocking ratios for the different fish species, according to their importance were: silver barb, 26-34%; common carp, 19-26%; tilapia, 13-21%; *Macrobrachium*, 1-16% (high in AR systems and low in AC/VAC systems); silver carp, 4-8% (high in AR systems, low in A systems); catfish, 1-5% (high in AC/VAC systems; and low in A systems), other species 11-21% (Table 6.4). The dominance of omnivorous species and the low ratio of filter feeding fish species, however, led to insufficient utilisation of the available natural fish food resources, low fish growth and low survival rates.

Table 6.4. Stocking structure (%) in the surveyed fish farming systems.

Stocked species	A[1] (n = 86)	AC/VAC[2] (n = 89)	AR[3] (n = 87)
Silver barb	34.3 ± 29.5	25.5 ± 22.5	31.3 ± 27.6
Common carp	20.7 ± 18.2	18.8 ± 15.7	25.5 ± 24.8
Tilapia	19.4 ± 1.5	21.4 ± 20.7	12.5 ± 8.3
Silver carp	4.1 ± 2.6	5.6 ± 3.3	8.4 ± 7.3
Catfish (*Pangasius*)	1.7 ± 2.3	5.4 ± 3.9	0.2 ± 0.1
Macrobrachium	5.5 ± 2.3	0.9 ± 0.8	16.4 ± 15.4
Other species	14.3 ± 14.1	21.4 ± 20.1	11.1 ± 10.7

[1] Fish culture in ponds and/or garden canals - non-integrated fish farming systems.
[2] Fishpond-livestock with/without vegetable and/or fruit-tree garden - integrated fish farming systems.
[3] Fish-cum-rice culture - integrated fish farming systems. ± = standard deviation.

Fish survival rates were generally low in all systems ranging between 2 and 36% (Table 6.5). Fish growth was not satisfactory, in spite of the substantially decreased density by the end of the culture period. However, recorded fish yields were rather high, 4.7, 11.6 and 0.7t ha^{-1} as averages for A, AC/VAC and AR systems, respectively, due to the extremely high stocking densities (Table 6.6) (WES Project Report, 1996-1997). The low fish survival and growth rates may be explained by the low quality of the stocking material, low nutrient inputs and insufficient fish feeding.

Direct application of inorganic and manure nutrients

Direct application of N and P nutrients stimulates phytoplankton growth and increases the rate of primary production, which is coupled with the life-supporting daily oxygen supply for the fish-producing processes. If we consider that the theoretical upper limit of primary production is around 15g C m^{-2} day^{-1}, this is an important natural resource to explore. However, in the surveyed 262 fish ponds, the rate of inorganic fertilisation was extremely low and ranged only between 9 and 82kg ha^{-1} $year^{-1}$. Only one farm applied superphosphate and one other farm applied diammonium monophosphate. Urea application was the most common practice, but at very low dosage. The yearly application of N fertiliser was only around 3-30kg N ha^{-1}. In the temperate zone with a 140 day production cycle, the optimal dose of inorganic N fertilisation with high-density polyculture stocking is 150kg N ha^{-1} (Olah *et al.*, 1986). Thus, inorganic fertilisation of fish ponds was not widely applied in the Mekong Delta. This could, however, be a sustainable practice, if the inorganic nutrients were supplied to ponds with organic application, either with manuring or direct integration of animal husbandry.

Table 6.5. Fish survival rates (%) in the surveyed fish farming systems.

Stocked species	A system[1]		AC/VAC system[2]		AR system[3]	
	Cantho (n = 43)	Vinhlong (n = 43)	Cantho (n = 42)	Vinhlong (n = 47)	Cantho (n = 41)	Vinhlong (n = 46)
Silver barb	23.4 ± 28.0	20.4 ± 13.0	23.6 ± 26.8	26.7 ± 23.4	36.4 ± 59.8	24.7 ± 24.8
Common carp	31.1 ± 48.9	17.5 ± 17.5	22.0 ± 26.8	21.2 ± 26.5	33.7 ± 81.2	14.3 ± 17.5
Tilapia	-	-	-	-	-	-
Silver carp	36.4 ± 45.4	25.1 ± 10.4	24.9 ± 24.7	19.2 ± 17.6	15.0 ± 1.6	23.4 ± 27.0
Pangasius	4.8	2.0	37.1 ± 21.1	24.4	-	-
Macrobrachium	10.9 ± 3.3	9.8 ± 3.9	33.3	-	18.1 ± 13.5	25.4 ± 26.7
Other species	33.8 ± 27.8	37.8 ± 45.1	43.3 ± 30.6	71.6 ± 155	44.5 ± 77.5	24.6 ± 26.7

[1] Fish culture in ponds and/or garden canals - non-integrated fish farming systems.
[2] Fishpond-livestock with/without vegetable and/or fruit-tree garden - integrated fish farming systems.
[3] Fish-cum-rice culture - integrated fish farming systems. ± = standard deviation.

Organic enrichment of the pond bottom with animal manure or with any other organic waste was also not common practice in the Delta, either during pond preparation or later. In the surveyed ponds it was almost negligible. The reported range of 12 to 310kg ha^{-1} had no significant influence on the energy flow of the ponds, nor economic importance in the production of fish food organisms. The farmers did not practice organic fertilisation, especially manure, for either pond or terrestrial crops in the Mekong Delta.

Manuring is an important management parameter in integrated farming systems. The primary question of how much manure could these ponds process and convert into fish growth, has almost been answered already (Olah and Pekar, 1995). In the last two decades, there have been a large number of pond experiments to determine the optimal dose, and reviews have clearly documented that around 5g C m^{-2} day^{-1} is processable, if the stocking density is high enough to properly disturb the sediment-water interface (Moav *et al.*, 1977; Schroeder and Hepher, 1979; Burns and Stickney, 1980; Olah, 1986). These experiments have outlined the limits and established the potential of organic waste aquaculture in integrated farming systems. Fish yields obtained from manure-fed systems varied between 4-40kg ha^{-1} day^{-1}, with 25-35kg ha^{-1} day^{-1} averaged maximum values for the whole growing season (Moav *et al.*, 1977; Buck *et al.*, 1978; Schroeder, 1978, 1987; Hopkins and Cruz, 1982; Shan *et al.*, 1985). It has also been documented that large-scale commercial farming, applying much less than 1g C m^{-2} day^{-1}, is far from utilising the full potential of this natural resource.

Table 6.6. Fish harvesting practice and fish yields at the surveyed farms.

Fish harvesting practice and fish yields	Farming system		
	A[1]	AC/VAC[2]	AR[3]
Partial harvest (% of households)	51.2	41.6	40.2
Total harvest (% of household)	48.8	58.4	59.8
First harvest after stocking (months)	7.1 ± 2.5	6.8 ± 2.1	7.0 ± 1.9
Final harvest after stocking (months)	9.4 ± 3.5	9.4 ± 2.9	8.5 ± 2.5
Fish yields (kg ha^{-1} per crop)	4,680 ± 4,310	11,600 ± 10,200	677 ± 653

[1] Fish culture in ponds and/or garden canals - non-integrated fish farming systems.
[2] Fishpond-livestock with/without vegetable and/or fruit-tree garden - integrated fish farming systems.
[3] Fish-cum-rice culture - integrated fish farming systems. ± = standard deviation.

Nutrient supply by pig and poultry integration

Although direct application of both inorganic and organic nutrients was a very minor source of C, N and P for aquaculture in the Mekong Delta, animal integration supplied a major nutrient flow. Most households had between one and four pigs, which were kept for a number of reasons. Pig was a traditional source of meat, a source of cash income, a means of saving and a source of manure. The raising period of a pig was about 8 months, reaching a final weight of 100-120kg per pig. Most households raised chickens which free ranged but were sometimes fed with rice. After 4-5 months they reached an average body weight of 1.5 to 2.0kg per head. Local poultry varieties were popular because they were easy to sell in the local market, tolerant to diseases and less demanding in terms of feed. Chicken was an important source of small cash income for many households. Chicken manure was also an important by-product. Ducks were commonly raised on both small and large scale for eggs. These were usually penned but were let out to forage on paddy fallow and water areas. Average numbers of poultry per household were between 11 and 40, with a very high standard deviation.

Pigs integrated with fish ponds represented a significant source of nutrients for fish production, although fish ponds were very small, having an average size of only a few hundred square metres. The average extrapolated number of pigs reached a high value of about 200 pigs ha^{-1}, but again with a very high standard deviation (Table 6.7). Integration of pigs with ponds varied from keeping pigsties directly over or near the pond to an indirect manual introduction of collected pig manure to the ponds. Although integration of pigs and ponds was commonly believed to be almost 100%, it was only partial since a significant part of the pig manure was used also for crops.

Table 6.7. Animal integration (number ha^{-1}) and fish feeding (kg ha^{-1} year^{-1}).

Farming systems province	Animal integration			Feeding		
	Pig	Duck	Chicken	Rice bran	Vegetable	Pellet
A						
Cantho	-	-	-	19,415 ± 36,042	9,858 ± 5,980	-
Vinhlong	-	-	-	11,913 ± 18,763	9,755 ± 7,548	1,941 ± 2,168
VACR/VAC						
Cantho	173 ± 284	-	28,464 ± 47,885	9,312 ± 15,357	12,467 ± 11,379	-
Vinhlong	202 ± 340	910 ± 747	1,009 ± 1,049	18,605 ± 39,995	3,911 ± 2,390	-
AR						
Cantho	-	-	-	7,970 ± 11,051	6,691 ± 7,603	18,263 ± 28,601
Vinhlong	34 ± 16	-	-	3,860 ± 4,334	-	-

According to a study carried out in India and Hungary, the average density of pigs integrated with fish was much less, reaching a maximum of 60 pigs ha^{-1} (Sharma and Olah, 1986). It was also concluded that raising 40-60 pigs ha^{-1} was sufficient to supply an average of 1-2g organic C m^{-2} day^{-1}. Although it is lower than the maximal daily 5g organic C m^{-2}, it is high enough to maintain an acceptable rate of daily fish production. As a result, no fish feed or inorganic fertilisers need to be applied in this type of farming system. The pig manure acts as a substitute for fish feed and pond fertiliser, which generally account for 50% of the input costs in conventional fish culture. These systems not only increase fish production but also reduce the cost of inputs for fish culture operations considerably. The organic C load resulting from the application of raw and fresh pig manure to the ponds enhances the photosynthetic activity and maintains high natural fish food production, resulting in increased fish yield.

Ducks in the Mekong Delta were raised penned and let out to forage on paddy fallow and water areas, including fish ponds. The average density of around 1,000 ducks ha^{-1} was just above the commonly used stocking and around the optimal stocking rate (Table 6.7). Until recently the duck stocking density was accepted to be 600-700ha^{-1} (Woynarovich, 1980; Barash *et al.*, 1982; Mukherjee *et al.*, 1991) both in temperate and tropical countries. This is the density of duck kept on the fish pond at one time. However, depending on the length of the growing season and temperate or tropical climate, several groups can be raised consecutively on the pond. In special density testing and optimising experiments, it has been demonstrated with both culture and ecosystem parameters that as many as 2,000 ducks ha^{-1} can be raised effectively at a time together with a high fish stocking rate (Pekar *et al.*, 1993; Pekar and Olah, 1998). In fish-cum-duck ponds, the rate of organic C decomposition was greater than in fish ponds with no ducks (Olah *et al.*, 1994), due to bioturbation by ducks in the fish pond ecosystem. Pond sediments with fish lost more N when ducks were present, in spite of an additional organic C load introduced through duck feed. In fish-cum-duck culture, the duck bioturbation effect has not been considered as an accelerator of community respiration. On the contrary, duck culture in fish ponds is widely discussed as a harmful N and organic C enriching process, deteriorating the chemical environment for fish, in spite of the positive effect of manuring. In recent experiments, effective bioturbation of duck diving, feeding and swimming activities were observed (Szabo, 1991). The above experiments have proven the rational use of duck integration, both in temperate and tropical farming systems, and explained the ecosystem processes in bioturbation, maintaining a safe and profitable culture with intensive stocking of fish and duck.

Beside pig and poultry manure, fish ponds received organic and inorganic nutrients indirectly with the introduced fish feed (Table 6.7). A wide variety of feed sources were used in fish culture in the Delta. Agricultural by-products like rice bran, broken rice, waste vegetables and animal wastes were the main pond inputs. However, the dominant feeds, rice bran and vegetables, are not N-rich foods and mainly function through cellulolytic processing. Pellets were used very occasionally in the Mekong Delta.

Recommendations for Future Research in Small-scale Freshwater Aquaculture in the Mekong Delta of Vietnam

- To study and determine the most beneficial water exchange rate for the different farming systems. Optimal water exchange would bring enough nutrients adsorbed on silt particles into the system and would allow sedimentation of silt to create high transparency for light penetration.
- To optimize stocking density and stocking structure for each farming system with the priority order of AR, AC, VAC and VAC-R systems. Optimal density and ratio of fish species with different feeding habits would ensure more effective utilization of the natural fish food resources and higher fish yield.

- To optimize nutrient inputs to the culture systems by inorganic fertilisation, organic manuring and direct integration of animals with the priority of using farm-origin organic wastes. Optimal nutrient supply would lead to higher production of natural fish food organisms and higher fish production.
- To develop various pre-treatment technologies for increasing the nutritive value of locally available agricultural by-products and farm-made fish feeds. Higher protein content and better digestibility of feed ingredients would improve the efficiency of supplementary feeding of fish.
- To study the effectiveness of the aquatic food web and nutrient cycles in processing and utilisation of organic wastes and other types of organic enrichment applied in culture systems. Better knowledge of the functioning of aquacultural ecosystems would help to improve the management of culture systems to sustain higher pond productivity.
- To study the sanitary aspects of existing aquaculture technologies utilising animal manure and human excreta with regard to both producers and consumers.
- To study the market potential of various indigenous fish species appropriate for culture in small-scale farming systems.

References

Barash, H., Plavnik, I. and Moav, R. (1982) Integration of duck and fish farming: experimental results. *Aquaculture* 27, 129-140.

Buck, D.H., Baur, R.J. and Rose, C.R. (1978) Polyculture of Chinese carps in ponds with swine wastes. In: Smitherman, R., Shelton, W. and Grover, K.T. (eds) *Culture of Exotic Fishes Symposium Proceedings*. American Fisheries Society, Auburn, Alabama, pp. 144-155.

Burns, R.P. and Stickney, R.R. (1980) Growth of *Tilapia aurea* in ponds receiving poultry wastes. *Aquaculture* 20, 117-121.

Hopkins, K.D. and Cruz, E.M. (1982) *The ICLARM-CLSU Integrated Animal-Fish Farming Project: Final Report.* ICLARM Technical Reports 5. ICLARM, Manila and the Freshwater Aquaculture Center, Central Luzon State University, Nueva Ecija. 96 pp.

Martin, J.M. and Meybeck, M. (1979) Elemental mass-balance of material carried by major world rivers. *Marine Chemistry* 7, 173-206.

Mitsch, J.W. and Jorgensen, S.E. (1989) *Ecological Engineering. An Introduction to Ecotechnology.* Wiley, New York. 472 pp.

Moav, F.T., Wohlfarth, G.W., Schroeder, G.L., Hulata, G. and Barash, H. (1977) Intensive polyculture of fish in freshwater ponds. *Aquaculture* 10, 25-43.

Mukherjee, T.K., Geeta, S., Rohani, A. and Phang, S.M. (1991). A study on integrated duck-fish and goat-fish production system. *Proceedings of an FAO/IPT International Workshop on Integrated Livestock-Fish Production Systems.* 16-20 December 1991, Kuala Lumpur, Malaysia. pp. 41-48.

Olah, J. (1986) Carp production in manured pond. In: Billard, R. and Marcel, K.T. (eds) *Aquaculture of Cyprinids*. INRA, Paris, pp. 295-303.

Olah, J., Ayyappan, S. and Purushotaman, C.S. (1990) Structure, functioning and management in rural undrainable fishponds. *Aquacultura Hungarica* 6, 77-95.

Olah, J. and Pekar, F. (1995) Transfer of energy and nitrogen in fish farming integrated ecosystems. In: Symoens, J.J. and Micha, J.C. (eds) *The Management of Integrated Freshwater Agro-piscicultural Ecosystems in Tropical Areas*. Royal Academy of Overseas Sciences, Brussels, pp. 187-201.

Olah, J., Sinha, V.R.P., Ayyappan, S., Purushothaman, C.S. and Radheyshyam, S. (1986) Primary production and fish yields in fish ponds under different management practices. *Aquaculture* 58, 111-122.

Olah, J., Szabo, P., Esteky, A.A. and Nezami, S.A. (1994) Nitrogen processing and retention in Hungarian carp farms. *Journal of Applied Ichthyology* 10, 335-340.

Pantulu, V.R. (1986). The Mekong River system. In: Davies, B.R. and Walker, K.F. (eds) *The Ecology of River Systems*. Dr W. Junk Publisher, Dordrecht, pp. 695-719.

Pekar, F., Kiss, L., Szabo, P. and Olah, J. (1993) Pond processing of high organic load in a fish-cum-duck culture system in Hungary. World Aquaculture'93, World Conference and Exhibition. 26-28 May 1993, Torremolinos, Spain.

Pekar, F. and Olah, J. (1998) Fish pond manuring studies in Hungary. In: Mathias, J.A., Charles, A.T. and Baotong, H. (eds) *Integrated Fish Farming*. CRC Press, Boca Raton, pp. 163-177.

Schroeder, G.L. (1978) Autotrophic and heterotrophic production of microorganisms in intensely-manured fish ponds and related fish yields. *Aquaculture* 14, 303-325.

Schroeder, G.L. (1987) Carbon pathways in aquatic detrital systems. In: Moriarty, D.J.W. and Pullin, R.S.V. (eds) *Detritus and Microbial Ecology in Aquaculture*. ICLARM Conference Proceedings 14, ICLARM, Manila, pp. 217-236.

Schroeder, G.L. and Hepher, B. (1979) Use of agricultural and urban wastes in fish culture. In: Pillay, T.V.R. and Dill, W.A. (eds) *Advances in Aquaculture*. Fishing News Books Ltd, Farnham, pp. 487-489.

Shan, K.T., Chang, L., Gua, X., Fang, Y., Zhu, Y., Chan, X., Zhou, F. and Schroeder, G.L. (1985) Observations on feeding habits of fish in ponds receiving green and animal manures in Wuxi, People's Republic of China. *Aquaculture* 46, 111-117.

Sharma, B.K. and Olah, J. (1986) Integrated fish-pig farming in India and Hungary. *Aquaculture* 54, 135-139.

Szabo, P. (1991) The effects of integrated fish-cum-duck culture on the water quality of fish ponds. Aquaculture Europe '91. International Conference, Dublin, Ireland.

Uhera, G., Nishima, M.S. and Tsuji, G.Y. (1974) The composition of Mekong river silt and its possible role as a source of plant nutrient in delta soils. A Mekong Committee Report. Department of Agronomy and Soil Science, College of Tropical Agriculture, University of Hawaii, Honolulu. 109 pp.

WES Project Report (1996-1997) Eco-technological and socio-economic analysis of fish farming systems in the freshwater area of the Mekong Delta. West-East-South Programme. College of Agriculture, Cantho University, Cantho, Vietnam. 124 pp.

Woynarovich, E. (1980) Raising ducks on fish ponds. In: Pullin, R.S.V. and Shehadeh, Z.H. (eds) *Integrated Agriculture-Aquaculture Farming System.* ICLARM Conference Proceedings 4, ICLARM, Manila, pp. 129-134.

Chapter 7

Aquaculture for Diversification of Small Farms within Forest Buffer Zone Management: an Example from the Uplands of Quirino Province, Philippines

M. Prein, R. Oficial, M.A. Bimbao and T. Lopez
International Center for Living Aquatic Resources Management (ICLARM), GPO Box 500, 10670 Penang, Malaysia

Abstract

Upland farming systems can have severe negative effects through intensified slash-and-burn activities, erosion-enhancing farm practices on moderate to steep slopes, and further encroachment into forest areas. ICLARM studied, in a 2-year applied research activity, opportunities for the improvement of existing pond management practices in an upland forest buffer zone environment, mainly through integration with other farm enterprises and through the introduction of polyculture. This chapter presents the results of aquaculture integration and production which are discussed in relation to farmer perceptions, other enterprises on the farm, opportunities for nutrient recycling, and in the broader context of sustainable natural resource management.

Introduction

Upland farming systems can have severe negative effects through intensified slash-and-burn activities, erosion-enhancing farm practices on moderate to steep slopes and further encroachment into forest areas. It is important that the farming systems subsequently introduced into these forest buffer zone areas are a sustainable combination of traditional and newly adopted enterprises which suit the nutritional, economic and socio-cultural needs of the farm households. They should further ensure the conservation and simultaneous wise utilisation of the adjacent forest areas. Recent experiences indicate that communities which have been granted long-term stewardship rights over clearly defined areas can sustainably manage these systems through co-management arrangements.

ICLARM conducted a study, from 1996 to 1998, in two communities within a forest buffer zone management scheme in the upland areas of Quirino province, Philippines, which was supported by an ongoing bilateral development assistance   programme   (DENR-Philippines   and   BMZ/GTZ-Germany: 'Community Forestry Project Quirino, CFPQ'). A People's Organisation (PO) of approximately 60 farm households in each community implements an agreed-upon management strategy of forested and farmed areas as part of project arrangements. The farmers are migrants displaced from their original homesteads due to the construction of a dam in a neighbouring province. They entered the area 15 to 25 years ago and have established farming enterprises for subsistence (upland rice, irrigated paddy rice terraces, vegetables and livestock) and for cash generation (bananas and ginger). Supported by the CFPQ, the communities through their POs have recently introduced activities geared towards more sustainable resource management and diversification of farm enterprises such as agroforestry plots, forest replanting, citrus varieties, lemon grass, rice-fish and pond aquaculture.

The CFPQ promoted the establishment of small farm reservoirs above terraced ricefields in 1993 which were widely adopted (Fig. 7.1). These reservoirs range in size from 50 to 1,200m^2 and provide opportunities for aquaculture. Subsequently, initial attempts at tilapia culture were made and some households established a reliance on this source of fish for their home consumption, with 'give-aways' to neighbours. Few nutrients were added to ponds, and only in the form of direct feed to the fish such as rice bran, rice hulls, taro leaves and ripe bananas.

Fig. 7.1. Reservoir ponds constructed at the top of a cascade of rice terraces provide an opportunity for aquaculture. Don Mariano Perez community, Quirino province.

Within this context, ICLARM studied opportunities for the improvement of existing pond management practices, mainly through integration with other farm enterprises and through the introduction of polyculture. This chapter presents some of the results in respect to aquaculture integration and production. Further details and results on socio-economics and farm sustainability are presented in Prein *et al.* (1999) and Prein *et al.* (in preparation).

Methods and Activities

Activities were conducted in two communities (*barangays*) in Diffun municipality, Quirino province: Don Mariano Perez (DMP), a community in a remote location between 600 and 1000m elevation without access to roads or electricity; and Baguio Village (BV), an adjacent community at lower elevation, the main settlement (*sitio*) of which is serviced by a road and has electricity.

Farmers normally did not use inorganic fertilisers for their crops (mainly rice), as these were difficult to transport. Only in the third year during the El Niño event did some farmers use small amounts for their rice crops. Most farmers grew azolla in their ricefields.

In the course of the ICLARM activity, farms were visited and their operations and pond management were studied. Furthermore, support was provided to the farmers and to the CFPQ through provision of technical advice in the form of training workshops, regular monthly farm site visits, the facilitation of fingerling supply and further formalised training through government services and facilities, e.g. the Bureau of Fisheries and Aquatic Resources (BFAR) (Fig. 7.2).

Fig. 7.2. Farmers planning for new recycling opportunities, including pond fish culture. Don Mariano Perez community, Quirino province.

All the *barangay* residents were invited to workshops on integrated aquaculture-agriculture conducted in different *sitios*. Those farmers who indicated an interest in aquaculture agreed to have closer interaction in the coming months of the Project and allowed their farm production to be monitored, became farmer-cooperators of the CFPQ-ICLARM collaborative Project. In DMP and BV, 50 and 17% of households had ponds, which was the result of an earlier national campaign in 1989-1993 promoting the establishment of on-farm reservoirs which could also serve for fish rearing. Only one third of all these households in each *barangay* which already had ponds also used them for fish culture. All these registered as cooperators which amounted to the total number of farmer-cooperators per *barangay*. The other two thirds of those with ponds *i.e.* small farm reservoirs but which did not practice fish culture, did not become cooperators.

Results and Discussion

Assessment and Intervention

The average area devoted to fish ponds by farm households in DMP and BV is 300m^2. Generally, the fishpond area accounts for the smallest portion (< 1%) of the land use within a farm. Reasons given by farmers as to why they do not devote more area to fish ponds are: (i) having mainly steep-sloped land; (ii) not having a long hose to tap spring water to divert it into ponds; and (iii) a lack of technical expertise on more productive aquaculture methods.

Only one farmer-cooperator in DMP reported a pond water retention of 8 months while the rest claimed that their ponds do not dry up even in April, the driest and hottest month in the area. However, farmers noted that compared to recent years before the Project, the volume of water from natural springs is now diminished. Water was still abundant in the 1980s but in two recent years (1997 and 1998), the local community noted that there was less rain. Moreover, rainfall did not come regularly in the expected months as it did previously when there was high rainfall in May; lately, rains have started 2 months later. Additionally, the 1997/98 season was an 'El Niño' event with widespread drought which also affected the project sites and caused ponds to dry up early, leading to low production or even total loss of the fish crop.

In the initial site surveys of pond operations in 1996, it was found that farmers in the area maintained constant water levels in their ponds by having a continuous inflow and flow-through of water. This was necessary as the manner of dike construction is according to that of their rice terraces which consist of thin high dikes without a clay core and much compaction, and therefore have high leakage and seepage rates (which is not a problem in the cascading arrangement of rice terraces). This resulted in low water temperatures and clear water in the pond, usually void of phytoplankton. It was also found that few operators added any form of fertilisers to their ponds. The project encouraged the regulation of water inflow towards more stagnant water to elevate temperature and facilitate the establishment of standing phytoplankton stocks *i.e.* green colour of pond water. Additionally, the input of on-farm biomaterials as fertilisers was encouraged, as other inputs were not available in these remote locations.

Fish Culture Management Following Project Intervention

The CFPQ-ICLARM collaboration provided technical advice in fish culture, i.e. appropriate fish species and stocking densities, feeding regimes, pond water management and culture systems. Farmer-researcher discussions used a broader perspective beyond only fish culture towards a view of the entire farming system (Lightfoot *et al.*, 1993a, b). Moreover, the orientation widened further from that of benefits derived from fish culture such as food and income contribution to family welfare, to issues of ecological benefits resulting from good natural resource management. Additionally, whole farm management, production and socio-economics were monitored utilising an ICLARM-developed tool (Lightfoot *et al.,* 1996), the results of which are presented in Prein *et al.* (1999).

Fish stocks

Farmers' usual practice was to have fish stocks in their ponds of mixed ages and sexes with continuous reproduction and partial harvesting of small amounts. Throughout the Project, farmers were given assistance in obtaining new fingerlings, i.e. adequate hatcheries were identified and contacted (notably in the area of Tanay on the northern side of lake Laguna de Bay), species were ordered and plastic bags with oxygen organised, and transport was provided for the 10-hour drive up to Quirino province. These costs were born by the Project, as well as the costs of the fingerlings in the first year. In the second and third years, transport costs *etc.* were still covered by the Project, but the farmers had to pay for the fingerlings themselves. Since 1996, around 25,000 fingerlings were distributed to around 50 farmers in the two *barangays*. Farmers were also introduced to fish nurseries in Quirino province.

Fish species

Tilapias were almost the only cultured fish species in two project sites, Nile tilapia (*Oreochromis niloticus*) were obtained in 1992 from Central Luzon State University (CLSU) in Muñoz, some of which were GMT strain (Graham Mair, personal communication); others were obtained from hatcheries and farms near Santiago and Cauayan and were blue tilapia (*O. aureus*). The farmers called the latter 'native' tilapia and liked their 'sweet' taste and cold tolerance. Some farmers kept 'native carp' in their rice fields. These were identified as *Carassius carassius*, a slender goldfish of silver-white colour which attained 8 to 12cm total length in the ponds and were also consumed by farm households. The project introduced common carp (*Cyprinus carpio*), grass carp (*Ctenopharyngodon idella*) and silver carp (*Hypophthalmichthys molitrix*). Some farmer-cooperators requested catfish (*Clarias gariepinus*) fingerlings.

Farmers were delighted with the newly-introduced species as these displayed rapid growth.

Fish stocking density

Farmers were advised to stock 2 fish m^{-2}. Initially, fish stocking densities followed by the farmers depended on the availability of fish fingerlings. However, with continuous, i.e. year-round fish rearing (without total harvest and complete draining of the pond), farmers were advised that fish densities should be maintained at the recommended level to avoid fish overcrowding and competition for pond nutrients so that potential fish growth could be attained.

Fish species composition

The project also suggested that farmers move from tilapia monoculture to polyculture in which separate natural food niches are exploited by different species, thereby producing more fish from a given pond with the inputs it is receiving. Fish species composition is a concern in polyculture systems and in the first year (1996/97) farmers were advised to use a 60 : 30 : 10 species composition in ponds stocked with tilapia, common carp and catfish (since there was considerable request for the latter), and 60 : 40 species composition for a tilapia and common carp combination. In the second year (1997/98) it was suggested that grass carp and silver carp should be added to the polyculture (Table 7.1).

Fish growth across species

Initial fish sampling showed that tilapia grew slowly at the Project sites (Table 7.1) as the low temperature (ranging from 18 to 32°C, averaging in the low 20s) in the upland area was not conducive for higher tilapia productivity.

The carp showed greater growth potential among stocked species at both Project sites. In the 1 year culture period, carp stocked at an initial weight of approximately 5g obtained an average weight that ranged from 129 to 300g, while the tilapias' average weight ranged from 44 to 95g. Farmers claimed to observe common carp feeding on clams, snails and tadpoles so that they concluded that the population of these aquatic species decreased after stocking this species. Although farmers raised fears of losing some aquatic life with continuous culture of common carp, it is likely that this predation was caused by catfish and not common carp. Farmers observed their own ponds and the aquatic life in them also in comparison to other farmers' ponds. Farmers had a strong feeling for the value of naturally occurring species, also, as these are not perceived to harm the fish and affect production.

Farmers observed that pond water in ponds stocked with common carp became turbid, i.e. milky brown in colour, and expressed fear that the dikes could easily be damaged or collapse as a result of the continuous burrowing activities of the fish. Suspended soil and other particles in turbid pond water inhibited phytoplankton development.

Table 7.1. Average fish size at sampling in farmer-cooperators' ponds, *barangays* Don Mariano Perez and Baguio Village, Diffun Municipality, Quirino Province. 1995/96 to 1997/98.

Species	Culture period (months)	Weight (g)			Length (cm)		
		Min.	Max.	Mean	Min.	Max.	Mean
Don Mariano Perez							
Polyculture Type I							
Tilapia	12	19.6	91.7	44.3	9.3	17.1	13.0
	24	32.9	107.6	66.3	11.3	17.7	14.6
Common carp	12	52.6	348.4	128.6	13.9	26.3	19.6
	24	110.0	517.3	229.8	19.1	32.7	24.3
Catfish	12	133.7	603.8	260.1	25.0	43.3	31.2
	24	500.0	827.0	625.7	31.0	46.5	39.5
Polyculture Type II							
Grass carp	12	167.5	479.1	300.0	21.3	34.2	26.3
Common carp	12	87.5	307.3	167.2	17.3	26.3	21.9
Silver carp	12	144.0	269.3	210.4	22.0	30.9	27.8
Baguio Village							
Polyculture Type I							
Tilapia	12	25.4	212.0	94.8	11.0	21.4	16.4
	24	36.0	136.0	77.9	12.1	20.0	16.3
Common carp	12	62.1	394.4	139.6	15.2	28.1	19.5
	24	104.0	316.0	184.2	19.0	27.8	22.3

Note: Polyculture Type I = tilapia, common carp and catfish stocked in October 1996 in DMP; Polyculture Type II = grass, common and silver carp stocked in September 1997 in DMP; Polyculture Type III = tilapia and common carp stocked in October 1996 in BV.

Another constraint for culturing carp in the area is that the fingerling supply from hatcheries is not always assured. Moreover, farmers may find it very difficult to breed grass carp and silver carp themselves so that appropriate small-scale hatchery technology would have to be introduced once demand increases and stabilises among farmers in the area.

Catfish also displayed good growth. However, the project did not encourage culture of this species because of potential negative effects on biodiversity in natural streams and water bodies. After the first stocking resulted in low recovery rates but very high growth rates, farmers decided not to stock catfish again.

Fish production across species

In 1995/96, tilapia was dominant among stocked fish in farmer-cooperators' ponds in Don Mariano Perez (Fig. 7.3). Tilapia production was estimated at 644kg ha^{-1} (Table 7.2). In the following years, farmers were able to stock some carp and catfish. Common carp gave the highest estimated production at 784kg ha^{-1} in 1996/97. The relatively higher fish production estimates in BV than DMP in 1996/97 were due to the application of commercial feeds by the farmer-cooperators in BV during that year. The drop in fish production for all species at the two project sites in 1997/98 from the 1996/97 culture year was caused by the water shortage due to the 'El Niño' event. While the fish ponds of nine farmer-cooperators completely dried up by April 1998, about half of the farmer-cooperators scaled down their fishpond operations because of limited water supply. Others were able to sustain their fishpond activities by accessing water from other sources, i.e. farmers who had hoses transferred them from other areas.

Table 7.2. Estimated annual fish production in farmer-cooperators' ponds, *barangays* Don Mariano Perez and Baguio Village, Diffun Municipality, Quirino Province. 1995/96 to 1997/98. Note: Figures in parentheses are the number of ponds sampled. Dash (-) signifies no stocking.

Barangay	Pond system (number of households)		Pond size (m²)		Production (kg ha⁻¹)					
	Individual pond	Pond in rice terraces	Average	Range	Tilapia	Common carp	Grass carp	Silver carp	Cat-fish	Total
Don Mariano Perez										
1995/96	5	8	310	47-942	644 (13)	-	-	-	-	644 (13)
1996/97	12	8	228	47-942	296 (20)	784 (11)	-	-	259 (7)	818 (20)
1997/98	12	8	225	37-942	124 (20)	220 (5)	349 (4)	73 (2)	106 (2)	267 (20)
Baguio Village										
1995/96	4	0	238	150-400	440 (4)	-	-	-	-	440 (4)
1996/97	4	0	238	150-400	733 (4)	1200 (4)	-	-	-	1933 (4)
1997/98	3	0	275	150-400	56 (2)	267 (2)	-	-	-	323 (2)

Pond fertilisation and nutrient cycling

Farmers cultured fish for home consumption (Fig. 7.4) and previously only used conventional on-farm resources, e.g. rice bran. The project encouraged the use of other on-farm resources as inputs from other farm activities besides rice cultivation such as spoiled fruit, fruit peelings, leaves, animal manure and kitchen leftovers.

In the pre-project year, i.e. 1995/96, less than half (47%) of the farmer-cooperators recycled bioresource materials into their fish ponds. The initial integrated aquaculture-agriculture workshops conducted with the farmers in mid-1996 convinced them of the merits of recycling although some farmers were already re-using some of their farm by-products, e.g. as livestock fodder. Thus, in 1996/97 all farmer-cooperators increased their recycling. However, seven (23%) farmer-cooperators discontinued their fishpond operations in 1997/98 due to the drought. The total number of bioresource flows combined across all farms reported by the farmer-cooperators at the two Project sites registered an increase from 44 to 136 and to 142 for 1995/96, 1996/97 and 1997/98, respectively (Table 7.3). This was a large increase of 209% in 1996/97 and a modest increase of 4% in 1997/98. Moreover, in this 3 year period, the number of different types of on-farm materials recycled in the farm increased from 8 to 12 and to 18, starting in 1995/96. The most common of these on-farm materials were vegetables (mostly giant taro), cereal by-products (mostly rice bran) and fruits (mostly banana) which represented around 30, 29 and 16%, respectively, of the total bioresource inflows.

In terms of inputs to the fish ponds, the total number of bioresource flows increased from 11 (1995/96), to 53 (1996/97) and to 65 (1997/98) (Table 7.4).

The significant increases in the second and third years indicated farmers' adoption of the integrated aquaculture-agriculture practice and the internalisation of its principles. The majority of the inflows to the fish ponds came from their irrigated rice and upland/fallow natural resource types (NRTs) representing around 48 and 34% of the total number of bioresource inflows, respectively (Table 7.4). The bioresource inflows from the upland/fallow NRT comprised seven material types with bananas accounting for the largest proportion (50%) among the counts of material types within this natural resource type (Table 7.4). On the other hand, there were only three material types used as bioresource inflows to the fish pond emanating from the irrigated rice NRT with rice bran accounting for the largest proportion (90%) among the counts of material types within this NRT.

Fig. 7.3. Harvesting tilapia, common carp and catfish from a pond at 800m elevation. Don Mariano Perez community, Quirino province.

Table 7.3. Bioresource flow types and counts per year on monitored farmer-cooperator farms, *barangays* Don Mariano Perez (DMP) and Baguio Village (BV), Diffun Municipality, Quirino Province 1995/96 to 1997/98. In parenthesis is the total number of IAA-households/respondents in the *barangay*.

Species category/ bioresource flow type	1995/96			1996/97			1997/98		
	DMP (24)	BV (6)	All (30)	DMP (24)	BV (6)	All (30)	DMP (24)	BV (6)	All (30)
Cereals	22	5	27	36	4	40	54	10	64
Corn bran	1	0	1	0	0	0	1	0	1
Corn grain	6	2	8	0	0	0	10	2	12
Rice bran	12	3	15	36	4	40	34	6	40
Rice grain	3	0	3	0	0	0	8	2	10
Rice straw	0	0	0	0	0	0	1	0	1
Fruits	5	0	5	27	7	34	30	3	33
Avocado	0	0	0	0	0	0	4	0	4
Banana	5	0	5	27	7	34	23	1	24
Papaya	0	0	0	0	0	0	3	2	5
Trees	0	0	0	1	0	1	1	0	1
Leucaena	0	0	0	1	0	1	1	0	1
Vegetables	8	3	11	30	6	36	23	6	29
Chayote	0	0	0	1	0	1	1	0	1
Giant taro	7	3	10	25	6	31	19	6	25
String beans	1	0	1	0	0	0	0	0	0
Squash	0	0	0	1	0	1	0	0	0
Sweet potato	0	0	0	1	0	1	2	0	2
Water spinach	0	0	0	2	0	2	1	0	1
Weeds[1]	0	0	0	12	0	12	3	0	3
Animals	1	0	1	7	2	9	8	0	8
Chicken meat	0	0	0	0	0	0	3	0	3
Manure[2]	0	0	0	2	0	2	2	0	2
Termites	1	0	1	5	2	7	3	0	3
Aquatic plants	0	0	0	4	0	4	4	0	4
Azolla	0	0	0	4	0	4	4	0	4
Total flows	36	8	44	117	19	136	123	19	142
Average flows	1.5	1.3	1.5	4.9	3.2	4.5	5.1	3.2	4.7
	Material type total (number)								
Cereals	4	2	4	1	1	1	5	3	5
Fruits	1	0	1	1	1	1	3	2	3
Trees	0	0	0	1	0	1	1	0	1
Vegetables	2	1	2	5	1	5	4	1	4
Weeds	0	0	0	1	0	1	1	0	1
Animals	1	0	1	2	1	2	3	0	3
Aquatic plants	0	0	0	1	0	1	1	0	1
Total	8	3	8	12	4	12	18	6	18

[1] Includes tiger grass, itch grass, etc.
[2] Includes chicken, pig and water buffalo (*carabao*).

Non-conventional bioresource materials that were used as inputs to the fishpond included weeds, animal manure and azolla (Table 7.4). Except for animal manure, all on-farm materials thrown into the fishpond are directly utilised by the fish as food. While fish also feed on animal manure, animal manure normally serves for pond fertilisation. However, the practice of using animal manure as a fishpond input is not yet accepted by the majority of farmers in the area because of their apprehension that the quality of the fish may be affected when the fish feed on animal manure. Farmers fear that fish cultured in manured ponds are not safe to eat, especially when they prepare a local special fish dish (*pinapaitan*) for which smaller fish (about 2-5cm) are harvested, washed and cooked ungutted.

Fig. 7.4. A meal of tilapia of various sizes prepared for a group of work-sharing farmers. Don Mariano Perez community, Quirino province.

Conclusions

After the 3-year activity, it can be concluded that farmers have internalised the benefits of bioresource recycling. The total recycling counts of the 30 farmer-cooperators increased from 44 to 142 over the 3-year period. Before the Project, less than half of the farmers-cooperators were practising recycling. Moreover, in the last year farmers utilised as many as 18 types of on-farm materials for recycling compared to only eight types before the project.

The ecological benefit of recycling was also accompanied by cash savings for the purchase of commercial fertilisers particularly in BV. In fish culture, farmers utilised on-farm materials such as rice bran/straw/hull, spoiled fruits/vegetables, chopped leaves instead of formulated fish feeds, and animal manure for pond fertilisation. As farmers used on-farm material inputs in fish culture, the proportion of cash cost to total costs decreased from 81 to 18% (Prein *et al.*, 1999).

Table 7.4. Bioresource inflows to fishponds by natural resource type (NRT) source and species/enterprise category, combined for all 30 farmer-cooperators from both *barangays* Don Mariano Perez and Baguio Village, Diffun Municipality, Quirino Province, 1995/96 to 1997/98.

NRT origin of flow	Species category/Bio-resource flow type	Flows (number)		
		1995/96	1996/97	1997/98
Forest reserve	Total	1	2	2
	Animals			
	Termites	1	2	2
Upland/fallow	Total	2	26	22
	Animals			
	Termites	0	5	1
	Cereals			
	Corn bran	1	0	1
	Rice bran	0	1	0
	Fruits			
	Avocado	0	0	2
	Banana	1	14	10
	Papaya	0	0	1
	Vegetables			
	Giant taro (leaves/tubers)	0	6	4
	Weeds	0	0	3
Orchard	Total	0	1	4
	Fruits			
	Avocado	0	0	2
	Papaya	0	0	1
	Trees			
	Leucaena	0	1	0
	Vegetables			
	Sweet potato	0	0	1
Homestead	Total	1	2	4
	Animals			
	Chicken meat	0	0	3
	Manure (various)	0	2	1
	Vegetables			
	String beans	1	0	0
Irrigated rice	Total	7	21	27
	Aquatic plants			
	Azolla	0	4	0
	Cereals			
	Corn grain	0	0	1
	Rice bran	7	16	25
	Rice straw	0	0	1
	Vegetables			
	Water spinach	0	1	0
Fishpond	Total	0	1	6
	Aquatic plants			
	Azolla	0	0	4
	Trees			
	Leucaena	0	0	1
	Vegetables			
	Water spinach	0	1	1
Total		11	53	65

Thus, the upland communities in DMP and BV experienced and benefited from some economic, social and ecological services of the practice of integrated aquaculture-agriculture farming systems. Fish culture, although currently a relatively small component of their upland farming systems, generated great interest among the local communities. If farmers put in more resources and time into fish culture, with continued external technical advisory support, integrating fish culture in their existing farming systems may significantly increase the nutritional quality of food and cash income of the farm households. Moreover, on a broader perspective of integrated resources management for sustainable agriculture, there may be an increased appreciation of the communities' role as resource stewards of their forestry resources due to sustainable livelihoods.

The CFPQ-ICLARM project has established farmer desire for improved management of fish culture and has demonstrated its benefits for the farm family and local communities. A modestly and appropriately designed and managed hatchery within one of the *barangays* will ease the difficulty of securing fingerlings of desired species and required numbers from external sources, and reduce the high fingerling mortality rate due to stress during transport. This will enable the community to be almost self-reliant in their fingerling supply. According to previous experience, one or several farmers usually specialise in seed production after mastering pond culture practices. This further transition should be encouraged and supported.

References

Lightfoot, C., Bimbao, M.P., Dalsgaard, J.P.T. and Pullin, R.S.V. (1993a) Aquaculture and sustainability through integrated resources management. *Outlook on Agriculture* 22(3), 143-150.

Lightfoot, C., Bimbao, M.P., Lopez, T.S., Villanueva, F.D., Orencia, E.A., Dalsgaard, J.P.T., Gayanilo, F.C., Jr, Prein, M. and McArthur, H.J. (1996) *Research Tool for Natural Resource Management, Monitoring and Evaluation (RESTORE)*, Vol. 1. field guide., Vol. 2. software manual. ICLARM, Manila.

Lightfoot, C., Dalsgaard, P., Bimbao, M.P. and Fermin, F. (1993b) Farmer participatory procedures for managing and monitoring sustainable farming systems. *Journal of the Asian Farming Systems Association* 2(1), 67-87.

Prein, M., Bimbao, M.A.P., Lopez, T.S. and Oficial, R. (1999) Upland integrated aquaculture-agriculture systems in forest buffer zone management: final report to BMZ-GTZ for the Philippine-German 'Community Forestry Project Quirino', June 1999. ICLARM, Manila. 91 pp.

Prein, M., Lopez, T.S., Bimbao, M.A.P. and Oficial, R. (in preparation) Fish farming in upland forest buffer zones: a case study in the Philippines. ICLARM Technical Report, Penang.

Chapter 8

A Description of the Rice-Prawn-Fish Systems of Southwest Bangladesh

G. Chapman and J. Abedin

GOLDA Project, CARE–Bangladesh, GPO Box 226, Dhaka 1000, Bangladesh

Abstract

Between the late 1970s and the mid 1980s a small number of farmers in Southwest Bangladesh began to test the stocking of giant freshwater prawn (*Macrobrachium rosenbergii*) in rice fields. Extremely rapid expansion has occurred since 1990, mainly because this prawn is sold to the export market and is thus of very high value compared with traditional crops. The modified rice fields, locally referred to as *gher*, have large peripheral trenches. Consequently, the surrounding dikes are much larger than those seen in typical rice-finfish operations. Although most farmers operating *gher* systems also grow finfish and rice, few utilise the dikes for crops, and the farm components are generally not well integrated. Except for the systems found in one specific area, the focus of production has been on the prawns. Despite the potential realization of high profits, farmers who operate *gher* are often in a state of considerable vulnerability. Farmers use high levels of inputs for which they require large loans that represent considerable risk. Additionally, environmental impacts associated with *gher* systems may be key determinants of these systems' sustainability in the medium and long term. The *gher* systems appear to be unique among rice-fish systems in that small-scale farms in an impoverished area are producing a high value product for an export market. However, because of the current constraints, the sustainability of the systems is questionable. Diversifying crops and decreasing costs, as opposed to increasing prawn production, is possibly the most effective strategy.

Introduction

Rice-prawn-fish culture in Southwest Bangladesh is an indigenous technology developed solely by farmers. Between the late 1970s and the mid 1980s, a small number of farmers began to test the feasibility of stocking giant freshwater prawn (*Macrobrachium rosenbergii*) in rice fields. These specially modified rice fields are locally referred to as *gher*. Since 1990, the number of households that have adopted *gher* farming systems has expanded very rapidly, primarily because prawns are sold to the export market and are thus of very high value compared with traditional crops. Clearly, farmers have focused on the prawn component.

The objective of this paper is to describe the *gher* systems of Southwest Bangladesh and examine the associated constraints.

Methods

Information gathered during the course of project activities was used in this description of *gher* systems. Surveys were conducted with project participants to obtain data regarding costs and returns relating to the main *gher* management practices observed in the area.

Results and Discussion

The predominant *gher* system of Southwest Bangladesh is characterized by trenches that occupy up to half the area of the rice field, and dikes with areas much greater than those seen in typical rice-fish fields. Wild postlarvae are stocked directly into rice fields at high densities (usually 3.5-5, but up to 10 postlarvae m^{-2}, although with postlarvae survival rates of about 35% actual stocking densities may be much lower than indicated) from late April to May, near the beginning of the wet season. Commencing in July, prawns are mainly fed the meat of a native snail, *Pila globosa*, over a 12 to 14 week period (home-made feed mixtures are often provided as a supplement to snail meat). Feeding from 30 to 60kg ha^{-1} day^{-1} is common, although much higher rates are observed. The main harvest is from November to December with small prawns being held over for the succeeding production cycle.

Because of the high value of prawns, the other components of *gher* farming systems, namely, rice, fish (typically silver carp, *Hypophthalmichthys molitrix*; catla, *Catla catla*; and silver barb, *Barbodes gonionotus*), and a variety of dike crops (fruit, vegetable, and tree) have been largely but understandably neglected (Fig. 8.1). The cropping system is essentially rotational since most farmers engaged in the *gher* system grow only dry season rice, with prawns and a small number of fish constituting the wet season crops. Where there is a wet season rice crop, prawns are generally restricted to the trenches as farmers believe that the rice crop inhibits prawn growth. Consequently, prawns cannot take advantage of the richer natural forage associated with a planted rice field. Furthermore, as there are no rice plants for protection, cannibalistic attacks during moulting are likely to be high. Prawn yields of 250 to 450kg ha^{-1} crop^{-1} are typical. Where water is brackish before the rainy season begins, some farmers raise black tiger shrimp (*Penaeus monodon*) in rice fields for a period of

90 days. Harvesting of this species occurs at the onset of the rains, around May, and thus coincides with the stocking of prawn postlarvae.

Fig. 8.1. A typical *gher* system in Southwest Bangladesh characterized by few dike crops, limited finfish culture, and often no wet-season rice crop with the focus on prawn production.

Currently, a fundamental constraint to the *gher* system is high production costs. Three elements typically account for 80% or more of the recurrent production cost of prawns: postlarvae, feed (primarily snail meat), and the cost of loans (mainly required for snail meat and postlarvae purchases). These elements represent from about 15-20, 30-44, and 20-34% of this cost, respectively. The large loans (averaging about Taka9,700, US$1 = Tk47, with interest rates of up to 10% month^{-1}) secured by farmers represent a major risk factor in *gher* operations.

Moreover, the likely considerable environmental costs associated with the *gher* system function as important constraints. Many *gher* have been constructed in low-lying, seasonally or permanently flooded, water bodies that have traditionally supported important native fish populations. Also, because of the heavy reliance on snail as a feed, local populations in at least one district appear to have been drastically reduced. As the ecological role of this snail is not known, predicting the types and severity of consequences related to its removal from the ecosystem is difficult. The disposal of snail shells in ditches impedes drainage. The availability of fodder for livestock has also decreased since the rapid expansion of the *gher* farming system. Likewise, important social concerns including livelihood changes of some groups within communities, management of large debt, impacts on women, and conflicts between *gher* operators and those without *gher*, have emerged.

Very recently, a growing number of farmers have placed more emphasis on dike crops and fish, and decreased the focus on the prawn component (Fig. 8.2).

This change has been partly the result of external factors that have strongly affected prawn production in 1997/98, and which increased the vulnerability of *gher* operators. A threatened import ban on prawns from Bangladesh in 1997 led to lower selling prices for producers. In 1998, the cost of postlarvae was three to four times that of 1997 due to a poor wild catch of postlarvae. Furthermore, the 1998 flood caused huge prawn losses through escape.

Fig. 8.2. A diversified *gher* system in Southwest Bangladesh places increased importance on finfish, and especially dike crop production.

The prawn component easily accounts for the highest net profits and associated costs in typical *gher* practices (Table 8.1). An atypical *gher* system is found in Pinguria village where farmers have used a more balanced approach that places more importance on rice, fish, and dike crops. In this diversified *gher* system, dike crops (due to trees that produce fruit, fuelwood, and lumber) yield net profits much higher those from the prawn component of the typical *gher* system (Table 8.2). Since the large costs and loans related to the prawn component are greatly reduced in the more diversified system (Tables 8.1 and 8.2) because relatively few prawns are stocked and snail meat is not used, the high risk associated with the typical *gher* system is lowered considerably.

Conclusions

The *gher* systems are unique among rice-fish systems in that small-scale farms produce a high value product for an export market. The potential for deriving high income is large but so are the risks associated with large loans, the relative inexperience of farmers, increasing or unstable input costs, uncertain postlarvae availability, and environmental impacts. These risks, which act as

Table 8.1. Mean production cost and return (Tk ha^{-1}) of the typical *gher* system with emphasis on prawns, subdivided into three stocking and feeding practices, in Bagerhat district, Bangladesh.

Practice	Item	Cost[1]	%	Return	%	Net profit	Benefit : cost ratio
1) Stock post-	Prawn[2]	67,367	91	85,186	80	17,819	1.26
larvae (PL) and use	Fish	900[3]	1	7,482	7	6,582	8.31
homemade feed	Rice	4,560	6	10,687	10	6,127	2.34
	Dike crops	1,121	2	2,921	3	1,800	2.61
	Total	73,948	100	106,276	100	32,328	1.44
2) Stock (PL) and	Prawn	85,027	93	105,773	83	20,746	1.24
use snail meat as	Fish	957	1	7,482	6	6,525	7.82
main feed (prevalent	Rice	4,560	5	10,687	9	6,127	2.34
practice)	Dike crops	1,121	1	2,921	2	1,800	2.61
	Total	91,665	100	126,863	100	35,198	1.38
3) Stocking	Prawn	108,133	94	150,138	88	42,005	1.39
juveniles and use	Fish	1,126[1]	1	7,482	4	6,365	6.64
snail meat as main	Rice	4,560	4	10,687	6	6,127	2.34
feed (least prevalent	Dike crops	1,121	1	2,921	2	1,800	2.61
practice)	Total	114,940	100	171,228	100	56,288	1.49

[1] Non-cash costs were not included in the analysis; US$1 = Tk47.
[2] For the prawn component, $n = 30$ for practices 1 and 2 and $n = 12$ for practice 3; data were pooled across the three practices for fish, rice, and dike crops due to the small number of farmers having these components: $n = 48$ (fish), $n = 11$ (rice), $n = 12$ (dike crops). Though data were pooled across practices, this analysis provides a realistic view of the cost and return of these three components for each practice.
[3] For each practice, estimated costs were used for the fish component.

Table 8.2. Mean production cost and return (Tk ha^{-1}) of a diversified *gher* system in Pinguria village, Bagerhat district, Bangladesh ($n = 12$).

Practice	Item	Cost[1]	%	Return	%	Net profit	Benefit : cost ratio
Diversified[2]	Prawn	11,871	47	23,417	20	11,546	1.97
gher	Fish	7,571	30	18,454	16	10,883	2.44
system	Rice	4,418	17	13,604	1	9,186	3.08
	Dike crops[3]	1,415	6	63,103	53	61,688	44.60
	Total	25,275	100	118,578	100	93,303	4.69

[1] Non-cash costs were not included in the analysis. US$1 = Tk47.
[2] This diversified system places more emphasis on rice, fish, dike crops (fruits, vegetables, and up to eight tree species producing fruit, lumber, and fuel), and less emphasis on the prawn component, compared with the typical *gher* system. Little feed of any kind is provided to prawns, which are stocked at lower rates than in typical *gher*, as juveniles.
[3] Return from tree crops was calculated as net present value.

severe constraints, will undoubtedly be strong determinants of the sustainability of these farming systems. A critical issue is the identification of an appropriate feeding strategy, a subject that farmers have begun to study themselves. A focus on diversification and reducing costs, rather than increasing prawn production, may be the most profitable, long-term approach.

Acknowledgements

We wish to thank the farmer participants and field staff of the GOLDA project who supplied the information for this paper. The GOLDA project is financially supported by the Department for International Development, UK.

Chapter 9

Fertilisation of Ponds with Inorganic Fertilisers: Low Cost Technologies for Small-scale Farmers

J. Pant, P. Promthong, C.K. Lin and H. Demaine

Aquaculture and Aquatic Resources Management, School of Environment, Resources and Development, Asian Institute of Technology, PO Box 4, Klong Luang, Pathumthani 12120, Thailand

Abstract

A farmer-managed trial of pond fertilisation with synthetic fertilisers, urea and triple superphosphate (TSP) was carried out with 12 farmers in Northeast Thailand. Project farmers were recommended to fertilise their ponds at a rate of 28kg nitrogen (N) and 7kg phosphorus (P) ha^{-1} week^{-1} as urea and TSP, respectively. Farmers applied fertilisers on average close to the recommended rate for N but at nearly double the rate of P. Fish were stocked at 3 fish m^{-2} and cultured for an average of nearly 8 months (range from 4 to 11 months). All but one farmer continued the trial until harvest. Almost all Project farmers experienced a substantial increase in fish yield which was noted to be associated with a change in water colour from clear/turbid to phytoplankton-rich green/dark-green after fertilisation. The average fish yield was nearly 6t ha^{-1}, nearly three times higher than the previous yield with traditional input. Economically, the fish price in the local market was at least Baht30kg^{-1} (Baht40 = US$1 approximately) compared with an average operational cost estimated to be only Baht10kg^{-1} of fish production. Intensification of farm ponds through fertilisation with urea and TSP was a relatively low-cost method to increase fish yields in Northeast Thailand using off-farm inputs.

Introduction

Fish is the major, traditional source of animal protein in the diet among even the poorest people in Northeast Thailand. Whilst natural fish stocks have markedly declined due to population pressure and environmental degradation (Edwards *et al.*, 1991), demand for fish and fish products has greatly exceeded local supply in Northeast Thailand in recent years. A need to intensify productivity of small-scale aquaculture systems to produce sufficient supplies for local household consumption is well recognised (Edwards *et al.*, 1997).

Although fertilisation of ponds with ruminant livestock manure and other on-farm by-products can produce fish at minimal cost, fish yields are correspondingly low. Edwards *et al.* (1991) reported that farmers had to collect and load about 4t of fresh buffalo manure to obtain only 20kg of fish from an average of a 200m^2 experimental farm pond as it is a relatively ineffective pond fertiliser. Besides, there is a limited scope for livestock to be integrated into small-scale aquaculture in Northeast Thailand as manure is traditionally used for crop production in the resource-poor farming systems in the region.

One relatively low-cost strategy to increase fish yield is through fertilisation of ponds with chemical fertilisers. Pond fertilisation has been researched on-station by the Pond Dynamics/Aquaculture CRSP Project at the Asian Institute of Technology (AIT), Thailand for a number of years. On-station experiments at AIT have demonstrated that fish yields in ponds fertilised with a modest input of urea and TSP exceed 4t ha^{-1} year^{-1} (Lin *et al.*, 1997). Similarly, on-station experiments carried out in Northeast Thailand produced an extrapolated yield of about 5t ha^{-1} year^{-1} of sex-reversed tilapia by fertilising ponds with urea and TSP (AOP, 1992).

A problem of low survival of stocked fry due to the presence of predators (snakehead fish, walking catfish and snakes) is well known in Northeast Thailand. However, nursing the relatively small and commonly available fry of only 2.5cm up to fingerlings of 5-6cm (5g) in *hapas* (nylon net cages) suspended in farmers' ponds has been found to be an effective measure to reduce predation by fish (Edwards *et al.*, 1996; Demaine *et al.*, 1999). In Thailand, tilapia plays an important role in providing low-cost animal protein to poorer rural and urban people, and ranks as the most commonly cultured freshwater fish. Although prolific fecundity has been a major drawback in tilapia culture (Ufodike and Madhu, 1986; Szyper *et al.*, 1995), stocking mono-sex male fish controls unwanted population and increases growth rate (Mair and Little, 1991).

This paper presents the results of on-farm trials, introduced to farmers by AIT Aqua Outreach in 1995/96, on fertilisation of fishponds with synthetic fertilisers, urea and TSP, in relatively small-scale, resource-poor farming systems in Northeast Thailand.

Methodology

Selection and Recruitment of Project Farmers

Farmer managed on-farm trials were carried out with 12 small and medium-scale farmers who volunteered to take part in trials in Udorn Thani, Nakorn Phanom and Sakorn Nakorn provinces of Northeast Thailand. The Department of Fisheries (DoF) of the Royal Thai Government (RTG) facilitated the process of farmers' selection and recruitment. For farmers' selection, a number of farmers were shortlisted based on a set of criteria: a relatively small-scale farmer, having aquaculture as one subsystem, and farming as the main occupation. The shortlisted farm households were briefed on the nature of the trial and asked whether they were willing to participate. Finally, the farm households willing to volunteer to participate in the trials were recruited. Upon recruitment, pond size and phytoplankton richness as indicated by water-colour in ponds were assessed prior to commencing the trial.

Project Recommendations

Project farmers were recommended to fertilise ponds at a rate of 28kg N ha^{-1} week^{-1} as urea and 7kg P ha^{-1} week^{-1} as TSP (Table 9.1). They were also instructed to dissolve fertilisers in water overnight before pouring the solution into the pond. Project farmers were recommended to nurse fry in a *hapa* suspended in the pond prior to releasing fish into the pond. A mixture of pig feed concentrate (2 parts) and fine rice bran (1 part) was recommended as feed for fry throughout nursing. Participant farmers were advised to stock the pond at a rate of 2-3 fish m^{-2} of sex-reversed, all male Nile tilapia (*Oreochromis niloticus*).

As sex-reversed tilapia seed was still scarce in the region, Project farmers were supplied with fry free of charge. Project farmers were also supported with some amount of urea and TSP, and supplementary feed for fry nursing in *hapas*. The recommended culture period for these trials was 6 months with an expected extrapolated fish yield of around 3-4t ha^{-1}.

Farm Visits and Monitoring of Trials

Project staff visited farms fortnightly throughout the trial period to record the activities carried out by participant farmers. However, Project staff did not intervene to recommend changes to farmers' production system practices such as stocking, feeding, and harvesting fish. A partial economic analysis of fish production in these trials was done based on input costs and total output estimated on the basis of total harvest (cumulative or single) and the fish price (Baht30kg^{-1}) in the local market. Assessment of labour use for different day to day activities such as feeding was not practicable as they involved negligible time and farmers were not able to maintain a record of such activities. At the end of the trial, a short questionnaire survey was conducted among the farmers who completed the trial to harvest.

Table 9.1. Technical recommendations for inorganic fertilisation on-farm trials.

Pond inputs		
Input	Input rate	Considerations
Urea (46 : 0 : 0)	61kg ha^{-1} week^{-1}	Apply after dissolving in water
TSP (0 : 46 : 0)	35kg ha^{-1} week^{-1}	Soak overnight in water and mix thoroughly before applying to the pond
Chicken manure	Depending upon availability	Amount of urea and TSP should be adjusted with the amount of manure applied (N: 28kg ha^{-1} week^{-1} and P: 7kg ha^{-1} week^{-1}

Stocking recommendation			
Stocking density	Species	Stocking period	Considerations
2-3 fingerlings m^{-2} (20,000 - 30,000 fingerlings ha^{-1})	Sex reversed all male tilapia	6 months	Nursing fry in *hapa* prior to stocking [recommended size to stock: 5g (6-8cm)]

Results and Discussion

Aquaculture Subsystem of the Project Farmers

Fish culture was found to be a relatively new activity for most of the Project farmers. Five Project farmers were new entrants with 1 year experience while the others had 2 to 6 years experience in fish culture, with the exception of Mr. Phai who was the only farmer with long experience, over 30 years (Table 9.2).

Pond size varied greatly among Project farms ranging from 300 to 1,440m^2 (Table 9.2). The water quality assessed prior to the trial indicated inadequate nutrient inputs in most fish ponds of Project farmers as water colour was turbid or clear. However, plankton-rich green water was found in Mr. Phai's pond, probably reflecting his much greater experience in fish culture.

Fry Nursing

All Project participants nursed fry in *hapas* prior to stocking. Details of farmers' nursing practice are presented in Table 9.3. Although size of *hapa* ranged from 5 to 20m^2, it was only around 5m^2 in two thirds of the total Project farms. The stocking density of fry in the nursing *hapas* ranged from 93 to 556 fish m^{-2}. In general, variation in the total number of fry nursed by Project farmers was closely related to the size of pond since the Outreach staff had estimated the number of fry required to maintain the recommended stocking density (Tables 9.1 and 9.2) while distributing fry to the farmers.

Table 9.2. Pond size, length of involvement in fish culture, stocking density and stocking period. SD denotes standard deviation.

Farmer	Pond size (m^2)	Involvement in fish culture (years)	Culture period (months)	Stocking density (fish m^{-2})
Tongsoon	1,440	1	5	2.7
Phai	800	30	4	3.3
Sopar	1,300	1	11	2.4
Boonsong	400	3	9	1.1
Kampong	350	1	9	2.2
Sunan	345	1	9	3.1
Prasobchai	494	2	5	1.9
Wantamid	896	6	4	3.1
Sanguansak	885	4	9	3.1
Surin	380	4	9	7.9
Tawatchai	300	1	9	4.3
Mean	729.0	5.3	7.5	3.2
SD	404.0	8.8	2.5	1.8

Table 9.3. Nursing details. SD denotes standard deviation.

Farmer	Hapa size (m^2)	Fry size at start (cm)	Fingerling size at end (cm)	Days nursed	Number at start of nursing	Density at nursing (fry m^{-2})	Survival at end of nursing (%)
Tongsoon	20.3	2	8	35	4,000	198	99.5
Phai	20.3	2	6	56	3,200	158	84.4
Sopar	20.3	1	8	31	3,200	158	96.9
Boonsong	4.9	1	7	41	800	165	68.8
Kampong	4.9	1	7	57	1,000	206	77.5
Sunan	4.9	1	6	30	1,200	247	1.6
Prasobchai	5.4	1	6	35	1,000	185	95.5
Wantamid	20.3	1	7	43	4,000	198	75.0
Sanguansak	5.4	1	7	30	3,000	556	96.7
Surin	32.4	1	7	43	3,000	93	100.0
Tawatchai	4.9	1	8	51	2,000	412	65.0
Mean	13.1	1.2	7.0	41.1	2,400	234.2	78.3
SD	9.8	0.4	0.8	10.0	1,233	133.2	28.5

The nursing period ranged from 30 to 57 days. Despite large variations in the fry density in the *hapa* and the nursing period, their effect on the growth of fingerlings was not obvious as size of fingerlings at the end of nursing ranged from 6 to 8cm. Farmers fed an average of 130g of pig feed concentrate and 76g of fine rice bran to 1,000 fry day^{-1} which, in general, followed the recommended ratio.

The survival rate of fry at the end of nursing ranged widely from 2 to 100% but was estimated to be 75% or above among three quarters of the Project farms. Relatively low survival rates of fry while nursing may have been due to limited experience of farmers in nursing and feeding fry. For example, Mr. Sunan applied pig fat rather than the recommended feed, which killed almost all the fry in his *hapa*. However, subsequently, he stocked other locally available seed to continue the trial.

Stocking Density, Period of Stocking and Species Stocked

The average stocking density in the trials of 3 fish m^{-2} was within the recommended range in all farmers' ponds, except for Mr. Surin and Mr. Boonsong — the former stocked 8 fish m^{-2}, the highest stocking density among Project farmers, whereas the latter stocked only 1 fish m^{-2}, the lowest (Table 9.2).

All but one farmer continued the trial until harvest. The culture period ranged from 4 to 11 months with an average of nearly 8 months (Table 9.2). Two farmers had to terminate the trial after only 4 months as the water level in their ponds dropped rapidly after the cessation of the rainy season. In contrast, a number of farmers, whose ponds retained sufficient water, extended the stocking period to 8 to 11 months. The farmers' preference for larger sized fish was the main reason for extending the stocking period.

Despite Outreach recommendations and project support for monoculture of sex-reversed tilapia for the on-farm trial, a number of farmers also added a few other fish species. However, these additional species did not have any noticeable effect on stocking density and yield in most of the farm ponds.

Fertilisation of Ponds

The average fertilisation rate applied by farmers was quite close to the recommended rate for N but it varied among Project farmers (Fig. 9.1). Although all Project farmers applied urea, only two farmers, Mr. Tongsoon and Mr. Tawatchai, reported using the recommended rate. Mr. Phai applied the highest amount of urea, estimated at an extrapolated rate of around 147kg urea ha^{-1} week^{-1}; as he also applied organic manure to his pond, he fertilised at an estimated excessively high rate of N (73kg N ha^{-1} week^{-1}) — more than twice the recommended rate. Two other farmers, Mr. Sunan and Mr. Sanguansak, also applied an excessive amount of urea at 50% higher than the recommended N fertilisation rate. The remainder of the farmers applied urea at a rate of 14-28kg N ha^{-1} week^{-1} except Mr. Wantamid, the only farmer who applied urea at far below the recommended rate.

Most farmers applied P at a higher than recommended rate with the average rate of application of TSP being nearly double the recommended rate (Fig. 9.1). Mr. Phai also applied the highest amount of TSP (35kg P ha^{-1} week^{1}). In contrast, Mr. Surin and Mr. Wantamid did not apply TSP at all. Mr. Prasobchai was the only farmer to apply the recommended rate of TSP.

All farmers reported that they had applied livestock manure in their ponds in addition to urea and TSP. Therefore, both inorganic and organic sources were considered to estimate the total N and P supply to the pond (Fig. 9.1).

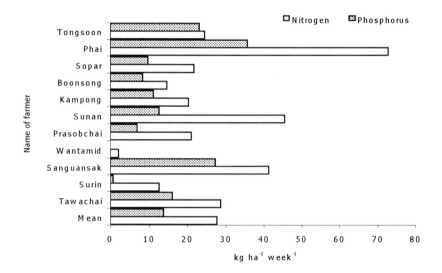

Fig. 9.1. Average nitrogen and phosphorus (kg ha^{-1} week^{-1}) applied during stocking by the Project.

Although most farm ponds had clear or turbid water prior to the start of the trial, the colour of the water in all ponds had changed to green to dark green when water quality was re-assessed at the end of the trial (Fig. 9.2). The change in water colour was noted to be associated with the growth of plankton as a result of fertilisation.

Fish Production

Extrapolated fish yields ranged from nearly 2 to over 8, with an average of nearly 6t ha^{-1}. All but one farmer experienced a substantial increase in fish yield (Fig. 9.3). Despite very little control over the trial, the average yield obtained in these on-farm trials was apparently higher than that expected from a farmer-managed on-farm trial. In contrast, fish yields obtained in other on-farm trials were significantly lower than that obtained from on-station experiments with similar pond inputs. The average extrapolated fish yields obtained using fresh buffalo manure in an on-station experiment on the AIT campus was 2.6t ha^{-1} year^{-1}, while it was only 1.7t ha^{-1} year^{-1} in an adaptive field trial with corresponding pond inputs in Northeast Thailand (AIT, 1986). Whilst substantial increase in fish yield in these trials appeared to be associated with high *in-situ* density of phytoplankton after fertilisation, there was no significant relationship between the amount of fertilisers applied and fish yield. However, fertilisation of ponds with inorganic fertilisers produced nearly three times higher fish yields than yields with traditional on-farm inputs (Edwards *et al.*, 1996).

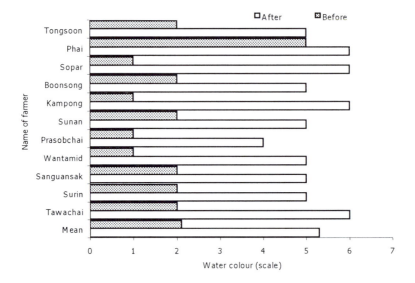

Fig. 9.2. Water colour (before and after trial) of the participant farmers' ponds. Scaling for water colour: 1 (clear), 2 (turbid), 3 (greenish + turbid), 4 (light green), 5 (green) and 6 (dark green).

The Project farmers followed different stocking, feeding and harvesting strategies. Correspondingly, total fish production also varied. In a number of cases, the yield resulted from multiple harvests, in others it was from a single harvest at the end of the season.

Although varying stocking and harvesting strategies by participant farmers might have affected fish yields, it was not feasible to assess such differences in these on-farm trials. However, among the various factors, the difference between the recommended (6 months) and practised (7.5 months) culture period may have contributed significantly to the relatively higher average yield estimated from these trials than the expected based on on-station, single harvest experiments. There was a positive correlation coefficient of 0.56 between yield and the culture period. For four farmers who had stocked fish for 6 months or less, Messrs Phai, Prasobchai, Tongsoon and Wantamid, the extrapolated yield was estimated to be about 5t ha^{-1}. In contrast, four other farmers who had stocked fish for over 9 months, Messrs Kampoong, Sunan, Surin and Tawatchai, harvested an extrapolated average fish yield of around 8t ha^{-1}.

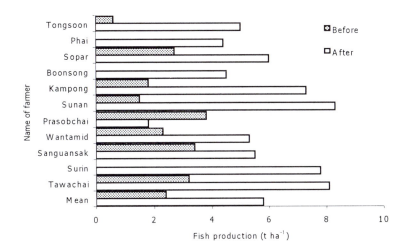

Fig. 9.3. Fish production (t ha^{-1}) before and after fertilisation of the pond with urea and TSP.

Social and Economic Considerations

Although farmers in Northeast Thailand are often reluctant to fertilise ponds with manure, particularly pig manure and night soil (Demaine *et al.,* 1999), fertilisation with inorganic fertilisers (urea and TSP) was acceptable to all Project farmers. Farmers were not hesitant in practising this newly introduced technology to fulfil their household consumption need of fish. Moreover, farmers also reported that they could easily sell surplus fish production over household consumption in the local market.

Looking at the economics of the enterprise, fish price in the local market was at least Baht30kg^{-1}, while the average cost of fish production was estimated to be only Baht10kg^{-1} (Fig. 9.4). Cost of production was mainly associated with the quantity of the input applied, particularly urea and TSP. Cost of production was quite high in Mr. Phai's case, estimated to be Baht15kg^{-1} fish, although he obtained a high total yield. Likewise, cost of production was relatively high in other Project farm ponds, namely of Messrs Prasobchai, Sanguansak, Surin and Tawatchai. For the rest, the input cost was less than Baht10kg^{-1} fish production. Mr. Wantamid had the lowest cost of production, only around Baht4kg^{-1} fish production. Mr. Wantamid had the lowest cost of production, only around Baht4kg[-1] fish production but total production was also low due to inadequate input supply. Messrs Kampoong and Sunan obtained a higher yield with relatively low investment as they had applied fertilisers close to the recommended rate; their extrapolated yields were 7 and 8t ha[-1], respectively, with cost of production of about Baht8 and 7kg[-1] fish, respectively, rather less than the average cost of production (Figs 9.3 and 9.4).

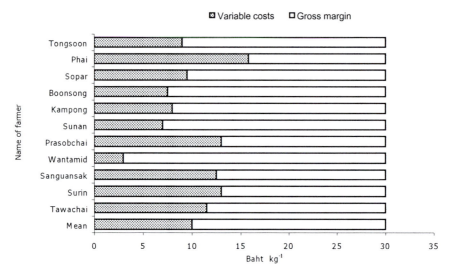

Fig. 9.4. Total variable costs and gross margin (Baht kg[-1]) of fish production.

Almost all the project farmers were very satisfied with the significant increase in fish production from their pond at the end of the trial. Most farmers had also decided to continue this practice in the coming years as well. Mr. Kampoong, though he was satisfied with the production, expressed difficulty in purchasing fertilisers, in terms of both cost and availability in the local market. This applied particularly to TSP. All the participant farmers reported that the recommendations were easy to follow and they were confident that they could continue the practice without technical assistance in the years to come. Moreover, farmer-to-farmer dissemination of technology was also evident. Almost all Project farmers reported that they had shared the recommendations with their relatives and/or neighbours. Most of the farmers were also satisfied with the size of fish obtained at the end of the trial. However, three farmers, Messrs Sunan, Tawatchai and Wantamid, expressed a desire for larger sized fish (2-3 fish kg[-1]) than that they had obtained (4-5 fish kg[-1]). Mr. Surin was the only farmer to express dissatisfaction with the size of individual fish obtained at the end of the trial but he had stocked at an excessively high density of 8 fish m^{-2}. Hence, project farmers' overall perception of fertilisation of ponds with urea and TSP was highly satisfactory.

The pond on small-scale farms in Northeast Thailand is not merely for stocking fish but has multiple purposes (Edwards *et al.*, 1996). In particular, pond water is widely used for irrigating fruits and vegetables; and it is also a source of water for farm animals for many households. In some cases it is the only source of water for household purposes such as washing and cooking. Thus, a farmer with only a single pond in a relatively dry area is always concerned about the multiple use of pond water and may not wish to heavily fertilise water. Farming households may only consider pond fertilisation if they have an alternative source of domestic water. In addition, environmental impact of the fertilisation of farm ponds with inorganic fertilisers, which was beyond the scope of these on-farm trials, needs to be thoroughly assessed. Hence, future research work should thoroughly address these issues for a given location prior to disseminating the pond fertilisation technology using inorganic fertilisers in the wider context in the Northeast region of Thailand.

References

AIT (1986) *Bufalo/Fish and Duck/Fish Integrated Systems for Small-scale Farmers at the Family Level.* AIT Research Report No. 198. Asian Institute of Technology, Bangkok.

AOP (1992) *Inorganic Fertilisation of Fishponds: CRSP Experiments at Huay Luang.* Aqua Outreach Working Paper No. 21. Asian Institute of Technology, Bangkok.

Demaine, H., Innes-Taylor, N.L., Turuongruang, D., Edwards, P., Little, D.C. and Pant, J. (1999) *Small-scale Aquaculture in Northeast Thailand: a Case Study from Udorn Thani.* AARM Program, Asian Institute of Technology, Bangkok. 86 pp.

Edwards, P., Demaine, H., Komolmarl, S., Little, D.C., Innes-Taylor, N.L., Turuongruang, D., Yakupitiyage, A. and Warren, T.J. (1991) Towards the improvement of fish culture by small-scale farmers in Northeast Thailand. *Journal of the Asian Farming Systems Association* 1, 287-302.

Edwards, P., Demaine, H., Innes-Taylor, N. and Turuongruang, D. (1996) Sustainable aquaculture for small-scale farmers: need for a balanced model. *Outlook on Agriculture* 25(1), 19-26.

Edwards, P., Little, D.C. and Yakupitiyage, A. (1997) A comparison of traditional and modified inland artisanal aquaculture systems. *Aquaculture Research* 28, 777-788.

Lin, C.K., Teichert-Coddington, D.R., Green, B.W. and Veverica, K.L. (1997) Fertilisation Regimes. In: Egna, H.S. and Boyd, C.E. (eds) *Dynamics of Pond Aquaculture.* CRC Press, Boca Raton. pp. 73-107.

Mair, G.C. and Little, D.C. (1991) Population control in farmed tilapias. *Naga* 4(2), 8-13.

Szyper, J.P., Lin, C.K., Little, D.C., Setboonsarang, S., Yakupitiyage, A., Edwards, P. and Demaine, H. (1995) *Techniques for Efficient and Sustainable Mass Production of Tilapia in Thailand. Proceedings of Sustainable Aquaculture 95*. Pacon International, Hawaii. pp. 349-356.

Ufodike, E.B.C. and Madhu, C.T. (1986) Effects of methyltestosterone on food utilization and growth of *Sarotherodon niloticus* fry. *Bulletin of the Japanese Society of Scientific Fisheries* 52(11), 1919-1922.

Chapter 10

Improved Management of Small-scale Tropical Cage Culture Systems in Bangladesh: Potential Benefits of an Alliance between an NGO and a Western Research Institute

K.I. McAndrew[2], C. Brugere[1], M.C.M. Beveridge[1], M.J. Ireland[2], T.K. Roy[2] and K. Yesmin[2]

[1]*Institute of Aquaculture, University of Stirling, Stirling FK9 4LA, UK*
[2]*CARE Bangladesh, Dhaka 1209, Bangladesh*

Abstract

Unlike in most Asian countries, cage aquaculture has no tradition in Bangladesh. Although cage culture has been shown in some cases to be technically viable, it has previously failed when an attempt was made to involve rural farming and fishing households. The CARE-CAGES Project introduced cage aquaculture to poor rural communities throughout Bangladesh, and was supported by a Research Project, funded under the DFID RNR Aquaculture Research Programme. The chapter describes the activities and outputs of the Research Project, which has been integrated into the larger development Project (CARE-CAGES), with mutual benefits to both.

Introduction

A major constraint to the development of cage fish culture by poor communities is the lack of technologies appropriate to their social, institutional, resource and environmental context. A 2-year, DFID-funded Research Project (R7100) to improve the management of small-scale cage culture was recently awarded to the Institute of Aquaculture (IoA), University of Stirling, and the Asian Institute of Technology (AIT), Bangkok. AIT are collaborating with the University of Fisheries, Nha Trang, Vietnam, while IoA is working primarily with a DFID-funded non-governmental organisation (NGO) project in Bangladesh — the

CARE-CAGES (Cage Aquaculture for Greater Economic Security) Project. The chapter details the structure, aims, activities and outputs of the Bangladesh component of Project R7100.

Why Cage Aquaculture?

Cage aquaculture has certain advantages over other aquaculture systems that are potentially important in terms of uptake by rural poor and landless people. The integrity of the cage unit means that large, communal water bodies can be used and, crucially, the ability to culture fish is not reliant on the ownership or leasing of land. Hence, where access to a water body can be achieved, landless people can grow fish in cages and obtain nutrition and/or income from the fish produced. In addition, a common problem encountered with ponds in Bangladesh is multiple ownership. This can result in conflicts when determining the ownership of fish produced in traditional pond aquaculture, resulting in the under-utilisation of the water resource for fish production. In cage aquaculture the ownership issue is simple: the owner(s) of the cage being the owner(s) of the fish within the cage.

Cages have other advantages over traditional aquaculture (Beveridge, 1996; Beveridge and Stewart, 1998). In the monsoon of 1998 the water in many pond farms increased to a level where fish were able to swim out of the ponds and were subsequently lost. Fish in cages cannot escape, provided a top net is present and the net is not damaged. Cages also exclude predators. A farmer in Sylhet region noted that while snakes and birds took large numbers of fish from his stocked pond, fish within cages were unharmed. A further advantage of the cage systems is that caged fish are easily caught and harvested, a principal reason why one NGO officer in Barisal kept fish in cages. When he had guests, a few large fish could readily be chosen to impress his guests, avoiding the expense of having to kill livestock.

The CARE-CAGES Project

The culture of fish in cages is a promising aquaculture technology already proven in many other Asian countries (Beveridge, 1996; Beveridge and Stewart, 1998; Beveridge and Muir, in press). Cage aquaculture technology has been tested since the late 1970s in Bangladesh research establishments. Although results have been variable, they demonstrated that it was technically viable (Hossain *et al.*, 1986; Haque, 1978). However, failure of projects in Kaptai Lake (Mollah *et al.*, 1992; Felix, 1986), Dhanmondi Lake (Karim, 1993) and, despite early encouraging results, in Parbatipur (Johnson, 1993) have left a strong impression in the aquaculture and fisheries agencies that cage aquaculture was not feasible in Bangladesh.

The CAGES project run by CARE Bangladesh is a pioneer in developing and promoting this system of fish culture throughout Bangladesh, and the first aquaculture development Project in the country to focus on cage aquaculture systems. Since September 1995 this DFID-supported Project worked in a number of sites across the country, building on lessons learnt by the Northwest Fisheries Extension Project (NFEP) and CARE during their earlier collaboration (Gregory and Kamp, 1995). The project is currently in a phase of rapid expansion and plans to work with 1,770 households through 44 NGOs in 1999, compared with 632 households and 22 NGOs in 1998 (Stewart *et al.,* 1999) (Fig. 10.1).

Fig. 10.1. Cages in an irrigation canal, Sylhet.

Being a technology new to Bangladesh and with only limited information on cage culture systems on which to base designs, the first three years of the project have focused heavily on the development of basic aspects of the technology and on establishing the appropriate approach to development. The activities of participant farmers and partner NGOs have greatly contributed to this process.

In the first season of the Project in 1996, CAGES staff worked directly with farmers and were able to offer advice based on the resources and wants of individual farmers. Since 1997 the Project has worked through local NGOs, with Project staff providing training in all aspects of cage culture to NGO workers, who pass this information on to their farmer groups. Early research trials carried out at the CAGES Research Farm were important in providing staff with knowledge and experience. These early trials, as well as experiences from earlier attempts at cage aquaculture in Bangladesh and the wider literature, formed the basis of advice, which was disseminated to NGO/farmer beneficiaries. There is, however, no substitute for experience. As a new production system, cage aquaculture had mixed results during these early years, with many farmers failing to grow fish successfully. Although not financially at risk — as CAGES provided nets, fingerlings and in some cases feed — many farmers stopped cage farming, due to inconsistent returns. The project memorandum (ODA, 1995) did note that 'the basic technology would require small changes to fit the environment and economic conditions of Bangladesh'. In practice these adjustments were considerable, especially considering the limited resources available to Bangladeshi farmers. It is hoped that the strengthening of the research component of the CARE Project, through the DFID-funded Research Project, will lead to reduced risks to farmers and increase the ability of rural poor farmers to successfully farm fish in cages as one component of their livelihood strategy.

Although improving each year, the consistency of results from the cage systems continues to be a concern. Approximately 70% of households achieved a profit from cage production operations in 1998, compared to only 30% in 1997. The current average net profit per household for 1998, based on 9% of households, is Taka116 (US$1 = Tk47). Although this is a significant improvement on 1997, it is still well short of the project target of Tk2,400 per household year[-1]. The most common constraints faced were high post-stocking mortalities, poor feeding strategies and stock losses due to storms or theft. The inexperience of the households involved and deficiencies in support to households during this first year of work meant that technical and social constraints were not overcome. An improved understanding of the system gained through experience has been the principal reason for the improved performance.

The CAGES Project has previously collaborated with a range of research institutes, the Bangladesh Department of Fisheries (DoF), and in-country aquaculture projects. The DFID aquaculture research programme (ARP) had earlier supported an MSc project from the Institute of Aquaculture (IoA), University of Stirling, on the use of periphyton to enhance tilapia production in cages (Huchette, 1997). This institutional relationship is currently being further developed through the present Research Project, which began in February 1998, and will run until March 2000. IoA is supporting the on-going research of the CAGES Project, and new linkages are being forged with AIT and the University of Fisheries, Vietnam, through sharing of experiences, information and reciprocal visits.

The Research Project - Rationale, Aims and Synergies

The present research project, Project R7100, was established to help develop sustainable small-scale, cage fish culture in inland (Bangladesh) and coastal (Vietnam) waters, which are presently not well developed. It is hypothesised that a major constraint to the development of small-scale, cage fish culture by poor communities has been the lack of technologies appropriate to the social, institutional, resource and environmental context. By examining a number of current cage culture activities in Bangladesh and Vietnam, the project aims to improve understanding of the context, and develop guidelines for small-scale, cage culture development at both structural/strategic levels and at the level of the small producer, through improved management. The CARE-CAGES link presented an excellent opportunity with many potential synergies, and brings added value to both the research and development programmes of DFID. It should be noted that the development project had a small research element that proved essential in the early stages of the project. However, the CAGES Project is now benefiting from additional and more focused inputs available from the IoA. Furthermore, the integration of Project R7100 with the CAGES Project fully exploits the promotional pathways through which research results can be transformed into products for dissemination and adoption.

In a recent CAGES Project review (Stewart *et al.*, 1999), it was noted that the Research Project offers considerable value for money in that the research design and delivery are integrated within an on-going Project which can provide backup and support. The reviewers also commented that participatory research passes some of the costs of research management to collaborating institutions and end-users. The fact that local participants justify their costs against potential benefits derived (an element of the 'beneficiary pays' principle) encourages a culture of value for money (Stewart *et al.*, 1999). It is acknowledged that the success of the Research Project is reliant on the continued goodwill and commitment of CAGES staff, but it is hoped that the projects will be mutually beneficial.

Initial Activities

Preliminary activities of the Research Project involved visits to each region where the CAGES Project operates to develop an understanding of the issues involved and to identify the needs of cage operators. In close collaboration with CAGES staff, the following activities are currently being undertaken.

Literature Review of Cage Culture in Bangladesh

This will be based on existing published information, including project documentation of partner institutions and interviews with secondary stakeholders (including project staff, government departments, university researchers and relevant NGOs) in study regions. The review will define the multidisciplinary context, key parameters and performance indicators, and identify issues of importance to cage aquaculture in Bangladesh. A similar country literature review of Vietnam is currently being coordinated by AIT researchers, with both outputs intended to be submitted for publication in a peer-reviewed journal.

Documenting NGO/Farmer Experiences

An important role of the Research Project is to support the documentation by CAGES staff and NGOs of the many farmer/NGO trials that have taken place to better focus current research priorities and support future trials. An output of this activity will be a document produced in both English and Bangla. It is hoped that other cage trials and experiences in Bangladesh from outside the CARE-CAGES Project will also be included. The output will have several benefits to NGOs and farmers, including:

- Informing the farmers and NGOs of trials that have already been conducted and the key results — it is hoped that this will prevent the repetition of experimental trials and the 'reinventing of the wheel' which at present is a major problem with farmer/NGO research in the CAGES project.
- Conclusions and recommendations drawn from the trials will allow farmers/NGOs to focus trials on areas in which further knowledge is important to the success of cage aquaculture in Bangladesh.
- All farmers/NGOs whose trials appear in the booklet will be fully acknowledged — it is hoped that this will encourage and motivate further research by farmers/NGOs.
- Farmers/NGOs will learn from the process of documenting their trials in a formal research format and encouragement and advice will be given on experimental design, data collection and analysis, and writing up of trials, if so desired — it is fully acknowledged that the trials are the intellectual property rights of the farmers and advice will only be given if requested.
- Many farmer/NGO trials, although not scientifically rigorous, are nevertheless of interest to persons outside CARE Bangladesh, and may also be of benefit to researchers and farmers beyond the life of the present Project.

Publishing and Disseminating Experimental Research Results from the CARE-CAGES Research Farm

Over the last two years much interesting experimental work has been carried out on the CAGES research farm, and written up in internal CARE reports. The Research Project is currently supporting the preparation of several papers for publication in peer-reviewed, international journals. Others will be published in the farmer/NGO trial publication detailed above. The publishing strategy is designed to:

- Encourage and motivate staff.
- Strengthen research design, implementation, data analysis and documentation skills among CAGES staff.
- Allow valuable information to be available to the wider research community.

The Research Project is also engaged in focusing and supporting the design of future research at the CAGES research farm. Support includes assistance with literature reviews, statistical methods, and writing up research outputs. All trials being undertaken will come from direct contact with cage farmers in Bangladesh through the network established by CARE-CAGES. In this way all research should be highly focused on the needs of cage operators.

Key Social Issues

Much of the present research has been carried out to improve the technical feasibility of cage culture in the targeted areas and to help farmers develop appropriate methods of culture to suit their needs. However, all cage culture is being carried out within the social context of Bangladesh and the uptake of technology is dependent upon it being acceptable to potential beneficiaries and the communities in which they live. Three social issues were identified by CARE-CAGES staff as being important to the continued success of the cage culture development project. A social survey was designed jointly by IoA and CARE staff to provide an understanding of the issues involved through participatory and culturally sensitive methods of investigation (Fig. 10.2). The survey dealt with the following issues:

- Understanding the reasons behind the success and failure of cage operators and determining expectations — this will enable the key factors that determine success and failure of cage culture to be recognised, and used to advise other farmers.
- Understanding the decision-making process within cage operator households regarding the management of cages and the division of labour and opportunity costs related to the activity — it is hoped this will enable CARE-CAGES staff to support the participation of women and avoid the burden of any negative effects of cage aquaculture to fall disproportionately on women and children.
- Understanding the complexities of the implementation of cage culture at the community level and to identify possible areas of conflict that might arise between various water user groups such as fishers, women, bathers, clothes washers and cage operators, as well as mechanisms of conflict self-regulation.

The results of the surveys are being processed in order that the results may be used by CAGES staff to refine and develop extension strategies for CAGES and partner organisations. Information derived from the social surveys will also be used to develop guidelines for policy makers and planners. Preliminary results suggest that, although cage culture has created some changes in the use of water bodies by villagers, the changes have been accepted by community members and have not led to conflicts between the various groups of water users. However, if the changes were to be exacerbated by increases in the number of cages, there is a potential for animosity and conflict between water users. At present, cage culture is accepted by villagers and often triggers curiosity and interest from neighbouring community members, thereby contributing to an increase in the social importance of those who practice it (both male and female cage operators). The cultural and religious context of

Bangladesh is predominant in determining the extent of women's participation in cage aquaculture, often restricting their contribution to auxiliary house-based tasks, particularly in the more fundamentalist areas of the country.

Fig. 10.2. Children explaining their understanding of cage culture, Jessore.

A similar social survey will be conducted with the same farmers 1 year later to document changes in the perception of cage aquaculture by farmers and the wider community and to assess the impacts of cage aquaculture. This is especially important in light of the expected increase in production which could potentially result in increased resource use conflict and environmental degradation.

Environmental Issues

Environmental issues were highlighted in the project memorandum as integral to the successful development of cage culture. However, as noted in the CARE-CAGES mid-term review, the implementation of scientifically sound environmental monitoring schemes for each site is impossible, given resources available in Bangladesh (Stewart *et al.,* 1997). Although harvesting biomass has been up to 100kg m^{-3} in some 1m^3 cages, the small average cage size of 8-9m^3, and limited numbers of cages in each water body (typically two to eight in a group) has meant that at present environmental degradation by cage aquaculture is not an issue as demonstrated in preliminary social survey results. Local eutrophication of some water bodies (usually small in size) has occurred, although none of these has led to fish kills or serious problems to other water resource users. Several cage operators, however, intend to substantially increase production in 1999, with one operator in Comilla intending to farm 70 cages of pangas (*Pangasius sutchi*). In such situations the carrying capacity of the water body may be exceeded, resulting in environmental degradation which could result in poor growth rates or fish kills, as well as resource conflicts with other water users. Operators planning such dramatic expansion are being advised on the potential negative consequences, and their situation is being carefully monitored.

In cage aquaculture escaped (feral) fish are inevitable (Beveridge, 1984, 1996; Beveridge *et al.*, 1994). Production figures from 1997 show that in the five regions, 19-35% of all fish stocked escaped from the cages into surrounding water bodies (Mallick, 1997). The feral fish will then compete with — and in the case of carnivorous species predate upon — local indigenous fish. The mid-term review team for the CAGES Project recommended that a research priority should be use of local species in order to minimise the impact of escapees. NGO/farmer trials on several native candidates for cage culture took place last year and results are currently being analysed. However, exotic tilapia species are the most commonly cultured species in cages in Bangladesh. A recent survey of tilapia in Bangladesh found only limited evidence of breeding in the wild, suggesting that their presence was having only minimal impact on fisheries (Ireland *et al.*, 1996). Nevertheless, as with other omnivorous and herbivorous fish, the escaped fish will compete with native species for food. Exotic carnivores, especially the African catfish (*Clarias gariepinus*), are not recommended for stocking; not only because of potential environmental problems but also because the species is uneconomic to farm due to its low market price.

In previous years there have been serious problems with fish mortalities, especially during stocking, with some farmers having to re-stock fry up to four times before sufficient numbers survived to make culture worthwhile (Mallick, 1997). Fish were often underfed, leaving the fish vulnerable to infection by opportunistic pathogens. The CAGES Technical Officer encouraged farmers in Barisal to increase the protein level of feeds, and this more expensive feed led to much better growth and survival of fish, and subsequent profitability. Growth and survival of fish is improving each year as farmers become more experienced and knowledgeable about cage culture practices. Production figures for 1997 showed that on average 52% of fish stocked were lost during the culture period, although only around 34% occurred during the grow-out phase (Mallick, 1997). With improved advice being offered and experience gained by cage farmers to date, Technical Officers expect that mortalities in subsequent years will be greatly reduced.

One pathogen that can easily be observed without the need for fish health laboratories is an isopod parasite that is found in the Naohata river, Jessore region. The parasite was first reported by the Jessore Technical Officer and a local NGO who reported that infected fish were dead and anaemic within 2 hours of infestation. These Bangladesh researchers have already carried out a preliminary study of the parasite which showed the parasites to be causing economic loss to cage farmers on the river as fish stocking has to be delayed until the monsoon arrives and 'the parasites disappear'. An MSc student from IoS has recently successfully completed his dissertation on the parasite with field work taking place during April/May 1999. All findings were disseminated to local cage farmers immediately and a copy of the thesis given to the CAGES Project.

Information Transfer to CAGES and Associated Staff

The Research Project aims to create effective information transfer among all participating groups of researchers. Bangladeshi staff have access to information and expertise through the IoA link to both literature and knowledge of a diverse disciplinary team of aquaculture specialists from IoA. The Bangladesh project has already benefited from advice on *hapa* breeding, parasitology, database analysis and social issues.

Workshops

Workshops are used as a means of transferring information and stimulating discussions on important topics. Recent/planned workshops include those on:

- *Social issues.* In December 1998 a social scientist from IoA and the CAGES Project Co-ordinator conducted a workshop on the importance of socio-economic issues in the success of technical programmes. The workshop was attended by all CAGES staff and staff from other CARE programmes. Workshop proceedings have been produced and are available through CARE-Bangladesh.
- *Experimental design and statistical methods.* Three 1-day workshops on experimental design and statistical methods are planned between now and the Research Project end. These will be conducted jointly by IoA and specialist CAGES staff. The aim is to develop skills in experimental design, data collection and storage, data manipulation and data presentation skills among CAGES staff, resulting in better research outputs from the project.
- *Hapa breeding of tilapia.* A major problem faced by cage farmers in Bangladesh is the availability and poor quality of fry. One way to ensure timely and good quality supplies of fingerlings is to breed tilapia in *hapas*. A workshop on *hapa* breeding of tilapia took place in March 1999, with researchers from NFEP, IoA, focal NGOs and CAGES staff present. The workshop focused on the *hapa* breeding that has been carried out in Bangladesh and elsewhere in Asia, with the aim of documenting attempts and making recommendations to potential *hapa* breeders in Bangladesh.
- *Key issues in Bangladesh and Vietnam cage aquaculture.* As described above, Project R7100 has a Vietnamese component, managed through AIT.

An end of Project workshop is planned for February 2000, which will involve researchers and development workers from Bangladesh and Vietnam, and IoA and AIT. Key issues in cage aquaculture will be discussed in these strikingly different social and physical environments (Bangladesh freshwater and Vietnam coastal aquaculture) and compared and contrasted. Outputs will include:

- Guidelines for the production of low/medium input, low/medium output cage culture extension and training materials.
- Guidelines for planners, policy makers and development specialists to assess the potential for cage culture within social, economic and institutional environments and improve the selection of technological options.

Improvement of CAGES database

There was concern in the recent annual review about the usefulness of the database currently used by the CARE-CAGES Project (Stewart *et al.*, 1999). An IoA researcher visited Bangladesh and worked on the database in conjunction with CAGES staff for 3 weeks in March/April 1999. Work was aimed at improving the database and the skills of CAGES staff. The 1998 production data has recently been entered into the database by CARE staff. Current priorities are to analyse the database to investigate key technical parameters for success, to allow improved advice to beneficiaries. Data in 1998 are also being linked with previous years' production figures, as well as with previously collected social and cultural information on households and on their success or failure. It is expected that the now functional database will be a powerful information source for project planning.

Recommended Practices for Cage Farmers in Bangladesh

A further output of the research project is a compilation of information on recommended practices for successful cage culture in Bangladesh. The output will incorporate:

- The knowledge and experience of CAGES field staff, NGOs and farmers involved in cage aquaculture in Bangladesh.
- Information from Cage Production Reports, detailing success and failure of cage farmers, produced by CARE-CAGES.
- Data from trials conducted by university and research facilities in Bangladesh on cage aquaculture.
- Experience derived from the Bangladesh DoF at Kaptai and the Oxbow Lakes projects.
- The wider literature.

This output will provide information on recommend practices to minimise risks and maximise returns on investment, with contents including:

- The biology of commonly cultured fish in Bangladesh.
- Recommended technical options, including details on feeds (type, amount) and feeding systems (feeding basket, fine mesh cage bottom), stocking density, size of fry at stocking, species-specific cage designs.
- An analysis of costs associated with inputs of all technical options and a realistic assessment of the returns that can be expected.
- A summary of findings from the social surveys currently being carried out, and their practical implications for the potential of cage aquaculture in Bangladesh.

The output will be made available to CAGES staff, all contributors, large NGOs and appropriate government departments, as well as other interested parties. As information is gained and developed, the booklet will be updated and thus will be published in a loose-leaf format.

Conclusions

It is hoped that the wide range of support provided by Research Project R7100 will enable the CARE-CAGES development Project to better achieve its aims, allowing the successful uptake of cage culture technologies appropriate to the social, institutional, resource and environmental context of Bangladesh. Working in partnership with the CARE-CAGES network gives IoA researchers access to large numbers of cage operators, enabling research to be demand-led and allowing pathways for data collection. The synergies achieved represent excellent value for money. The outputs of the Research Project will be useful not only to the CAGES Project but information, experiences and recommendations from the Bangladesh experience will be made available to the wider research and development communities.

References

Beveridge, M.C.M. (1984) *The Environmental Impact of Freshwater Cage and Pen Fish Farming and the Use of Simple Models to Predict Carrying Capacity*. FAO Technical Paper 255. FAO, Rome. 131 pp.

Beveridge, M.C.M. (1996) *Cage Aquaculture*, 2nd edn. Fishing News Books, Oxford. 346 pp.

Beveridge, M.C.M. and Muir, J.F. (in press) Environmental impacts and sustainability of cage culture in Southeast Asian lakes and reservoirs. In: van Densen, W.L.T. and Morris, M.J. (eds) *Fish and Fisheries of Lakes and Reservoirs in Southeast Asia and Africa*. Westbury Publishing, Otley, UK.

Beveridge, M.C.M. and Stewart, J.A. (1998) Cage culture: limitations in lakes and reservoirs. In: Petr, T. (ed.) *Inland Fishery Enhancements*. FAO Technical Paper 374. FAO, Rome, pp. 263-279.

Beveridge, M.C.M., Ross, L.G. and Kelly, L.A. (1994) Aquaculture and biodiversity. *Ambio* 23, 497-502.

Felix, S.S.M. (1986) *Terminal Report on Bangladesh Aquaculture Development Project (ADB)*. International Agro-Fisheries System, Manila. 62 pp.

Gregory, R. and Kamp, K. (1995) Fish culture in cages by woman. In: BAFRU (eds) *Aquaculture Extension in Bangladesh: Experiences for the Northwest Fisheries Extension Project 1989-1992*. Institute of Aquaculture, University of Stirling.

Haque, A.A.K.M. (1978) On the use of floating pond for fish-farming under Bangladesh condition. *Bangladesh Journal of Fisheries* 1(2), 155-157.

Hossain, A., Islam, A., Ali, S. and Hossain, A. (1986) Culture of some fishes in cages. *Bangladesh Journal of Aquaculture* 6-7(1), 65-69.

Huchette, S. (1997) Technical and economical evaluation of periphyton-based cage culture of tilapia (*Oreochromis niloticus*) in Bangladesh. MSc Thesis. Institute of Aquaculture, University of Stirling. 92 pp.

Ireland, M.J., Tapash, K.R., Nuran Nabi, S.M., Rahman, M.A., Huque, S.M.Z. and Aleem, N.A. (1996) Are tilapia breeding in the open waters of Bangladesh? The results of a preliminary countrywide survey. CARE Bangladesh ANR Workshop, 2-3 February, 1997, Jessore.

Johnson, K. (1993) Cage aquaculture trials in Rangpur, findings and recommendations.

Karim, M. (1993) Status and prospects for pen and cage culture in Bangladesh. FAO/UNDP TA-Project BUD/83/010. IDA/WD Agriculture Research II project. Fisheries Research Institute, Ministry of Fisheries and Livestock, Dhaka. 58 pp.

Mallick, P.S. (1997) Cage production reports 1997. CAGES Project, CARE Bangladesh. Dhaka, Bangladesh. 20 pp.

Mollah, M.F.A., Haque, M.K.I., Halder, G.C. and Ahmed, K.K. (1992) Cage culture of Indian major carp and silver carp at different stocking rates in Kaptai lake. *Bangladesh Journal of Agricultural Science* 19(2), 257-264.

Overseas Development Administration (1995) Cage Aquaculture for Greater Economic Security (CAGES). Project Memorandum (August 1995-July 2000). Aid Management Office, Overseas Development Administration (ODA), British High Commission, Dhaka, Bangladesh.

Stewart, J.A., Ireland, M.J. and Townsley, P. (1999) DFID/CARE CAGES Project. Annual Review. January 1999. Unpublished Report. 44 pp.

Stewart, J.A., Rusenow, T., Townsley, P. and Beveridge, M.C.M. (1997) DFID/CARE CAGES Project. Mid-Term Review. 24 November-10 December, 1997. Institute of Aquaculture, University of Stirling, Stirling, Scotland. 22 pp.

Chapter 11

Towards Sustainable Development of Floating Net Cage Culture for Income Security in Rural Indonesia: a Case Study of Common Carp Production at Lake Maninjau, Indonesia

A. Munzir and F. Heidhues

Department for Agricultural Economics and Social Sciences in the Tropics and Subtropics, University of Hohenheim (490), D-70593 Stuttgart, Germany

Abstract

Commercial floating net cage culture is a rapidly growing industry in rural Indonesia. It has become the economic base for income generation in poverty alleviation programmes of the Government of West Sumatra province in two villages at Lake Maninjau. The increasing interest in cage development raises questions of its determining factors. This study investigated the influence of different factors on common carp production in cages at Lake Maninjau. Although production function analysis showed that higher fish production could be attained by increasing inputs, there was visual indication of the degradation of water quality. For sustainable development of cage culture, it is necessary to consider the carrying capacity of the lake, the ecological dynamics of the cage system, and the formulation and implementation of regulations to promote environmental preservation.

Introduction

Raising fish in floating cages is well known in Southeast Asia, e.g. in the Mekong river and in large rivers of Sumatra and Kalimantan (Christensen, 1989). There are also submerged cage systems in small rivers, which have been operated in Indonesia for a long time (Costa-Pierce and Effendi, 1988; Dahuri, 1997; Soewardi, 1997). According to Sedana (1997), cage culture had not started before 1978 in some Sumatran provinces such as Riau province and until the end of the 1980s, it was only a sideline business for fish farmers; production was mainly for family consumption. The cage system did not seem to be a profitable method of fish culture either in freshwater bodies in Sri Lanka (Galapitage, 1981).

Commercial floating net cage culture is widespread today in rural Indonesia, especially in West Java and West Sumatra, having been practised since the beginning of the 1990s. Soewardi (1997) reported that intensive floating net cage culture was first introduced in Lido Lake, West Java in the 1970s. Large-scale systems were developed in Saguling and Cirata hydropower reservoirs after 1986 and in Jatiluhur reservoir since 1990. Total production of the three reservoirs in 1996 was 48,703t, almost twice the total fish production of West Java Province in 1995. The floating net cage system becomes a rapidly growing industry. It has the potential for both income generation and food security improvement in rural Indonesia by increasing the purchasing power of fish farmers. The cage industry may play an important role in the future of Indonesia's aquaculture development based on opportunities to use reservoirs for its expansion in many parts of Indonesia (Husen, 1997).

The Government and residents in two villages, Pantai Barat and Pantai Panjang, West Sumatra province, decided to develop net cage culture as part of a poverty alleviation programme. As both villages are located on the slopes of Lake Maninjau in the Barisan mountains, there is only limited land for crop cultivation so the government launched a fund called *Dana Ingress Desa Tertinggal* (IDT Fund) for the poor to start operating and developing fish cages.

Research into the sustainable development of cage farms which was necessary to secure the rural community's income generation, remained to be carried out in the area. This became even more pressing after mass mortality of cultivated fish in January and October 1997, the year the El-Niño phenomenon occurred which brought environmental disaster and the economic and political crises erupted in Indonesia.

This chapter analyses the factors that determine common carp (*Cyprinus carpio*) production as the base for a strategy of future sustainable development of floating net cages. It was hypothesised that the net cage production system at Lake Maninjau was still in the condition of increasing returns to scale.

Methodology

The study was based on primary and secondary data. A total of 80 cage farmers who were rearing common carp in Pantai Barat and Pantai Panjang villages were randomly selected for interview using standardised questionnaires with open and closed questions. Ten explanatory variables, two of them being dummy variables, were selected to explain the cage production system. For this purpose, a Cobb-Douglas (C-D) production function model was specified as below.

$$Y = \alpha \, X_1^{\beta 1} \, X_2^{\beta 2} \, X_3^{\beta 3} \, X_4^{\beta 4} \, X_5^{\beta 5} \, X_6^{\beta 6} \, X_7^{\beta 7} \, X_8^{\beta 8} \, D_1 \, D_2 \, e^{U}$$

where:

Y	=	gross output of fish (kg culture per time)
X_1	=	total rearing area (m^3)
X_2	=	quantity of fingerlings (kg culture per time)
X_3	=	quantity of fish meal (kg culture per time)
X_4	=	rearing period (month culture per time)
X_5	=	labour used (man days culture per time)
X_6	=	capital investment (Rupiah, Rp.)
X_7	=	formal education of cage aquaculturist (year)
X_8	=	experience of farmer (year)
D_1	=	dummy variable = 1, for those who received *Dana Ingress Desa Tertinggal* (IDT Fund) from government or a member of an IDT group, 0 otherwise
D_2	=	dummy variable = 1, if cage farm located in Pantai Barat, 0 otherwise
α and β_i =		parameters (regression coefficients) to be estimated
U	=	random error or disturbance term
e	=	natural logarithm.

To evaluate the relative influence of each of the explanatory variables on output, the production function was estimated by using multiple regression techniques.

Results

Technical Aspects

The enclosure of the floating cages was made of polyethylene net. It consisted of an outer net to protect the fish from predators and an inner net as a rearing area. The enclosure was suspended from a frame construction made of bamboo, the simplest construction. A large and well-developed net cage fish farm was constructed of wood and drums as floats with a small hut constructed on the frame. The rearing area of one unit of the floating net cage, varied from 4 x 4 x (2.5-3) to 7 x 7 x $3m^3$. Some fish farmers used a water column of more than 3m in depth or between 2 and 5m depths. The most common depth was 3m.

The Adoption of Technology

Floating net cage technology was introduced into West Sumatra by a young well-educated aquaculturist in 1991. He tried to rear fish in one net cage of 4 x 4 x 3m in size at Lake Maninjau. Before 1991 the fishing activities of the rural community were limited to small-scale fishing and collecting small mussels from the lakeshore. The government supported the development of the net cage technology in 1992 by providing a fund for 17 fishers' households to run one cage each. Subsequently, the technology was widely adopted by the residents of Maninjau so that the number of units and households engaged in cage culture increased rapidly (Fig. 11.1). Most fish cultured were common carp (*Cyprinus carpio*). Some fish farmers also reared some Nile tilapia (*Oreochromis niloticus*).

Fig. 11.1. Floating net cage culture in Lake Maninjau, Indonesia.

Number of Households in Net Cage Culture

The number of households engaged in net cage culture increased rapidly, although between 1992 and 1994 there are no recorded data. The office of the government fisheries service started recording data on the development of cages in 1994 (Table 11.1).

Table 11.1. The development of the number of household net cage farmers and cage production. Source, Kantor Dinas Perikanan Cabang Kabupaten Agam (1997).

	Household		Cage units		Production		Average production (kg)	
Year	Number	Growth (%)	Number	Growth (%)	(kg)	Growth (%)	Per household	Per cage unit
1991	17	-	17	-	-	-	-	-
1992	NA	-	NA	-	-	-	-	-
1993	NA	-	NA	-	-	-	-	-
1994	NA	-	NA	-	450,000	-	-	-
1995	536	-	1,234	-	555,000	-	1,035.5	449.8
1996	570	6	1,886	53	2,467,765	345	4,329.4	1,308.5
1997	617	8	2,017	7	3,744,165	52	6,068.3	1,856.3

NA = Not available.

The increasing number of rural households engaged in cage culture as well as the increasing number of cages per fish farm caused a rapid increase in total fish production. In the year 1995-1996, 345% growth was reached (Table 11.1). The increase of cage unit numbers, total production and the number of household's farmers engaged in net cage culture are presented in Fig. 11.2.

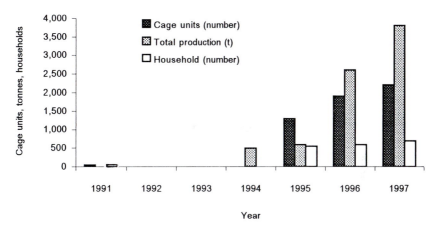

Fig. 11.2. The increase in cage unit number, total production and households engaged in net cage culture at Lake Maninjau. Data for 1992 and 1993 are not available.

The rate of increase of net cage production at Lake Maninjau slowed down in 1997 due to rising fish mortality as a result of degraded water quality. The office of the government fisheries service reported that 35.5t of fish died in cages and 58.5t were sold at half price between 10-12 January 1997. This disaster happened again in October 1997 when cage fish farmers lost 153.7t of fish.

Based on visual observation, it was assumed that there was an overturn of the lake's waters because of strong winds, which lifted detritus from the lake's bottom, causing a serious degradation of water quality. However, there was no comprehensive research on this matter.

Although the rate of increase of net cage production tended to slow down due to the mortality of cultured fish, the average production cage per unit and per household increased rapidly (Fig. 11.3). This indicated that running the cage culture farming system was a popular choice to secure income generation as well as to alleviate poverty in these rural areas.

Estimated Production Function

The estimated coefficients of the Cobb-Douglas production function are summarised in Table 11.2.

Table 11.2. Analysis of the estimated coefficients of Cobb-Douglas production function.

	Intercept	X1	X2	X3	X4	X5	X6	X7	X8	D1	D2
Production coefficients	-0.731	0.109	0.480***	0.217***	0.044	0.013	0.264**	0.242	0.042	0.012	-0.069*
Standard error	0.656	0.095	0.116	0.053	0.129	0277	0.127	0.151	0.079	0.025	0.042
T-Ratio	-1.114	1.143	4.124	4.066	0.342	0.048	2.071	1.610	0.536	-0.470	-1.629
Sig. T	0.269	0.257	0.000	0.0001	0.734	0.962	0.042	0.112	0.594	0.640	0.108
Mean value:											
- Inputs		2.087	2.399	3.325	0.374	1.973	6.084	0.810	0.402	0.625	0.625
Standard Deviation		0.248	0.282	0.246	0.090	0.073	0.276	0.122	0.134	0.487	0.487
- Output	3.176										
Standard Deviation	0.279										
Multiple R	0.959										
R^2	0.920										
Adjusted R^2	0.909										
Standard error	0.084										
Durbin-Watson test	1.230										

Analysis of variance:

	DF	Sum of squares	Mean square
Regression	10	5.679	0.568
Residual	69	0.490	0.007
F = 79.915		Significant F = 0.0000	

*** Highly significant at the level of 1%.
** Significant at the level of 5%.
* Significant at the level of 10%.

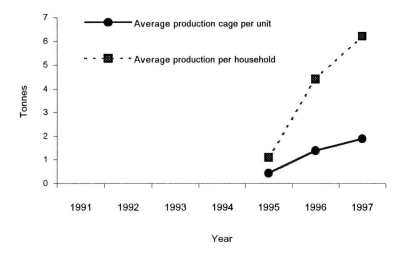

Fig. 11.3. The average production in tonnes cage per unit and tonnes per household of cage culture at Lake Maninjau.

Discussion

Despite some criticism of the C-D production function, the C-D model is widely used in farm and general economic production studies. It conforms to economic theory and provides a compromise between an adequate fit of data, computational feasibility and sufficient degrees of freedom to allow statistical testing (Douglas, 1976; Asnawi, 1981; Soekartawi, 1990). The C-D model was used in this study because it was adequate to reach the defined objective and to test the underlying hypotheses.

Olatubi (1991) tested four functional forms of the production function model to explain the input-output relation in rural aquaculture in Nigeria, i.e. double-log function, semi-log function, exponential function and linear function. He found that the linear form yielded the 'best fit'. Similar tests in Philippine milkfish aquaculture were done by Chong and Lizarondo (1981) for three models, i.e. linear, quadratic, and C-D; and the result was that the C-D model, which is linear in its logarithmic form, gave the best result.

The high R^2 and significance parameter (F-value) indicated that the estimated production function fitted the data well. The model explained 92% of the yield variation of the cage production system. Two explanatory variables in the model were highly significant, the quantity of fingerlings or stocking rate (X2) and the quantity of fishmeal (X3). There was also no problem with dominant variables or multicollinearity, as shown by the Durbin-Watson statistic (Table 11.2).

Saputra (1988) argued that the optimal stocking density of fingerlings in a fish cage is determined by environmental factors and fish species. As these differ from area to area no standard stocking densities have been recommended. The stocking rate applied by the cage fish farmers at Lake Maninjau varied in a range of 70 to 200kg of total fingerlings in one cage, which varied depending on cage size. The average stocking rate was 2.1kg fingerlings m^{-3} or 12.2kg m^{-2}, which produced fish at an average of 12.6kg m^{-3} or 72.3kg m^{-2}, respectively.

From the average weight of common carp fingerlings of 305 kg stocked into a cage production system at Lake Maninjau, the yield was 1,815kg fish or six times the initial biomass in one culture period of an average of 2.4 months. Based on several experiments, Saputra (1988) found multipliers of 8 to 9 for tilapia in 4 to 5 months. For common carp he found a multiplier of 6 times in 6 months after stocking. These findings, however, cannot be directly compared with the production condition of cages at Lake Maninjau due to differences in the species, initial size of fingerlings and environmental factors.

From the point of view of income security and poverty alleviation in rural areas, the cage production system could play an important role for fish farmers since fish can be harvested regularly in small quantities. It is more profitable than traditional aquaculture and rice cultivation. According to Soewardi (1997), cages tend to be inexpensive and easy to construct and manage, requiring only modest skill. The profit from one cage is equivalent to 2.5 times the profit of a 1ha rice field.

Referring to experiences and recorded data from several production cycles by the innovator of cage culture at Lake Maninjau, the food conversion ratios of common carp production were 1.5 for fish seed 2-5cm; and between 1.8 to 2.0 for fingerlings of size of 5-8cm. These food conversion ratios are considered to be adequate.

The capital investment (X6) was found to be a significant factor in yields of the whole production process of net cage culture. On average, the capital requirement amounted to US$586 (Rp.1,464,550 at an exchange rate of Rp. 2,500 per $US1 by the beginning of the economic crisis in the autumn of 1997). In the context of poverty alleviation, fish farmers had access to capital from the IDT Fund launched by the government. This fund was organised by small groups (IDT groups) of fish farmers in the form of a revolving credit system.

Education, although not significant, has a positive sign, indicating that it may contribute to agricultural production and productivity. In this regard, Duraisamy (1992) distinguishes worker and allocate effects. The 'worker effect' refers to technical efficiency — a more educated farmer's ability to produce more output from a given bundle of inputs. The worker effect arises because education improves the quality of labour. The 'allocate effect' refers to allocate efficiency — the ability of the educated farmers to obtain, analyse and understand economically useful information about inputs, production and commodity-mix. This enhances their ability to make optimal decisions with regard to input use and product-mix.

In this study, formal education of cage fish farmers (X7) and their membership in IDT groups (D1) had no significant effect on production. Previous studies in West Sumatra showed that West Sumatra's society was an open community in adopting profitable agriculture innovations quickly. This was an important reason in understanding why this area had attained self-sufficiency in rice production by 1968, 16 years earlier than at national level (Scholz, 1977; Krimmel *et al.,* 1990). This may explain why neither formal education nor membership in IDT groups had significant effects on net cage production.

The area of cage farms, which was included in the C-D model as D2, had a moderately significant influence on cage production. Although the environmental disaster relating to the impact of El Niño in 1997 included burning forests and an extended long dry season as well as ash from volcano activities which influenced the water quality of the lake, the analysis also indicated that excessive intensity could endanger the future sustainability of cage development. Mass mortality of fish, which happened in 1997 with water quality degradation at certain locations of high-populated cage fish farming, seems to confirm this statistical result.

Lessons from other areas, such as from Laguna de Bay in the Philippines, reinforce that concern. The yield of capture fisheries of Laguna de Bay has been declining since the introduction of aquaculture. Total fish production (capture fisheries plus aquaculture) has been declining since the early 1980s after rapid growth of aquaculture, which was introduced to the area in the early 1970s. The decline was attributed to the negative effects of aquaculture on the lake system as well as to a general deterioration of the aquatic environment due to the load of silt from the deforested hills, domestic and industrial wastewater loaded with nutrients and toxins from residential subdivisions and real estates set up in the vicinity of the lake over the past 20 years (Focken *et al.,* 1998).

The sum of all coefficients of the C-D production function (Σ β_i) was 1.3. This value indicated that the input-output relationship of cage culture at Lake Maninjau at this level of production exhibited increasing returns to scale. It means that if all inputs specified in the production function were increased by 1.0%, output would increase by 1.3%. This finding is relevant in connection with the estimation made by the government's office of fisheries service of Agam district. It has calculated that at an average size of 4 x 4 to 7 x 7m for each cage unit, more than 10,000 units of cages could be built at selected sites along the shore areas of Lake Maninjau under proper management (Kacab Diskan, 1996). Beside Lake Maninjau (9,950ha), West Sumatra province also has three other lakes, namely Lake Singkarak (13,011ha), Lake Diatas (3,150ha), and Lake Dibawah (1,450ha) which are not yet utilised for cage fish farming. At the same time, such intensification requires careful monitoring of water quality and the entire aquatic system to avoid harmful effects to the natural resource base.

Conclusions

The empirical result of the study based on a production function model showed that higher production of commercial cage culture at Lake Maninjau could be obtained through the use of more input. Additional inputs, especially in the form of higher rates of stocking and feeding for fish culture, should be balanced with the carrying capacity of the lake. Long-term sustainability of the cage fish farming system requires the maintenance of water quality. A comprehensive study on the carrying capacity of the lake, the ecological dynamics of the cage system and minimum standards of water quality are required to determine the ecological limits of cage expansion.

Since the net cage system is a relatively new development and is important for poverty alleviation and income security in rural Indonesia, formulating and implementing regulations to promote environmental preservation and ensure its sustainability are of vital importance.

References

Asnawi, S. (1981) Irrigation and the performance of the improved rice technology. A case study in West Sumatra, Indonesia. PhD dissertation at the Australian National University. 287 pp.

Christensen, M.S. (1989) *Techniques and Economics of Intensive Cultivation of Kelawat and Lempan Carp in Floating Cages. A Handbook for Extension Workers and Farmers.* Deutsche Gesellschaft für Technische Zusammenarbeit (GTZ) GmbH, Eschborn. 138 pp.

Chong, K. and Lizarondo, M.S. (1981) Input-output relationship of Philippine milkfish aquaculture. In: Hulse, J.H., Neal, R.A. and Steedman, D.W. (eds) *Proceedings of Aquaculture Economics Research in Asia.* IDRC, Singapore, pp. 35-44.

Costa-Pierce, B.A. and Effendi, P. (1988). Sewage fish cages of Kota Cianjur, Indonesia. NAGA, April, pp. 7-9.

Dahuri, R. (1997) The status of Indonesian fisheries trade and development. Paper presented in Seminar and Workshop on Supply/Demand in Fisheries Trade in O.I.C States. Organised by Islamic Centre for Development of Trade, the Ministry of Ocean Fisheries and Merchant Marine of the Kingdom of Morocco, and the Islamic Development Bank. Agadir, Morocco, June 25-27, 1997.

Douglas, P.H. (1976) The Cobb-Douglas production function once again: its history, its testing, and some new empirical values. *Journal of Political Economy* 84(5), 903-915.

Duraisamy, P. (1992) Effects of education and extension contacts on agricultural production. *Indian Journal of Agricultural Economics* 47(2), 205-213.

Focken, U., Batac-Catalan, Z. and Becker, K. (1998) Baseline data for sustainable management of capture fishery and aquaculture in Laguna de Bay, Philippines: a survey of artisanal fishery. In: Jarayabhand, P., Chaitanawisuti, N., Sophon, A., Kritsanapuntu, A. and Panichpol, A. (eds) *Book of Abstracts, the Fifth Asian Fisheries Forum and International Conference on Fisheries and Food Security Beyond the Year 2000, November 11-14, 1998, Chiang Mai.* Aquatic Resources Research Institute, Chulalongkorn University, Bangkok. p.56.

Galapitage, D.C. (1981) Economics of cage culture of tilapia in Sri Lanka. In: Hulse, J.H., Neal, R.A. and Steedman, D.W. (eds) *Proceedings of Aquaculture Economics Research in Asia.* IDRC, Singapore. IDRC, Singapore, pp. 82-89.

Husen, M. (1997) The region of community's fisheries industry as the model of an agribusiness development for freshwater fish culture. Proceeding industrialising, social planning and the role of government in agriculture development. Book I. Institute for Agriculture Research and Development of Indonesian Agriculture Department. CV Dewi Sri Jaya, Jakarta. 123 pp. (In Indonesian.)

Kacab Diskan (Kantor Cabang Dinas Perikanan) Kabupaten Agam (1996) Potential and opportunities of investment in the fisheries sector of Agam district. Fisheries Office of Agam District, West Sumatra. Unpublished. 10 pp. (In Indonesian.)

Krimmel, T.G., Duve, T., Fleischer, G., Ismal, G., Madjid, M., Piepho, H., Schnoor, A., Sommer, M. and Wentzel, S. (1990) Towards an institutionalisation of monitoring and evaluation of project impact. The example of projects in the small-scale irrigation sector in West Sumatra, Indonesia. Verlag Josef Margraf Scientific Book, Weikersheim. 221 pp.

Olatubi, W.O. (1991) The economics of aquaculture: a study of homestead fish production in concrete ponds in Lagos State, Nigeria. Special aspects of rural development in Nigeria. Materialien 19. Zentrum fuer regionale Entwicklungsforschung der Justus-Liebig-Universität Gießen. 75 pp.

Saputra, H. (1988) Fish culture in floating net cage. CV Simplex, Jakarta. 131 pp. (In Indonesian.)

Scholz, U. (1977) Minangkabau. Die Agrarstruktur in West-Sumatra und die Moeglichkeiten ihrer Entwicklung. Giessener Geographische Schriften 41, Gießen. 217 pp.

Sedana, I.P. (1997) Aquaculture in Sumatra. In: Takashima, F., Takeuchi, T., Arimoto, T. and Itosu, C. (eds) *Aquaculture in Asia. The Proceedings of the Second International Seminar on Fisheries Science in Tropical Area, Tokyo, Aug. 19-22, Japan, 1997.* Japan Society for Promotion of Science and Tokyo University of Fisheries, Tokyo, pp. 3-8.

Soekartawi (1990) Theory of production economics with special topic on the analysis of Cobb-Douglas function. CV. Rajawali, Jakarta. 257 pp. (In Indonesian.)

Soewardi, K. (1997) Aquaculture in West Java. In: Takashima, F., Takeuchi, T., Arimoto, T. and Itosu, C. (eds) *Aquaculture in Asia. The Proceedings of the Second International Seminar on Fisheries Science in Tropical Area, Tokyo, Aug. 19-22, Japan, 1997.* Japan Society for Promotion of Science and Tokyo University of Fisheries, Tokyo, pp. 9-13.

Chapter 12

Promoting Aquaculture by Building the Capacity of Local Institutions: Developing Fish Seed Supply Networks in the Lao PDR

D. Lithdamlong[1], E. Meusch[2] and N. Innes-Taylor[3]

[1]*Regional Development Committee, Livestock and Fisheries Section, Division of Agriculture and Forestry, PO Box 16, Savannakhet, Lao PDR*
[2]*AIT AARM – Cambodia, Aquaculture Office, Department of Fisheries, PO Box 835, Phnom Penh, Cambodia*
[3]*AIT AARM - Lao PDR, Livestock and Fisheries Section, Division of Agriculture and Forestry, PO Box 16, Savannakhet, Lao PDR*

Abstract

As AIT sought to expand its regional Aqua Outreach network to the Lao PDR in 1993, it was confronted with a novel approach to promoting aquaculture: a government request to support their existing systems without separate project staff. The resulting development of technologies has, as a result of this approach, probably proceeded more slowly that it would have through a project team based in-country. However, the experience of the past four years indicates that a focus on facilitating the development of operational systems within local institutions has been effective. Research showed that inadequate fish seed supply was the major limiting factor to the development of aquaculture in the province. Therefore, training and development initiatives were launched to improve the capacity of the Savannakhet Livestock and Fisheries Section of the Department of Livestock and Fisheries as a development institution at both the provincial and district level. Production of the provincial fish-breeding centre was boosted and a network of farmers was established to distribute fry throughout the province. By 1995 provincial staff were able to start their own research into low-cost, small-scale seed production at the farmer level. In 1997 over 30 families were members of a network of 'spawning farmers' that produced and sold over 1 million fish seed locally, more than the output of the provincial hatchery.

Introduction

AIT Aqua Outreach began as a research/development project working with the Thai Department of Fisheries in Udorn Thani province. The main thrust of the project was to investigate mechanisms to convey proven aquaculture technologies to small-scale farmers. When the programme entered its second phase, it expanded into a regional programme to include the Southeast Asian countries of Cambodia, the Lao PDR (Laos), and Vietnam. When making this transition, the programme retained its goal of improving rural livelihoods through promotion of appropriate aquatic resource management related technologies. The overall purpose of this enlarged programme was to define its purpose as strengthening national institutions' capacity to promote aquatic systems management on a long-term basis. This shift to institutional strengthening has been reflected in the methodology adopted by AIT Aqua Outreach-Lao PDR.

AIT Aqua Outreach began working with the Savannakhet Livestock and Fisheries Section (L & F) of the Department of Livestock and Fisheries (DLF), its partner institution in Laos, in early 1993. The programme sought to develop strategies for aquatic resource development within the context of its partner's institutional structure and capacity. This was consistent with the programme's commitment to sustainability and the Lao Government's request not to work through separate project staff. The concern of the Lao Government was to avoid what was referred to as a 'sunset project', i.e. beautiful, just before disappearing. They did not want a conventional project with the classical risk that 'development would disappear with the project'. The result has been that the programme has focused on developing operational systems within the L & F. This experience has contributed to greater levels of participation and ownership, and thus more sustainable development. The increased institutional capacity that has been fostered through the relationship between Aqua Outreach and the L & F has led to significant progress in addressing technical issues concerning aquatic resources development.

Addressing Shortages in Fish Seed Supply as an Entry Point to Development

It was determined early within the programme that fish seed supply was a major constraint to developing aquaculture. Savannakhet, located in one of the largest floodplain areas in Laos, had been targeted for aquacultural development since the mid-1970s. Potential for production and farmers' interest were both high, but the issue of fish seed availability had never been successfully addressed. Pak Bor, the provincial fish hatchery, was producing a relatively low number of fingerlings and, because of logistical problems that are common in Laos with its dispersed population and underdeveloped road system, distribution was quite limited. The majority of the fingerlings available in Savannakhet at the time were transported across the river from Northeast Thailand, originating from the numerous small-scale hatcheries there. The strategies adopted by the L & F in association with Outreach to address these problems were: (i) to increase seed production and distribution at Pak Bor in the short term; and (ii) decentralise seed production in the long term. The mechanisms developed to reach these ends were the 'nursing network', a network of nursing fish seed to fingerlings; and the 'spawning network', farmers spawning fish to produce hatchlings within the village.

Developing ways to address constraints related to fish seed supply shortages served as an entry point into other aquatic resources management activities including community managed fisheries and rice-fish culture. Other important issues being addressed include those concerning wild fish. Currently, most of the aquatic production in rural Laos comes from capture fisheries in small-to-medium sized water bodies. The conservation and possibly enhancement of these resources are an important part of the L & F's role in aquatic resources development.

Nursing Network – Focusing on the Rural Poor

The 'nursing network' was developed by the L & F to promote decentralised nursing and distribution of young fish, or fry. It involves district and provincial level government personnel who produce the fry, select and train farmers, and distribute the fry. Farmers who carry out the nursing and distribute fingerlings to their neighbours are the final step and key actors in the process. The provincial fish hatchery acts as the centre of the network and farmers act as nodes for distribution. The farmers who carry out the nursing also have the important function of 'adding value' to the fish seed, a highly perishable product. Provincial and district level L & F staff function as the link between the hatchery and the nursing farmers. The target group of this network, or beneficiaries, are the rural farmers who purchase the nursed fingerlings for grow-out in ponds.

The nursing network uses techniques modified from cage-based nursing technologies developed at AIT and later refined during the first phase of Aquaculture Outreach in Udorn Thani province, Thailand. In this case, nursing fry in *hapas* is used not only to increase the size and thus the value of fish fingerlings, but also to develop seed distribution points in remote areas. Fry (1-2cm) are transported to remote areas to be 'nursed' in net cages (*hapas*) to a larger size (5-7cm) before being sold to local farmers. There are several advantages to distributing fish seed through such a network. First, fry of this size can be produced cheaply by the central hatchery and the quick turnover of pond space allows them to increase production. Second, fry of this size are inexpensive and more easily transported to farmers than larger fish seed. Third, farmers nursing the fry to larger sizes are able to sell the larger fingerlings at a profit to generate income for their families. Finally, farmers purchasing the fingerlings are able to buy quality fish seed locally at a reasonable price (Fig. 12.1).

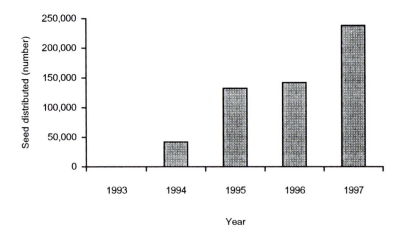

Fig. 12.1. Numbers of fish seed distributed through the nursing network.

Establishing the nursing network has been a process of institutional capacity building requiring many skills other than those concerning aquaculture technologies. Developing systems for promoting the nursing network has been a flexible process based on constant evaluation. Each year the policies of the network have changed based on the growing experience of the L & F in administering the system. An annual review and planning process has been used to assess impacts and make further refinements in areas such as farmer selection and incentives provided to farmers who participate. The result is that the network has been more effective each year in focusing on rural poverty and food security, which are important priorities of the government for addressing aquatic resource development issues.

After three years of successfully promoting the nursing network, the L & F is now specifically stressing the involvement of poor and landless villagers. The reason for this decision is to emphasise the potential to provide poorer families in the community with a source of income. Landless villagers are able to nurse fry by placing *hapas* in public water bodies or neighbours' ponds to generate income with a reasonably low investment. Ms. Boonmee is the single head of a household in Palansai district. She has several children and no steady income. The family has no agricultural land and there are very few employment opportunities in the village. Ms. Boonmee attended training on nursing fish fingerlings in *hapas* at the district office and joined the nursing network. She is nursing 4,000 fingerlings in two *hapas* placed in a school pond near her house (Fig. 12.2). She purchased the fry and has rented the *hapas* on credit from the local L & F office. Her main cash investment will be for rice bran to feed the fingerlings, an input that is relatively cheap and locally available. Because taking care of the fish is light work, her young daughters will be able to help with the nursing activities. At the end of the 8-10 week nursing period, Ms. Boonmee will have tripled her investment, making much needed cash for her family, while the farmers in her village will have easy access to good quality fingerlings.

Fig. 12.2. Ms. Boonmee's nursing *hapas* placed in the village school's fishpond.

Because fish seed are cheap and easily transported as early fry, the nursing network has been able to reach farmers in remote areas. Mr. Mee is an upland rice farmer in Dong Village in Sepone district. Dong Village lies on what was once the Ho Chi Minh Trail and Mr. Mee raises fish in the many bomb craters that cover his hillside farm. Last year Mr. Mee joined the nursing network and nursed 2,000 fingerlings to raise in his own ponds and to sell locally (Fig. 12.3). He sold fingerlings to farmers from remote villages, some travelling 2 days to purchase as few as 50 fingerlings. The fish were transported back to the villages in open clay jars and traditional water gourds. Earlier trials run by the L & F show that this method of transportation can be fairly effective when fish are moved in small numbers. Ms. Boonmee said that this year many more villagers from remote areas, most of whom are ethic minority 'Lao Tung', plan to buy fish fingerlings. Before the nursing network reached the area, fish seed was rarely available in the mountainous regions of the province.

Small-Scale Seed Producers Have an Impact

Although the nursing network solves many problems associated with providing seed to farmers in remote areas, this strategy has a number of limitations. First is the limit on the number of fry that the provincial hatchery will be able to produce, and second is the limit implied by the subsidised nature of the distribution methods. The most logical way to overcome these limitations is to facilitate the development of decentralised hatchery production. In the case of the activities in Savannakhet, this means family-run, small-scale hatcheries in remote areas providing fish seed to local farmers, in other words a spawning network. The nursing network is intended, therefore, to act as a bridging mechanism to this decentralised seed production scenario. By participating in the nursing network farmers become familiar with nursing fish and raising them in ponds. They become familiar with the fish species used in aquaculture and begin to raise broodstock. By making fingerlings available in remote areas, a market for fish fingerlings begins to develop among the farmers raising fish. As this nursing network develops, certain farmers begin to specialise in fish culture. It is these farmers that are recruited into the spawning network.

Fig. 12.3. Mr. Mee's nursing *hapa* in a bomb crater on his farm located in the uplands near the Ho Chi Minh Trail.

The spawning network consists of farmers who are producing fish seed using simple, *hapa*-based hatchery techniques developed by the Savannakhet L & F and AIT Aqua Outreach (Fig. 12.4). Member farmers in the spawning network, most of whom have participated in the nursing network for 2-3 years, already have a pond, broodstock, and familiarity with handling fish in *hapas*. They also have an established market of local farmers who previously purchased fingerlings through the nursing network. The farmers are provided with a simple set of spawning equipment (nylon mesh *hapa*, fine mesh incubation *hapa*, syringes for injecting hormone, premixed hormone, and a manual on fish breeding) and given a short, hands-on training in the basic techniques for spawning *Cyprinus carpio* (common carp) and *Barbodes gonionotus* (silver barb) in *hapas* at a local pond. As members of the network, farmers are provided with ongoing technical support in the form of advice from district and provincial L & F staff. Premixed fish breeding hormone is made available at a reasonable cost. Farmers may either purchase it directly or obtain it through district staff who make regular visits to the provincial office with coolers for transporting animal vaccines. By establishing itself as the only readily available source of hormone, the L & F is able to monitor the amounts and the quality of the hormone used (Fig. 12.5).

Fig. 12.4. A small scale hatchery based on *hapa*-based technologies.

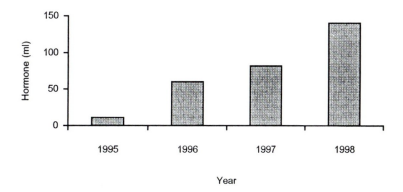

Fig.12.5. Amount of premixed fish spawning hormone distributed to the spawning network.

Because most of the farmers participating already have experience nursing and raising fish, the success rate of the spawning network members has been very high. In the first year, 1996, 19 farmers were able to produce and sell over 400,000 fish fingerlings. In 1997 the group had expanded to 29 farmers and sold over 1 million fingerlings (Fig. 12.6). This is substantial considering production at Pak Bor, the provincial government fish hatchery, has only been able to produce a maximum of slightly over 700,000 fingerlings annually (Fig. 12.7). In one district, a local small-scale hatchery producer has supplied the fry for the 1998 nursing network. This is a good example of how the decentralisation of seed production and distribution can be taken another step toward sustainability.

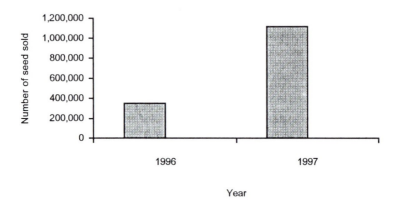

Fig. 12.6. Fish seed sales from private hatcheries.

In a similar way to the nursing network, the strategy of promoting the spawning network has evolved from year to year. The first year consisted of working very closely with hand-picked farmers who already had a close relationship with the L & F. In the years following there has increasingly been a shift away from this focused approach to that of broadcasting the information to more and more farmers. The result has been that a smaller percentage of farmers trained each year have been successful in producing fish seed (Fig. 12.8), but the overall number of seed produced continues to increase. Simply looking at production numbers does not, however, take into consideration problems such as drought that may temporarily reduce both supply and demand for fry, or the number of fingerlings that are stocked in farmers' own ponds rather than sold. Issues like these need to be carefully considered as the L & F continues to support the development of the network in the future.

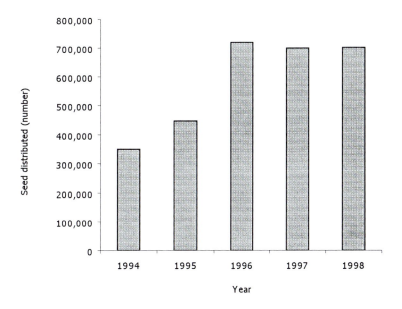

Fig. 12.7. Fish seed production at Pak Bor Provincial Fish Hatchery.

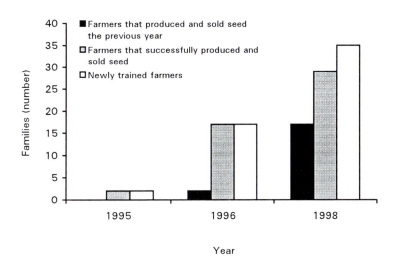

Fig.12.8. Families trained in spawning techniques and successfully producing and selling fish seed.

Opening Doors to More Remote Areas

Promoting fish culture in the mountainous districts of Savannakhet has always been problematic. Efforts of the L & F to work in these areas are hampered by logistical problems caused by long distances and bad roads which make fish seed distribution difficult. Even developing the nursing network has been difficult because some areas are impossible to reach during the monsoon season, the time of year when nursing network activities take place. In these remote areas, developing the capacity to produce fish seed locally is the only way to develop sustainable aquaculture activities.

Mr. Pok is a member of the spawning network in Sai Sam Pan village, Phine district, one of the four mountainous districts in eastern Savannakhet province. He was a successful member of the nursing network and joined the spawning network in 1996. In 1997 he sold over 17,000 fingerlings to 32 farmers in these mountainous areas. In 1998 Mr. Pok decided to invest further in fish seed production and replaced his net-based system with two small concrete tanks. By the end of July he had already sold 30,000 fingerlings and still expected to produce more. By making fish seed available in Phine district, Mr. Pok has made it possible for farmers to consider fish production as a part of their farming system. If the availability of fingerlings is in question, farmers can ill afford to take the risk of preparing to culture fish.

Developing Field-Level Capacity

Developing the capacity of district officers to effectively function as development agents is the key to the sustainable development of livestock and fisheries in the Lao PDR. However, many of these officers live and work in remote rural areas with limited access to new information. The personal development of such staff should be an on-going activity for any institution and in the case of the DLF, the systems to meet these needs should be designed to accommodate the rapidly changing needs of the organisation.

In an attempt to develop flexible and appropriate methods of staff development, the Savannakhet L & F has developed a series of modular learning packages targeted at district staff that facilitate learning at a distance. These 'distance-learning modules' allow district officers to access new information and study at their own pace without having to leave family and work responsibilities unattended. Based on the priorities of the district officers themselves, modules have been developed firstly in basic English language skills and fish culture techniques. These modules, and more importantly the distribution and support systems required for their management and administration, are currently being tested in selected districts of Savannakhet province. The modules have proved popular with district staff and additional modules on how to conduct training, the establishment and co-ordination of fry nursing networks and systems of field data management are currently being developed.

New Models for the South

The experience of capacity building within the Savannakhet L & F Section has become a model for other provinces in the Southern region of the Lao PDR. The DLF has recently established the country's first regional development organisation, called the Regional Development Committee (RDC). The RDC aims to further develop the capacity building models developed in Savannakhet for regional application through increasing the dialogue and co-ordination of development activities between provinces in the Southern region.

Although the RDC has its developmental focus on fisheries and livestock, it aims to assist provinces within the region to develop improved linkages with other development sectors, particularly within the fields of agriculture and forestry. The RDC is also an institutional experiment for the Lao Government. It is the Lao Government's first regional-level development agency and differs radically from other Government institutions. It has developed as an *alliance* of national and international institutions that brings together resources and expertise for development. Partners in this alliance work within a common framework that provides a structured and culturally appropriate forum for development. The RDC has an overall Co-ordinator appointed by the DLF who chairs a Provincial Co-ordination Committee that guides the development of a common management and administrative framework. This framework is used to develop partnerships within the alliance and maintain overall developmental focus and direction.

Although it is too early to judge the appropriateness of this institutional model for promoting the development of fisheries in Southern Laos, the capacity building efforts of Government officers in Savannakhet province demonstrate the effectiveness of this approach. Ultimately, local institutions have to take responsibility for development, and the success of the spawning and nursing networks in Savannakhet province demonstrate that a focus on developing local institutional capacity can produce relatively rapid and sustainable improvements to local farming systems.

Farmers were provided with a spawning set consisting of: (i) a small nylon *hapa* for holding and spawning broodstock of common carp and silver barb; (ii) a small cloth *hapa* for incubating spawn in ponds; (iii) syringes for the injection of fish hormone; (iv) fish breeding hormone; and (v) written guidelines for fish breeding. A key component of the technology is the fish breeding hormone which is distributed to the districts in the 'cold chain' established for livestock vaccines. In 1996 word spread of the simple and low-cost fish breeding techniques developed by the Savannakhet L & F section.

Distance learning modules (in English language; and fisheries) have been developed for district staff with the aim of developing a distance education system that can be sustained by provincial offices. More recent has been the establishment of the Regional Development Committee (RDC), with L & F representatives from three provinces in southern Lao PDR, to plan and obtain funding for promotion of seed production in Champasak and Kammouane provinces.

Chapter 13

Carp Seed Production for Rural Aquaculture at Sarakana Village in Orissa: a Case Study

Radheyshyam

KVK/TTC, Central Institute of Freshwater Aquaculture, Kausalyaganga, Bhubaneshwar-751002, India

Abstract

The case study deals with successful attempts made on seed production of carp by the rural poor for aquaculture development in the area. A participatory approach was resorted to for problem identification, constraints prioritisation and needs-based, problem-solving, technology implementation from the farmers' perceptions. Carp breeding was initiated with common carp (*Cyprinus carpio*) in 1987. In the following years induced breeding of rohu (*Labeo rohita*), mrigal (*Cirrhinus mrigala*) and catla (*Catla catla*) was introduced and a total of 239, 261 and 95 sets of species were bred, respectively. The successful experience in profitable carp breeding sustained the interest of the farmers so that they bred fish every year and this led to a rising trend in spawn production from 0.35 in 1987 to 21.3 million in 1997. A total quantity of 1,118.6kg of female brood fish could be utilised producing 70.9 million spawn in 11 years. Fry and fingerling production technologies were demonstrated in the rural area using small ponds of 0.02-0.10ha during 1987-1997, totalling 80 cases under different management practices. In 11 years, the production of 15.94 million fry and 1.88 million fingerlings made the neighbouring villages self-sufficient in terms of seed.

The estimated net income from spawn production was Rs.207,246 (US$4,820) and the return on expenditure was 274%, whereas from seed raising the net income was Rs.606,550 (US$14,106) and the average return on expenditure was 131%. The analysis of the present cases suggests that the adoption of carp seed production technology by the farmers was techno-economically viable, sustainable and employment generating. It had a visible impact on the bio-physical and socio-economic conditions of the village. Success motivated other farmers to undertake seed production and carp culture.

The response of the rural poor in adopting the technology of carp seed production and culture has been very positive. About 2,500 participants were trained in the technology following the 'Farmer-Led-Farmer Approach' at the village without jeopardising local ecosystems. The farm has become the focal point of training activities and a number of satellite units of seed production and carp culture have developed in the area.

Introduction

In labour surplus agricultural economies such as India, suitable diversification in livelihoods for the rural masses is most imperative in bringing about a positive change in the rural economic scene. In this regard allied food production technologies, particularly aquaculture, hold immense practical relevance for the rural poor. The primary requisite for aquaculture is the availability of a water resource which is reasonably abundant in the countryside, though varying in size, shape, seasonality and productivity, but still having high potential from a productivity point of view. Advancements and refinements in technologies over the years have made aquaculture eminently suitable and it has significantly contributed to the rural economy, both for poor farmers as well as for entrepreneurs, under defined sets of conditions. Even so, a large extension gap still persists which limits the reach of the technologies to the wide matrix of the rural community. In a situation like this, any random prescription of a technological package hardly provides a viable solution to the problems in the field unless location and resource-specific needs are emphasised and addressed. For sustained rural aquaculture development under diverse conditions, a concerted effort to disseminate appropriate technologies tailored to suit rural conditions and the exact needs of the farmers is required.

With priority attention to the factual information collected by survey, and analysis of the location and resource-specific needs of the farmers, the present studies on carp seed production for rural aquaculture development were carried out. Most earlier studies indicated that for rapid expansion and growth of rural aquaculture, paucity of carp spawn compelled village farmers to stock their ponds with riverine fish seed (Radheyshyam and Kumar, 1982). Records of the past several years reveal that due to lack of technical support and basic infrastructure facilities, carp breeding was rarely undertaken in the villages involving farmers (Radheyshyam et al., 1982). A few significant reported attempts on carp spawn production are those by Sarkar et al. (1984), Radheyshyam et al. (1985), Radheyshyam and Tripathy (1992), Radheyshyam and Sarkar (1998) and on carp seed raising by Selvaraj and Kanaujia (1979), Radheyshyam et al. (1982), Mohanty and Mohanty (1984), Tripathi et al. (1987), Radheyshyam et al. (1988), Sharma et al. (1988), Thakur et al. (1988), Patnaik et al. (1989), Radheyshyam and Tripathy (1992), Singh and Radheyshyam (1995), Radheyshyam (1997, 1998) and Jena et al. (1998). Despite pointed extension focus in this regard, the sustainability aspect of the production of carp spawn, fry and fingerlings by the farmers still remained a missing link.

The present case study was an endeavour to analyse events spanning over 11 years, leading to a village graduating into self sufficiency in spawn, fry and fingerling production by benefiting from technology demonstrations with a participatory approach in a down-top process. The most visible outcome was an assured supply of carp seed at reasonable price within the reach of farmers while simultaneously generating employment for the rural poor.

Materials and Methods

Mode of Approach

The study was conducted during 1987-1997 following an initial resource inventory survey carried out during 1986. The village situation, resource and livelihood were appraised by the representative farmers of the village through organising a meeting followed by individual interviews under researchers' observations. Based on the availability of minimum resources and interest, ten farmers were selected. A participatory approach was resorted to for problem identification, prioritisation and possible solutions from the farmers' perceptions. Concerted efforts were made to find a solution to the problems naturally resulting from need-based technological interventions. Aquaculture activities in Sarakana village were stimulated by a group of ten farmers trained in a participatory mode in 1986. These farmers later cultured food fish initially and gradually branched out into other activities such as carp seed raising, carp spawn production and integrated fish farming. However, the present case deals only with carp seed production as a commercial activity taken up in phases by farmers through a needs-based and problem-solving approach. The technology input into various aspects of carp seed production was analysed and evaluation of performance was made to assess the impact at the end of each production cycle. Although technological packages were made available to the farmers by the researchers, the farmers played a key role in planning, formulating and implementing the technologies throughout the study period under the guidance of researchers.

Background Information

The study was carried out in Sarakana village in the densely populated Khurda district of Orissa state, India, under the jurisdiction of which Central Institute of Freshwater Aquaculture (CIFA) is located near the capital city, Bhubaneswar. There were numerous rainfed and undrainable fish ponds located there. Extensive and semi-intensive culture were common practices and most fish production came from rural aquaculture.

Sarakana village, situated 12km northeast of CIFA, had 400 families with a population of 2,400. The male-female ratio was 0.98 and the adult-child ratio

1.71. Out of the total number of families, 73% belonged to the poor, 25% to the medium and 2% to the rich category. Per family annual income of poor, medium and rich families was Rs.3,000-6,000 (US$70-140), Rs.6,050-11,000 (US$141-256) and above Rs.11,000 (US$256), respectively, in 1986. The village was flood-prone and the main sources of drinking water were open wells and tube wells. Livelihood analysis of the village showed that the major sources of income of poor farmers were wage employment and share cropping whereas for middle class and rich farmers the major income came from their own agriculture and service (Fig. 13.1).

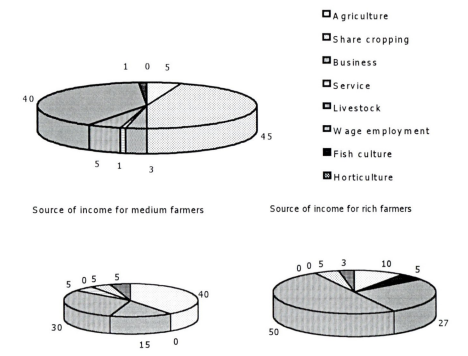

Fig. 13.1. Livelihood analysis of Sarakana village.

Fish Culture Resource Inventory

A survey conducted in 1986 indicated a rich potential for aquaculture development. The village possessed 3.4ha of *Gram panchayat* (village community) ponds and 2.0ha of privately owned ponds. The privately owned ponds were small, seasonal, flood-prone, shaded with large marginal trees and invariably they remained fallow. From the *Gram panchayat* ponds, 44% of the water area was feasible for fish culture but being infested with aquatic weeds, remained under-utilised. The remaining *Gram panchayat* ponds were completely covered with water hyacinth and other weeds and were invariably inundated with floodwater during the monsoon. There was a deep and lengthy drainage structure for floodwater flow which, however, used to remain covered with aquatic weeds after the recession of floodwater. The utilisation of these water areas was limited to crop irrigation, washing cattle, bathing and religious and domestic purposes. Sporadic instances of capture fishery occurred. According to the farmers, the fish yield averaged 60kg ha^{-1} year^{-1} from the flood-affected ponds and 300kg h^{-1} year^{-1} from flood-protected ponds.

Problems Identification

Before the initial implementation of the technological packages, meetings of participating farmers were organised and with the consensus view of the farmers a number of problems such as low fish yield in 1986, low survival of fry in nursery ponds in 1987, and low recovery of carp spawn in 1988 were identified. Problem-cause diagrams were prepared by participating farmers. Constraints were listed and prioritised by the farmers under the guidance of technical experts based on the criteria of their extent, severity, importance and frequency (Tables 13.1-13.3). Several appropriate and need-based technologies were implemented to remove the identified constraints. Following implementation of the technologies, adoption was recorded and the reasons for success were prioritised.

Technology Implementation

Carp breeding

Brood fish were sourced from an extensive culture fish pond of 1.5ha in the first phase of the study, except in 1989-1991 when farmers could not lease this pond from the *Gram panchayat*. During this period, broodfish were reared in a small pond of 0.10ha. However, during 1992-1997 mature fish were collected in the month of March for management in well-prepared, separate ponds on artificial feeds.

Spawn production was initiated in 1987 with controlled breeding of common carp. In subsequent years, Indian major carp breeding was gradually introduced and a total of 95, 239, 161 and 147 sets of *Catla catla, Labeo rohita, Cirrhinus mrigala* and *Cyprinus carpio,* respectively, were experimented with. Common carp breeding was undertaken twice a year without using any inducing agents with *Hydrilla* and *Najas* used as egg collectors. Breeding of *L. rohita* and *C. mrigala* was initiated in 1988, and *C. catla* in 1989, with practical demonstrations of the technology.

Table 13.1. Problem prioritisation for low fish yields in fish ponds at Sarakana village during 1986. Increase in degree of criteria from 1 to 5.

	Criteria					
Problems	Extent	Severity	Importance	Frequency	Total score	Rank
Lack of knowhow	5	5	5	5	20	
Lack of fish feed	5	4	5	5	19	II
Lack of manure and fertiliser	4	4	5	5	18	IIIa
Aquatic weed infestation	4	5	5	4	18	IIIb
Poor management of soil and water quality	4	4	5	3	17	IV
Poor pond management	4	4	4	4	16	V
Presence of predatory and weed fishes	3	4	4	4	15	VI
Lack of quality fingerlings	3	3	4	4	14	VII
Short duration lease	3	3	4	3	13	VIII
Flood prone	3	3	3	3	12	IX
Seasonality of ponds	3	3	2	3	11	X
Maximum ponds belong to *Gram panchyat*	4	4	1	1	10	XI
Lack of suitable piscicides	2	2	3	2	9	XII
Lack of money	2	1	3	2	8	XIII
Disease outbreak	2	1	2	2	7	XIV
Poaching	2	1	2	1	6	XV
Multiple-ownership	2	1	1	1	5	XVI

Indian major carp were injected with pituitary extract (PG), GnRH and ovaprim during June-August. PG was injected twice at 4-6 and 8-16mg kg^{-1} female. Males received one dose only of 5-6mg kg^{-1}. As a second alternative, GnRh was injected in double doses after dissolving one vial (400mg) in 20ml of distilled water at 0.4 and 0.9ml kg^{-1} female and 0.4mg kg^{-1} male. For a third alternative, ovaprim was used only for females in a single dose of 0.3-0.5ml kg^{-1} female. Males were injected with a glycerine extract of PG at 5mg kg^{-1} male in a single dose. Injected fish were released into breeding *hapas* fixed in a composite fish culture pond inside the net enclosure as described by Radheyshyam *et al.* (1985). Hatching was done using a *hapa*-hatching device during 1987-1995, whereas during 1996/97 cement concrete incubation pools were used.

Carp seed raising

The fry and fingerling technologies were implemented in the rural area using small ponds of 0.02-0.1ha during 1987-1997, totalling 80 cases (one case means one crop in a pond). In 56 cases eradication of unwanted fishes was done by dewatering the ponds. The pond bottom was exposed to sun. The ponds were dried during summer in 10 cases. The sun-dried ponds were allowed to accumulate rainwater. In 4 cases the ponds were treated with bleaching powder at 35-50mg l^{-1} while in another 10 cases mahua-oil cake was applied at 250-

300mg l^{-1} to eradicate unwanted fish from ponds. Manuring was done with cow dung at 5-25t ha^{-1} in instalments. While mahua-oil cake-treated ponds received a lower quantity of cow dung, bleaching powder treated ponds received more cow dung. Lime was applied by the farmers at 125-250kg ha^{-1} in all ponds other than the bleaching powder treated ponds. Kerosene oil was sprayed on the pond surface to control aquatic insects at 60-75l ha^{-1} a day before spawn was stocked.

Table 13.2. Problem prioritisation for low survival of fry in nursery ponds at Sarakana village during 1987. Increase in degree of criteria from 1 to 5.

	Criteria					
Problems	Extent	Severity	Importance	Frequency	Total score	Rank
Lack of knowhow	5	5	5	5	20	I
Presence of predatory fish	5	4	5	5	19	II
Poor pond management	5	4	4	5	18	III
Lack of carp spawn for stocking	5	4	4	4	17	IV
Stocking with riverine seed	4	4	4	4	16	V
Poor management of soil and water quality	5	4	3	3	15	VI
Lack of manure and fingerlings	4	3	4	3	14	VII
No insect control	3	3	4	3	13	VIIIa
Shaded with larger trees	4	3	3	3	13	VIIIb
Malnutrition	4	4	3	2	13	VIIIc
Poor plankton density	3	3	3	3	12	IXa
Late harvesting	3	3	3	3	12	IXb
Lack of funds	3	3	4	2	12	IXc
Lack of transport facilities	2	3	3	3	11	Xa
Lack of suitable piscicides	3	2	3	3	11	Xb
Seasonality of the ponds	3	2	2	3	10	XI
Lack of own ponds	2	2	2	3	9	XIIa
Aquatic weed infestation	2	2	2	3	9	XIIb
Disease outbreak	2	1	2	3	8	XIII

In 42 cases out of 80, ponds were treated with multiplex fertiliser (pre-mixed minerals and vitamin) at 3-5kg ha^{-1}, either separately or in combination with cow dung and de-oiled cake at 1,000 and 200kg ha^{-1}, respectively. In 19 cases a mixture of single superphosphate, groundnut-oil cake and cow dung in the ratio of 1 : 25 : 250 was applied. Rearing of *C. catla, L. rohita, C. mrigala* and *C. carpio* was carried out in monoculture to raise fry. The ponds were stocked with 2.5-10 million spawn ha^{-1}. Fry and fingerling rearing were done in succession. Mixed fry of *C. catla, L. rohita, C. mrigala, C. carpio, Hypophthalmichthys molitrix* and *Ctenopharyngodon idella* were stocked at 0.05-2.5 million ha^{-1} for fingerling rearing and were fed with rice bran and groundnut-oil cake at a conventional rate. Conventional feeds were fortified with micronutrients and vitamin C in two cases. The seasonal ponds were used for fish seed rearing during the monsoon and for paddy cultivation in the summer. Fry and fingerlings were harvested separately and were disposed of at farm sites in phases. The cost of various items was estimated for fry and fingerlings together. Average expenditure on individual items was computed from 80 cases.

Table 13.3. Problem prioritisation for low yield of carp spawn at Sarakana village during 1988. Increase in degree of criteria from 1 to 5.

	Criteria					
Problems	Extent	Severity	Importance	Frequency	Total score	Rank
Lack of knowhow	5	5	5	5	20	I
Lack of training	5	5	4	5	19	II
Lack of hatchery	5	4	4	5	18	III
Lack of breeding tools	4	3	5	5	17	IV
Lack of brood fish	4	4	4	4	16	V
Less spawn demand	2	3	5	5	15	VI
Lack of brood fish management	4	4	3	3	14	VII
Lack of brood fish ponds	5	3	3	2	13	VIII
Lack of access to inducing agents	3	3	2	3	11	IX
Less interest of farmers	2	2	3	3	10	X
Poaching of brood fish	2	1	3	3	9	XI
Sale of brood fish	2	2	2	2	8	XII

Results and Discussion

The breeding was carried out in '*hapas*' fixed in a composite fish culture pond containing *C. carpio* stock. The presence of *C. carpio* was found to cause severe damage to carp eggs in breeding *hapas*. Hence to avoid hazards of loss of viable eggs, the breeding *hapas* were fixed inside a net enclosure as suggested by Radheyshyam *et al.* (1985). Spawn production was initiated in 1987 with captive breeding of *C. carpio,* with a yield of 0.35 million spawn. In subsequent years common carp were also bred. Within a period of 11 years, 299.6kg of female brood fish were used for breeding work and 12.58 million spawn were produced.

While implementing the induced breeding of carp to the farmers during 1988 to 1997, a total of 239 sets of *L. rohita* were experimented with, out of which 186 fully and 28 partially responded, yielding 24.9 million spawn. The demand for *C. catla* spawn in the rural area was high but the breeding response of this species is known to be relatively low. However, attempts were made to breed *C. catla* in 1989 with a total spawn production of 0.2 million but breeding could not be done in 1990 due to lack of brood fish. From 1991 to 1997 *C. catla* were bred every year. A total of 95 sets of *C. catla* were bred, of which 78 responded fully and 12 partially, with a total spawn production of 18.53 million. The same *C. catla* brood fish were used once every year since 1992, giving better performance. Out of 261 sets of injected *C. mrigala*, 69% responded with a production of 26.4 million viable eggs. Spawn production from 1988 to 1997 was 14.9 million. Year-wise spawn production of *C. carpio, C. mrigala, L. rohita* and *C. catla* is shown in Table 13.4.

Table 13.4. Fry and fingerling production and gross income from 1987 to 1997 at Sarakana village.

| Year | Total number of ponds | Area (ha) | Production (10^5) | | Income (Rs.) | | Gross income (Rs.) |
			Fry	Fingerling	Fry	Fingerling	
1987	1	0.08	2.20	0.40	11,000	4,000	15,000
1988	3	0.24	3.20	1.01	12,000	12,200	24,200
1989	4	0.20	2.84	0.63	15,135	17,365	32,500
1990	7	0.40	9.40	0.84	36,225	17,700	53,925
1991	5	0.23	10.00	3.30	33,900	44,701	78,601
1992	6	0.29	7.80	2.70	51,000	51,300	102,300
1993	10	0.72	19.41	1.67	109,800	34,550	144,350
1994	9	0.64	17.24	1.34	90,025	16,150	106,175
1995	9	0.64	23.32	2.59	88,000	40,100	128,100
1996	9	0.64	28.44	2.52	137,730	55,700	193,430
1997	9	0.64	35.50	1.82	128,000	64,200	192,200
Total		4.72	159.35	18.82	712,815	357,966	1,070,781

In 11 years, 1,118.6kg of female brood fish comprising 642 sets were experimented with, out of which 73% responded fully and 12% partially, resulting in production of 136.62 million viable eggs and 70.9 million spawn. The spawn recovery of *L. rohita , C. mrigala* and *C. catla* from viable eggs was 37.99 ± 2.27, 38.48 ± 5.12 and 38.6 ± 2.5%, respectively, in a *hapa* hatching device. This low recovery from a *hapa* hatching device could be due to a combination of factors such as cutting of *hapas* by crabs, entry of unwanted fish into the outer hatching *hapas*, presence of predatory cyclopoid copepods in hatching *hapas* (Mishra *et al.*, 1980; Sarkar *et al.,* 1984), and sudden change in water temperature and depletion of dissolved oxygen content and cyclonic weather (Radheyshyam and Sarkar, 1998). The farmers were motivated to overcome these problems by constructing two hatching pools made of cement and concrete in 1995 and three in 1996. Cement concrete incubation pools became operational from 1996. As a result, the spawn recovery improved to 78.68 ± 3.63, 73.91 ± 0.52 and 83.65 ± 4.05% in *C. catla, L. rohita* and *C. mrigala,* respectively.

The concept of carp spawn production introduced in Sarakana village in 1987 has taken root there in a flourishing way. The breeding success led to the rising trend in spawn production from 0.35 million in 1987 to 21.35 million in 1997. The village now supplies quality fish seed to local villages. Part of the spawn produced was reared by the farmers and rest they sold at Rs.3,000-6,000 million^{-1} spawn at the farm site.

Based on the present market rate the estimated value of the spawn produced was Rs.282,980 (US$6,581). Overall average percentage composition of total expenditure on various items in carp spawn production is depicted in Fig. 13.2. Maximum expenditure incurred in this venture was depreciation (22%), followed by purchase of inducing agents (20%) and brood fish maintenance (16%). The estimated net income was Rs.207,246 (annual being Rs.188,410). The percentage return on expenditure was 274% and the profitability index was 2.74, suggesting the economic profitability of the technology for participating rural farmers.

Ponds shaded by trees were rendered unproductive by reduced sunlight. Accumulation of leaf litter and an excessive organic load in the pond further deteriorated water quality, adversely affecting carp and carp food organisms (Radheyshyam *et al.*, 1991). At times, masses of foamy brown/white frog eggs, which tend to fall into ponds during rains, caused a proliferation of tadpoles. Therefore, marginal trees and bushes were cleared before launching the seed-raising programme. Mahua-oil cake is one of the best fish toxicants to eradicate unwanted species but, due to limited availability, only 10 cases were treated with it; in most cases ponds were dried and/or dewatered. A mixture of de-oiled cake, cow dung and single superphosphate or a multiplex pre-minerals mixture and vitamins were used for sustained production of natural fish food organisms (Shirgur, 1977; Radheyshyam *et al.*, 1993).

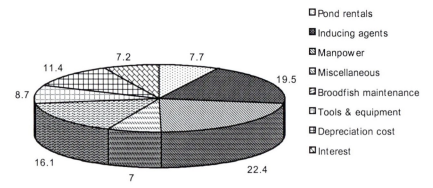

Fig. 13.2. Cost composition for carp spawn production.

A short-term rearing of 12-15 days with scientific management practices gives up to 94% fry survival (Tripathi *et al.*, 1991). Fry were harvested and/or thinned in phases in the present study according to the local demand, allowing an extended period of rearing (14-44 days). Prolonged retention of fry in nursery ponds adversely affected the survival of fry. After 32.84 ± 5.24 days rearing, the survival was 20-40% whereas after 17.4 ± 4 days it was as high as 61-75%. Overall average fry recovery was 50.06 ± 15.38% which was higher than reported rates of 20.3-22.4% (Radheyshyam *et al.*, 1982), 26.75% (Mohanty and Mohanty, 1984), 10-43% (Radheyshyam *et al.*, 1988) and 46.20% (Jena *et al.*,

1998) in rural Orissa. In 16 cases the nursery ponds were stocked at shallow water depth (35-45cm) followed by phased increase of water level at 4-day intervals, resulting in satisfactory fry recovery (50-70%).

The role of micro-nutrients in artificial feed on growth and economic profitability in carp seed production from nursery ponds was demonstrated during 1991 by fortification of feed with micro-nutrients and vitamin C, where the survival of fry was 60-70% (Singh and Radheyshyam, 1995). This developed confidence in village farmers who started using micro-nutrients and vitamin C in fish feed. A commercially available multiplex pre-minerals mixture with vitamins was introduced in 1992, which has accelerated plankton production and fry survival in nursery ponds without any adverse effect. It was subsequently used regularly every year by the farmers. The survival rate in this treatment was 30.8-72.3%.

The ponds in the village were rainfed and many became dry during the summer months and remained unutilised. Hence, for round year utilisation, seasonal ponds were used for fish seed production during the monsoon and later for paddy production in summer in succession for two calendar years. After one paddy crop, two crops of fry and one crop of fingerlings were taken. Fry survival for the first and second crop was 50% and 43%, respectively, against 39% before taking a paddy crop, suggesting the beneficial effects of paddy cultivation on the pond bottom for fish production. Fingerling recovery was 89% and 57% during the first and second year, respectively. The net income from rohu seed rearing was Rs.7,086 (US$165) in 1993 and Rs.15,000 (US$349) in 1994 from a 0.1ha pond. When paddy production on the pond bottom was compared with that from normal land, the yield was comparable at 4,375-5,000 and 4,313kg ha^{-1} crop^{-1}, in the pond bottom and on normal land, respectively. However, the profit percentage turnover was threefold higher in the pond paddy crop than the normal paddy crop (Radheyshyam *et al.*, 1991). Rotation of paddy farming during the summer and fish seed rearing in the monsoon has become a popular technology for seasonal ponds in rural areas.

Farmers involved were from middle-class families and started with extensive fish culture in a leased community pond. In 1986 the participating farmers were in possession of only a small 0.08ha pond shaded with large trees that remained unexploited. As part the participatory training, it was utilised for fish seed rearing. Thus, the year 1987 was an eye opener for the farmers when scientific fry rearing was initiated. This was the first year of making a profit of Rs.8,440 (US$196) from the 0.08ha small pond in 7 months. The hard-earned profits were ploughed back into a new and extended venture of fish seed rearing activities and the number of fish seed rearing ponds increased from 1 (0.08ha) in 1987 to 10 (0.72ha) in 1993. Fry and fingerling production were only 0.22 and 0.04 million, respectively, in 1987 but increased to 3.55 million fry and 0.18 million fingerlings in 1997 (Table 13.4). The cumulative fry and fingerling production in 11 years were 15.94 and 1.88 million, respectively. Gross earning ranged from Rs.15,000 (US$349) to Rs.193,430 (US$4,498) (averages of Rs.97,345 year^{-1}; US$2,264 year^{-1}). Net return ranged between Rs.8,440 (US$196) and Rs.111,560 (US$2,594) (average being Rs.55,140 year^{-1}; US$1,282 year^{-1}).

Average percentage return on the expenditure was 131% and the profitability index was 1.31, suggesting the economic profitability of fish seed raising for rural participating farmers. Percentage cost composition of fry and fingerlings is depicted in Fig. 13.3.

The total expenditure in the entire venture of fish seed raising amounted to Rs.464.24 x 10^5 (US$11 x 10^5) (Table 13.5). Supplementary feed has been reported as the major cost component at 50-70% of the total cost in aquaculture (Schroeder, 1978). However, in the present study the feed cost constituted 23% of total expenditure. Application of mixed fertiliser in phases might have accelerated natural fish food production which served as direct food for fish seed. Stocking materials which contributed 48% of the total expenditure were produced by the farmers themselves. The renovation of the ponds, removal of aquatic weeds, application of feed and fertiliser, watch and ward against poaching netting and harvesting of the seed which were done by the farmers contributed to 11% of the total cost.

The farmers were now self-confident because of their experience, with a realistic understanding of the scope of fry and fingerling rearing and choice of appropriate technologies. The adoption of the technologies in the rural area not only ensured an adequate quantity of quality fish seed but also generated income. Part of the earnings from fish seed production was used to develop the infrastructure of their farms and to implement the technologies and part was distributed amongst participatory farmers for their livelihoods.

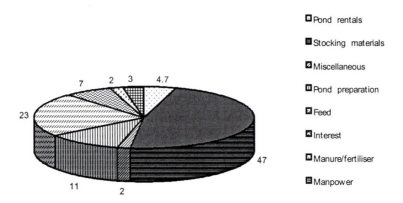

□ Pond rentals

▩ Stocking materials

▨ Miscellaneous

▥ Pond preparation

▨ Feed

◪ Interest

□ Manure/fertiliser

⊞ Manpower

Fig. 13.3. Cost composition for fry and fingerling production.

The fish seed production enterprise has been successful for several reasons. The most important among them was regular technical support rendered by the author and bottom-up participatory skill training followed by acquisition of their own infrastructure, increased profitability, quick return of money, sustainability, food security, interest and confidence of the farmers, easy adoptability of the technology and road facility for input and output transport (Table 13.6). The cumulative effect of the above reasons might be responsible for making Sarakana village a centre for carp seed production.

Expertise imparted to the farmers has made them self-reliant in their newly adopted avocation. Confidence in their workmanship and high returns have proved to be sufficient incentives for the farmers to invest more in creating new facilities for carp seed production. The adoption of carp seed production technology by the farmers was found to be techno-economically viable, employment generating and sustainable. It had a visible impact on the bio-physical and socio-economic conditions of the village. In 1986, there were only 20 private ponds (2.0ha) which increased to 28 (4.0ha) by 1997. Fish seed production was unknown to them in 1986 but spawn, fry and fingerling production in 1997 were 21.35, 3.55 and 0.182 million, respectively. Fish culture did not contribute to the family income of the farmers who adopted the technology in 1986 whereas in 1997 it contributed 45% of the total income (Fig. 13.4). Fish consumption by the farmers increased from 1% to 12% of their diet. Changing status of the families who adopted the technology is given in Table 13.7.

Table 13.5. Cost composition and average income generation through fry and fingerling rearing for 80 cases during 1987 to 1997.

Items	Expenditure (Rs. in 1000)	Composition on operational cost (%)
Rental value of the ponds	22.15	4.77
Dewatering and/or water filling	7.62	1.64
Bleaching powder	0.12	0.03
Mahua oil cake	2.64	0.57
Cowdung/gobar gas slurry	4.84	1.04
Lime	3.05	0.66
Single superphosphate	0.25	0.05
Multiplex	2.82	0.61
Kerosene oil	1.45	0.31
Spawn	108.58	23.39
Fry	114.02	24.56
Groundnut oil cake	82.05	17.67
Rice bran	23.51	5.06
Manpower	49.75	10.72
Miscellaneous items	9.00	1.94
Total working capital	431.84	-
Interest on working capital	32.39	6.98
Total operation cost	464.24	100.00

Returns:		
Gross income	1,070.79	
Net income	606.55	
Average annual net income	55.14	
Net return on expenditure (%)	130.65	
Profitability index	1.31	

Sarakana village employed 30 fish seed vendors to transport fish seed during July-November for 88 man-days for each vendor every year. These vendors each earned Rs.50-200 through fish seed supply to the farmers. Thus, the farmers of Sarakana not only supplied seed to end-users in remote villages but also provided income on generating employment opportunity for rural fish seed vendors.

Sarakana village has been a hub of freshwater aquaculture and actually a show window for technology dissemination in the rural area for the development of aquaculture in the region. According to the farmers, they have helped 2,500 rural fish farmers technically through skill demonstrations and discussions in pond preparation for food fish culture (76% of farmers), fingerling raising (12% of farmers), fry rearing (10% of farmers), carp spawn production (1% of farmers) and rotational paddy-fish farming (1% of farmers). This has had a visible impact on the development of rural aquaculture in a radius of over 25km. Out of these technically-helped farmers, 503 have adopted fish farming as a part-time profession in their villages through food fish culture (89%), carp fish seed raising (9%), carp spawn production (1%) and rotational paddy fish seed production (1%). This suggests that in addition to participating farmers, the rural farmers from local villages also benefited.

Table 13.6. Prioritisation of the reasons for success. Increase in degree of criteria from 1 to 5.

	Criteria					
Problems	Extent	Severity	Importance	Frequency	Total score	Rank
Regular technical support	5	5	5	5	20	I
Participatory training	5	5	4	5	19	II
Acquisition of own infrastructure	5	4	4	5	18	III
More profitable than other agriculture work	5	4	3	5	17	IVa
Sustainability of the technology	5	4	4	4	17	IVb
Providing food security	4	4	4	5	17	IVc
Interest of the farmers	4	4	5	3	16	V
Easy-to-adopt technology	4	3	4	4	15	VI
Transport facility	4	3	4	3	14	VII
Quick return of money	4	3	3	3	13	VIII
Co-ordination of the farmers	3	3	3	3	12	IX
Development of self confidence	3	3	3	2	11	X
Educated farmers	2	2	3	2	9	XI
Located at representative place	2	2	2	2	8	XII
Financially sound	2	2	2	1	7	XIII
Leasing of G.P. tank	2	1	1	2	6	XIV

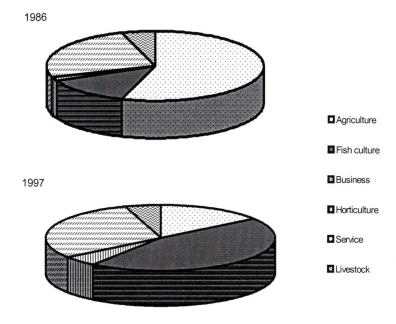

1986

1997

☐ Agriculture

▓ Fish culture

▨ Business

☐ Horticulture

☐ Service

◧ Livestock

Fig. 13.4. Livelihood analysis of the farmers of the target group in 1986 and 1997.

Conclusions

There is ample scope for both horizontal and vertical expansion of rural aquaculture. Special attention is required to supply quality fish seed and this requires techno-economical support. Fish seed production contributes to rural aquaculture development through increased food security of the rural people and through economic as well as social well being. Both subsistence rural fish farmers and entrepreneurs need to be recognised and a conducive environment provided for them to make an optimum contribution to economic and social welfare. Available advanced fish seed production technologies need to be tested and refined through a bottom-up participatory approach on the farms of innovative farmers. After ensuring the validity, workability, profitability and sustainability, these resource and location-specific, cost-effective, refined technologies should be disseminated to other farmers following a 'farmer-led-farmer approach'.

Table 13.7. Changing scenario of the target families during 1986-1997.

	Parameter	1986	1997
Number of families		7	7
Total population:	Male	14	18
	Female	12	16
	Total	26	34
Literacy (%):	Male	60	80
	Female	40	60
Per caput land (ha)		0.23	0.18
Livestock (number)		6	10
Own ponds:	Number	1	10
	Area (ha)	0.08	0.72
Leased ponds:	Number	1	1
	Area (ha)	1.50	1.50
Number of house:	Pucca	NIL	4
	Kutcha	4	3
	Cow shade	NIL	1
Household goods:	Chairs	NIL	6
	Cots	6	12
	Almirah	1	4
	Tables	NIL	4
	Radio	2	2
	Television	NIL	3
	Cycle	3	5
	Scooter	NIL	2
	Gas chulha	NIL	1
	Electric fan	2	5
Farm equipment:	Nets	NIL	2
	Hapa	NIL	12
	Cement hatchery	NIL	One hatchery complex with five hatching pools
	Oxygen cylinder	NIL	2
	Fry carriers	NIL	10
	Diesel water pump	NIL	2
	Electric operated pump	NIL	2
	Sprinkler	NIL	1 set
	Sprayer	1 set	1 set

Acknowledgements

I wish to express my gratitude to Dr S. Ayyappan, Director, Central Institute of Freshwater Aquaculture, for his encouragement and for providing facilities. Thanks are also due to Shri B.K. Sharma, Ex-CTO., S.K. Sarkar, CTO , KVK and TTC for their constant help in this work. Shri B.K., Sahoo, village farmer, and his team deserve special thanks for carrying out the participatory work on their farm.

References

Jena, J.K., Mohanty, S.N., Tripathi, S.D., Mohanty, A.N., Muduli, H.K. and Sahoo, S. (1998) Carp seed raising in small backyard and kitchen ponds: a profitable technology package for tribal women. In: Thomas, P.C. (ed.) *Current and Emerging Trends in Aquaculture*. Daya-Publishing House, New Delhi, pp. 94-98.

Mishra, R.S., Radheyshyam and Prasad, B. (1980) Predatory effect of cyclopoid copepods on fish spawn. Proceedings of the abstracts of National Academy of Science, held at Allahabad University during October 23-27. Published by National Academy of Science, India, pp. 7-9.

Mohanty, A.N. and Mohanty, S.N. (1984) *Rearing of Fry in Tribal Villages of Orissa*. Proceedings of the Souvenir and Seminar on Freshwater Fisheries and Rural Development, Rourkela, Published by Fisheries Department of Govt. of Orissa, pp. 9-12

Patnaik, S., Das, K.M. and Pani, K.C. (1989) Raising fish seed in weed cleared small rural ponds is profitable. *Journal of Zoological Research* 2(1&2), 67-70.

Radheyshyam (1997) *Rural Aquaculture-Sarakana Success Story*. CIFA Publications, Bhubaneswar. 7 pp.

Radheyshyam (1998) Carp seed production in rural area - a decade of experience. *Fishing Chimes* 18(4), 7-11.

Radheyshyam and Kumar, D. (1982). Fisheries extension for rural aquaculture. In: Proceedings of Fish Farmers Development Agency, Souvenir Balasore (Orissa). Published by Fisheries Department of Govt. of Orissa, India, pp. 95-105.

Radheyshyam, Safui, L. and Sarkar, S.K. (1998) Fish seed raising and paddy cultivation on rotation in seasonal pond of rural Orissa. In: Hameed, M.S. and Kurup, B.M. (eds) *Technological Advancements in Fisheries*. Cochin University of Science and Technology, Cochin, pp. 47-53.

Radheyshyam and Sarkar, S.K. (1998) Carp spawn production in Sarakana village - an emerging rural trade. In: Thomas, P.C. (ed.) *Current and Emerging Trends in Aquaculture*. Daya Publishing House, New Delhi, pp. 138-144.

Radheyshyam, Sarkar, S.K. and Singh, B.N. (1985) Observations on possible methods of averting hazards of loss of viable eggs in breeding hapas in rural ponds having common carp. *International Journal of Academy of Ichthyology (Proceedings V. AISI)* 6, 115-119.

Radheyshyam, Satapathy, B.B. and Selvaraj, C. (1982) *Utilization of Roadside Borrow Pits for Rearing of Carp Fry - a Case Study*. Proceedings of Fish Farmers Development Agency, Souvenir Balasore (Orissa). Published by Fisheries Department of Govt. of Orissa, India, pp. 3-6.

Radheyshyam, Satapathy, B.B., Singh, B.N., Verma, J.P., Sarkar, S.K., Kumar, K. and Dutta, B.R. (1988) Utilization of small backyard ponds for fish culture in rural area - a new perspective. *Journal of Zoological Research* 1(2), 129-139.

Radheyshyam, Sharma, B.K., Chatopadhyaya, D.N., Sarkar, S.K. and Satapathy, B.B. (1993) Effects of phased increase in water level on the survivability of *Cyprinus carpio* fry in nursery ponds. In: Mohan Joseph, M. (ed.)

Proceedings of the Third Indian Fisheries Forum, Pantnagar University India. Published by Asian Fisheries Forum Indian branch, pp. 59-63.

Radheyshyam, Singh, B.N., Satapathy, B.B., Sarkar, S.K., Kumar, K., Verma, J.P. and Dutta, B.R. (1991) Effects of shading on the productivity and economics of backyard ponds in rural Orissa. In: *Proceedings of the National Symposium on Freshwater Aquaculture, CIFA, Bhubaneswar India*. CIFA Publication, pp. 132-134.

Radheyshyam and Tripathy, N.K. (1992) Aquaculture as nucleus for integrated rural development - an experience. *Fishing Chimes* 12(9), 37-48.

Sarkar, S.K., Dutta, B.R., Kumar, K. and Singh, B.N. (1984) Preliminary observations on the possibility of adopting the simplified technique of induced fish breeding in rural Orissa - a case study. *International Journal of Academy of Ichthyology* 5, 177-180.

Selvaraj, C. and Kanaujia, B.R. (1979) Fish seed rearing in village pond. *Indian Farming* 29(2), 31-32.

Schroeder, G.L. (1978) Autotrophic and heterotrophic production of micro-organisms in intensively manured fish ponds and related fish yields. *Aquaculture* 14, 303-325.

Sharma, B.K., Thakur, N.K., Sarkar, S.K., Safui, L., Radheyshyam, Dutta, B.R. and Sarangi, N. (1988) Involvement of rural womenfolk in aquaculture under S & T program at KVK/TTC Kausalyaganga. In: *Proceedings of the All India workshop on Gainful Employment for Women in the Fisheries Field*. Published by Department of Science and Technology Govt. of India & Central Institute of Fisheries Technology, Cochin India, pp. 54-71.

Shirgur, G.S. (1977) Observations on rapid production of zooplankton in fish nurseries by intensive phased manuring. *Journal of Indian Fisheries Association* 1(2), 25-50.

Singh, B.N. and Radheyshyam (1995) A big role for micro-nutrients in artificial feed on growth, survival and economic profitability in carp seed production from nursery ponds. In: *Proceedings of the Tripura Fisheries Souvenir*. Published by Fisheries Department of Govt. of Tripura, India, pp. 16 - 12.

Thakur, N.K., Sarkar, S.K., Sarangi, N. and Sharma, B.K. (1988) Self employment of rural womenfolk through successional aquaculture in backyard ponds. In: *Proceedings of the All India Workshop on Gainful Employment for Women in Fisheries Field*. Published by Department of Science and Technology Govt. of India & Central Institute of Fisheries Technology, Cochin India, pp. 72 -81.

Tripathi, S.D., Kumar, D., Mohanty, S.N. and Muduli, H.K. (1987) Fish seed raising an important income component in rural aquaculture - a few case studies. Extension Education Bulletin, Directorate of Extension Education, Orissa University of Agriculture and Technology, Bhubaneswar, pp. 1-4.

Tripathi, S.D., Aravindakshan, P.K., Singh, R., Jena, S., Ayyappan, S., Muduli, H.K. and Mohanty, A.N. (1991) Carp seed rearing in ammonia treated nurseries. In: *Proceedings of the National Symposium on Freshwater Aquaculture*. Published by Association of Agriculturists & Central Institute of Freshwater Aquaculture Bhubaneswar, India, pp. 13-15.

Chapter 14

Freshwater Fish Seed Quality in Asia

D.C. Little[1,2], A. Satapornvanit[1] and P. Edwards[1]

[1]*Aquaculture and Aquatic Resources Management, School of Environment Resources and Development, Asian Institute of Technology, PO Box 4, Klong Luang, Pathumthani, 12120, Thailand*
[2]*Institute of Aquaculture, University of Stirling, Stirling FK9 4LA, UK*

Abstract

The rapid development of inland aquaculture in Asia has been based on the ready availability of fish seed to farmers. Although seed of the major cultured species is now produced in large quantities in hatcheries, poor quality is increasingly perceived as a major constraint to the success of fish culture, especially for new entrant farmers and poorer smallholders. However, clarification of the role of seed quality, as opposed to post-stocking management, as the cause of farmers' poor results is often lacking. Furthermore, poor seed quality can have a variety of causes relating to both genetics and management. A participatory methodology is described, 'state of the system' (SoS) reporting, that identifies probable causes of poor seed quality and important researchable issues with the people involved in seed production and distribution networks. SoS reporting may be a useful approach for accelerating accurate analysis and subsequent policy-making of such dynamic situations with, and for, the stakeholders involved. Characteristics of current fish seed networks, and promising strategies for the production and distribution of high quality seed to smallholders, are identified. Improved seed quality will depend both on greater awareness of the problem and on practical methods for its assessment at the hatchery or farm level. The role of institutions in improving, maintaining and monitoring fish seed availability and quality is discussed.

Introduction

The availability of quality fish seed is a prerequisite for adoption of sustainable aquaculture by smallholders. Hatchery development over the last 30 years has been essential for fish culture becoming part of livelihood systems in Asia. A wide range of species is now raised. Although carp, both indigenous and exotic species, and tilapia make up the bulk of fish raised in the region, several species of catfish and gourami are also important.

Techniques to mass-produce fish seed in the region were mainly introduced initially through the public sector but the availability of fry and fingerlings increased rapidly after adoption by the private sector. A variety of 'actors' now produce and distribute fish seed to rural areas, often as part of complex networks. Such entrepreneurs, who are typically concentrated in certain geographical areas, now produce the bulk of seed in countries such as Bangladesh, India, Thailand and Vietnam. Additionally, seed production by farmers themselves is now widely promoted in the region.

Whilst fish seed supply has exploded in many parts of Asia, there now appears to be a major problem of poor or erratic quality of seed reaching farmers. The possible underlying causes for the gradual deterioration in yields and individual size of many species of cultured fish are reviewed in the light of recent research. The quality and management of broodfish and hatchery practices that lead to inbreeding or contamination have been implicated, as has poor husbandry during nursing, handling or transportation. Infection with pathogens may also be an underlying cause of poor performance of fish seed. Schemes to upgrade stocks with wild or improved fish are now fashionable but these may be unsustainable unless the causes of quality deterioration are better understood and improved management strategies developed. Simple but reliable methods that can be used at the hatchery or farm gate to assess seed quality are required to improve the quality of fish stocked. The potential of challenge stress tests and the development of morphological keys as appropriate methods of quality control are reviewed below.

A framework for problem identification of fish seed quality is required which is both holistic and participatory. The emerging production and distribution systems for freshwater fish seed in Asia are complex and dynamic. Current policy making based on conventional data collection is problematic. Novel approaches are required that produce more up-to-date and reliable information and are feasible within the institutional constraints of the region.

Developmental Impact of Fish Seed Quality

High quality fish seed is important to both producers and consumers of farmed fish, but may be particularly critical if benefits are to be realised by poorer people. Fry or fingerlings that die or grow poorly after stocking might cause a temporary set-back to a well-resourced fish farmer but the experience may either discourage further investment by, or lead to a complete loss of interest of, a new entrant farmer. Improved quality fish seed, by enhancing the level and consistency of performance, should also lead to lower production costs of fish raised commercially, and lower prices for poorer consumers. The quality of fish mm seed also affects the returns and risks involved in production and marketing networks. Fish seed with high survival and healthy appearance benefit intermediaries in trading networks who are often from the poorest sectors of the community.

Why Does Fish Seed Quality Decline?

The basic causes of the gradual deterioration in yields and individual size of many cultured fish are unclear and may be obscured further by inadequate production and delivery of seed to farmers or poor management of fish seed by farmers once stocked (Fig. 14.1). The declines in fish production perceived by producers may reflect poor management following stocking of fish rather than intrinsic differences in seed quality. The perceptions of experienced commercial farmers, for whom fish culture is their main source of income, may be a more reliable indicator of possible changes in seed quality than farmers who have recently adopted fish culture. Ingthamjitr (1997) reported that hybrid *Clarias* catfish farmers in central Thailand, using fairly standard management practices, had suffered a decline in quality which was subsequently related to the quality and management of broodfish. Broodfish of fecund species for which relatively few fish are spawned may be particularly prone to mismanagement. Poor performance has been linked to inbreeding of carp in India (Eknath and Doyle, 1985). Deterioration in the quality of tilapia is frequently reported but probably has other causes. The 'stunting' observed in tilapia after several generations may be linked to negative selection or genetic introgression through contamination with feral fish (Macaranas *et al.*, 1986). Current schemes to upgrade stocks with wild or improved fish, e.g. GIFT, may be unsustainable unless the problems are better understood and improved management strategies are developed (Little and Edwards, 1997; Little, 1998).

Changes in the quality of fish seed may be related to changes in the management of brood and seed fish as producers react to dynamic opportunities and competition. In some cases, overproduction and reduced returns of hatchery operators have led to consolidation or specialisation of seed production that may negatively affect quality. Long established carp and mixed tilapia hatcheries in Thailand changed to more value-added products such as ornamental fish and black tiger shrimp as margins declined with the entry of many competitors. Many current producers of carp and tilapia are less experienced, which could result in poorer quality and less predictable quantities. Alternatively, quality might be expected to improve as producers begin to specialise in fewer species of fish or specific markets that require particular consistency of quantity and quality. This phenomenon is illustrated by the rapid rise of hatcheries producing monosex tilapia in Thailand.

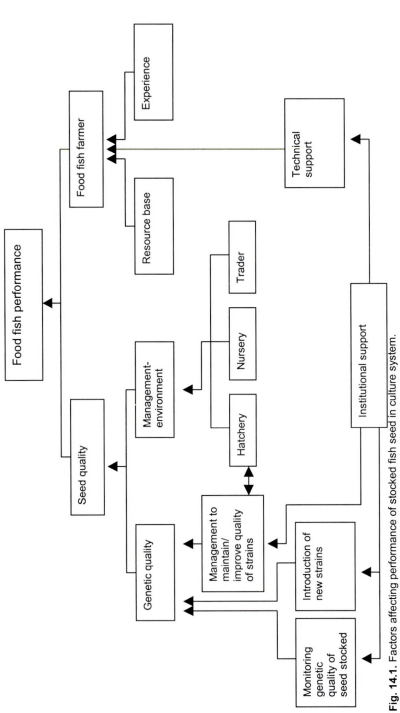

Fig. 14.1. Factors affecting performance of stocked fish seed in culture system.

Poor husbandry of fish seed during nursing, holding or transportation is believed to negatively affect later performance (Morrice, 1995) but this may not influence farmers to purchase locally produced seed. A survey in Northeast Thailand found that farmers preferred to buy tilapia seed from outside the village rather than from neighbours because of perceived quality differences (Sodsook, 1989). Competition between nursery producers and traders, and typically long distance transportation appear to result in poorer quality seed reaching farmers in the Red River Delta, Vietnam (Demaine, 1996).

The quality of fish seed stocked in community water bodies has also become a major issue as government and contracted private hatcheries now produce vast numbers of seed for stocking systems ranging from small community tanks or water bodies to open-water floodplains. The overall outcome of stocking is determined by a combination of natural, socio-economic and institutional conditions (Lorenzen and Garaway, 1998), as is the quality of the seed alone.

Fish Seed Availability

Local production and marketing of fish seed can stimulate and support local people to incorporate fish culture within their livelihood systems. The distance between the source of fish seed and the farmer plays a role in the farmer's decision of where to buy seed. This may explain the popularity of mobile traders who deliver seed to the farm in many parts of Asia (Lewis *et al.*, 1996; Ingthamjitr *et al.*, 1998). Clearly, species and techniques amenable to production of consistent quantity and quality of fish seed should be promoted. Most success has been achieved to date with small carp and tilapia in this respect (AIT Aqua Outreach, 1997; Gregory *et al.*, 1997). The AIT Aqua Outreach Project has promoted aquaculture through improving the availability of quality fish seed at the farm gate. The approach involved introducing advanced nursing in *hapas* of small fry purchased through well-developed trading networks (Little *et al.*, 1991) and institutionalising the development of simple hatchery, nursery and distribution systems in Lao PDR and Cambodia (Haitook, 1997; Little and Gregory, 1997). Initial analysis indicated a tendency in the study areas for concentration of operations to occur and yet a major developmental objective of institutions concerned with promoting aquaculture in rural areas should be to make fish seed available to very dispersed and small-scale producers (Box 14.1).

Approaches to Assessing Fish Seed Quality

Fish seed quality, or the perception of quality, might be expected to improve as awareness and knowledge among actors within trading networks grow. An improved understanding of quality by the final customers, i.e. farmers purchasing fish seed, is especially important. The large proportion of new entrants or recent adopters of fish culture is a challenge to accelerating this process.

Box 14.1. Are concentrations of fish seed producers good for improving the availability of quality fish seed?

Concentrations of seed producers:
- increase choice for customers, allowing them to cost-effectively select good value fish seed (low price, high quality)
- increase possibility of broodfish exchange to reduce problems of inbreeding of isolated stocks
- allow cost-effective monitoring of quality
- facilitate information exchange and rapid dissemination of knowledge and best practice among fish seed producers
- encourage development of a range of support services that provide additional livelihoods and in turn increase operators' knowledge and resources

But:
- concentrate income generation in a limited number of locations and people
- result in longer transport times of fish seed, increasing stress-related quality problems
- increase density of fish and problems of pathogen/parasite load
- increase competition and lower returns to individual operators leading to hatchery practices that negatively affect quality, e.g. use of small broodfish
- impact on availability of high quality water; close proximity of many hatcheries can

---

Simple techniques to assess seed quality that can be used at the hatchery or farm gate are required to improve the quality of stocked fish. Stress tests for quality control have been used mainly for shrimp post-larvae, as reviewed by Fegan (1992). Based on the concept of a physiological challenge (Wedemeyer, 1981), salinity stress tests have also been used to assess the quality of salmon smolts for stock enhancement and culture (Clarke and Blackburn, 1977; Duncan, 1996). A salinity challenge test has been developed recently for use with hybrid *Clarias* catfish hatchlings (Ingthamjitr, 1997). Stressors may also be applied as a management tool to try and reduce the number of weak, often pathogen-carrying seed. An example of this approach is the use of formalin to treat white-spot (baculovirus) infected, post-larval penaeids (Limsuwan, 1997).

Growth, condition, morphological changes and body composition have also been used to establish 'quality' of Atlantic salmon smolts (Dickhoff *et al.*, 1995). Fin quality (Davidson *et al.*, 1995), behaviour (Olla *et al.*, 1994) and colour changes (Cahill, 1989) are also useful indicators of quality. An analysis of survival of Atlantic salmon smolts in Canada found that 44-68% of the variation in survival rate could be attributed to individual smolt weight, quality and sea conditions (Farmer, 1994). Quality was related to an index that included fork length, weight, condition factor, incidence of fin rot and erosion, and physical abnormalities. Fin rot was differentiated from erosion and assessed for individual fins to produce a weighted score. Chowdhuri (1996) used scale loss and tail damage of carp after transportation as quality indicators in Bangladesh.

Whereas challenge tests are probably of most relevance for monitoring quality at the hatchery, simple keys that use visual assessment of species and quality could be used at the field level by extension staff and farmers. Availability of keys could educate fish seed purchasers and reduce cheating on species and quality that are prevalent in trading of the commonly cultured fish species in Asia.

Rapid improvements in fish seed quality are most likely to occur where there are clear and short-term benefits to both producer and customer. Competition between producers can stimulate the interest of hatcheries to improve quality. The hybrid *Clarias* catfish hatcheries of central Thailand compete to attract custom by replacing poorly performing seed, as well as competing on price (Ingthamjitr, 1997). In areas of concentrated hatcheries and sales outlets in Northeast Thailand, maintenance of fish seed in good condition prior to sale is essential. Strategies for mobile traders to purchase high quality seed from nursery operators and to maintain the quality vary with the method of transportation and infrastructure in different countries but are becoming increasingly complex.

Role of Institutions

The current dominance of private-sector fish seed production in those Asian countries with the most dynamic aquaculture sectors provides both opportunities and constraints for government institutions to play a role in maintaining or improving quality. Maintenance and improvement of high quality germplasm, and its multiplication to support the private sector, have been identified as key roles for public sector institutions. Success will depend on government institutions being able to develop and maintain close contacts with the diverse 'actors' in fish seed networks and being responsive to their changing needs. Consistent funding to secure and maintain the concentration of skills and resources necessary is also vital for this long-term role. The possibilities of *in situ* maintenance or improvement at the private hatchery level or farm, especially of indigenous species, should also be pursued as this promises to be a lower cost strategy to ensure genetic quality.

The likelihood of non-genetic factors influencing quality is a major issue requiring attention. Research on private-sector fish seed networks is clearly required to identify the major factors, and combinations of factors, leading to poor quality seed. Government institutions might initially develop a seed quality monitoring role, perhaps with certification of quality to raise awareness, but developing and disseminating knowledge to private producers, and a competitive and open market for fish seed must be the longer term objectives.

There are clear training and research needs within the institutions themselves if these objectives are to be met in the short-term.

Understanding Fish Quality Issues

The causes of poor seed quality and inconsistent performance of seed stocked in smallholder aquaculture are often multifactorial and are obscured by long and complex production and trading networks. Information about current practice and likely future trends in general is rarely available to institutions promoting rural aquaculture. Policy makers and implementers of rural development generally rely on conventional data collection, analysis and dissemination that can be both cumbersome and produces information of little value too late. Participatory methods to rapidly collect and synthesise information on current practice and opinion from a broad range of stakeholders have been brought together as 'state of the system' (SoS) reporting with local partner institutions (AIT Aqua Outreach, 1997). The methodology allows different perceptions about quality of the various actors in fish production networks to be rapidly collected and assessed. Field collection of strategic information is followed by analysis and a clear and simple presentation of summarised information to groups of the various stakeholders at a (SoS) workshop. A framework of questions is used to stimulate small group discussion, comment and criticism during the workshop. Small groups of different stakeholders are used to ensure that no single group dominates or unduly influences others and initial presentations of information are given in a neutral, uncritical manner. Key issues may be identified for further investigation, perhaps requiring selective additional fieldwork. Such SoS workshops allow triangulation of information collected and are a vital step in its synthesis into a balanced and useful document. They allow clarification of obscure or complex issues and ensure ownership of the information by the partner institutions and stakeholders involved. Ownership is further strengthened by the reports being written simultaneously, phrase by phrase, in both the local language and English by all participants rather than later translation by a smaller, select group of individuals. This also affects the type of 'language' used since stakeholders articulate their thoughts and comments both verbally and in writing to produce a narrative that is understandable by non-specialists.

Acknowledgements

The Fish Seed Quality Project involving partner institutions in four countries in Asia is supported financially through the Asian Institute of Technology by the Aquaculture Research Programme, DFID, UK. This article is based on the conceptual framework contained in the project document. Partner institutions are the Research Institute Number 1, Ha Bac and the Faculty of Fisheries, College of Agricultural and Forestry, Ho Chi Minh University, Vietnam; the Regional Development Committee (RDC) for Livestock and Fisheries Development in Southern Laos; the Department of Fisheries, Thailand; and the Department of Fisheries, Bangladesh. The AIT Aqua Outreach Programme through additional support provided by the Governments of Denmark and Sweden through Danida and Sida, respectively, has provided a facilitatory role in Laos, Thailand, and Vietnam and the DFIF Bangladesh Fisheries Programme has facilitated similarly in Bangladesh.

References

AIT Aqua Outreach (1997) Hatchery development in and around Ho Chi Minh City. State of the System Report. AIT Aqua Outreach, AIT, Bangkok.11 pp.

Cahill, D. (1989) The collection and analysis of smolt survey data leading to the development of a smolt quality scheme. MSc Dissertation, University of Stirling, Stirling. 85 pp.

Chowdhuri, A. (1996) Post-transportation survival of Indian major carps in Bangladesh. MSc Dissertation, Mymemsingh Agricultural University, Mymemsingh. 42 pp.

Clarke, W.C. and Blackburn, J. (1977) A seawater challenge test to measure smolting of juvenile salmon. *Fisheries and Marine Service Technical Report No. 705.* Environment Canada, Ottawa. 11 pp.

Davidson, K., Swan, P. and Hayward, J. (1995) The gulf region parr/smolt quality evaluation programme: 1989-1993. Canadian Data Report of Fisheries and Aquatic Science. 153 pp.

Demaine, H. (1996) Field work in the Red River Delta. Progress report. AARM Program, SERD, AIT, Bangkok. 10 pp.

Dickhoff, W.W., Beckman, B.R., Larsen, D.A., Mahnken, C.V.W., Schreck, C.B., Sharpe, C. and Zaugg, W.S. (1995) Quality assessment of hatchery-reared chinook salmon smolts in the Columbia River basin. In: Schramm, H.L. and Piper, R.G. (eds) *Uses and Effects of Cultured Fishes in Aquatic Ecosystems.* American Fisheries Society, pp. 292-302.

Duncan, N. (1996) Photoperiodic manipulation and its use in the all year round production of Atlantic salmon, *Salmo salar.* PhD Dissertation, University of Stirling, Stirling. 278 pp.

Eknath, A.E. and Doyle, R.W. (1985) Indirect selection for growth and life history traits in Indian carp aquaculture.1. Effects of broodfish management. *Aquaculture* 49, 73-84.

Farmer, G.J. (1994) Some factors which influence the survival of hatchery Atlantic salmon (*Salmo salar*) smolts utilised for enhancement purposes. *Aquaculture* 121, 223-233.

Fegan, D.F. (1992) Recent developments and issues in the penaeid shrimp hatchery industry. In: Wyban, J. (ed.) *Proceedings of the Special Session on Shrimp Farming*. World Aquaculture Society, Baton Rouge, pp. 55-70.

Gregory, R., Innes-Taylor, N.L., Guttman, H. and Little, D.C. (1997) Seed supply in aquaculture. Policy Paper Number 1. AIT Aqua Outreach, Asian Institute of Technology, Bangkok.

Haitook, T. (1997) Fish seed supply in Savannakhet, Lao PDR. MSc Dissertation, Asian Institute of Technology, Bangkok. 171 pp.

Ingthamjitr, S. (1997) Commercial hybrid *Clarias* catfish hatcheries in Central Thailand. PhD Dissertation, Asian Institute of Technology, Bangkok. 135 pp.

Ingthamjitr, S., Phromtong, P. and Little, D.C. (1998) Fish seed production and marketing in Northeast Thailand. *NAGA* 20(3-4), 24-27.

Lewis, D.J., Wood, G.D. and Gregory, R. (1996) *Trading the Silver Seed. Local Knowledge and Market Moralities in Aquacultural Development*. Intermediate Technology Publications, London. 199 pp.

Limsuwan, C. (1997) Reducing the effects of white spot baculovirus using PCR screening and stressors. *AARHI Newsletter* 6(1), 1-2.

Little, D.C. (1998) Options in the development of the 'aquatic chicken'. Fish Farmer, July/August, pp. 35-37.

Little, D.C. and Edwards, P. (1997) Contrasting strategies for inland fish and livestock production in Asia. In: Corbett, J.L., Choct, M., Nolan, J.V. and Rowe, J.B. (eds) *Recent Advances in Animal Nutrition in Australia*. University of New England, Armidale, pp. 75-87.

Little, D.C. and Gregory, R. (1997) A family affair in Cambodia. AASP *Newsletter* 1(3), 5.

Little, D.C., Innes-Taylor, N.L., Turongruang, D. and Komolmarl, S. (1991) Large seed for small scale, aquaculture. *Aquabyte* 4(2), 2-3.

Lorenzen, K. and Garaway, C. (1998) How predictable is the outcome of stocking? In: Petr, T. (ed.) *Inland Fishery Enhancement*. FAO Fisheries Technical Paper, No. 374, FAO, Rome. 463 pp.

Macaranas, J.M., Taniguchi, N., Pante, M.J.R., Capili, J.B. and Pullin, R.S.V. (1986) Electrophoretic evidence for extensive hybrid gene introgression into commercial (*Oreochromis niloticus*) stocks in the Philippines. *Aquaculture and Fisheries Management* 17, 249-258.

Morrice, C. (1995) NW Bangladesh aquaculture extension project annual report. NFEP 2, Parbatipur. 52 pp.

Olla, B.L., Davis, M.W. and Ryer, C.H. (1994) Behavioural deficits in hatchery-reared fish: potential effects on survival following release. *Aquaculture and Fisheries Management* 25(suppl), 19-34.

Sodsook, S. (1989) Nursing juvenile fish in nylon hapas suspended in earthen ponds. MSc Dissertation, Asian Institute of Technology, Bangkok. 100 pp.

Wedemeyer, G.A. (1981) The physiological response of fishes to the stress of intensive aquaculture in recirculation systems. Aquaculture in heated

Chapter 15

Genetic Technologies Focused on Poverty? A Case Study of Genetically Improved Tilapia (GMT) in the Philippines

G.C. Mair[1], G.J.C. Clarke[1], E.J. Morales[2] and R.C. Sevilleja[2]
[1]*University of Wales Swansea, Swansea SA2 8PP, UK*
[2]*Freshwater Aquaculture Center, Central Luzon State University, Nueva Ecija 3120, Philippines*

Abstract

This chapter examines the potential of genetics-based technologies to enhance the livelihoods of poor people in developing countries, in the context of current development paradigms. Issues related to impacts upon production and producers, fish consumers and upon employment opportunities are discussed using a case study of a technology for the production of genetically male tilapia (GMT), which has been developed and disseminated since 1995 in the Philippines. Small-scale tilapia culture in the Philippines is characterised and criteria by which small-scale, poor farmers can be identified are outlined. Analysis of current dissemination strategies revealed that on-going 'passive' mechanisms, whilst providing for sustainable dissemination, do not reach many small-scale farmers. However, small-scale, poor farmers do stand to benefit, in terms of improved livelihoods, from appropriate genetics-based technologies. Tilapia play an important role in the diet of both the urban and rural poor in the Philippines who stand to benefit greatly from contributions that genetic technologies can make to improved production efficiency. Evaluation of a number of alternative dissemination strategies indicates that strategic, supported partnerships with non-governmental organisations (NGOs) and people's organisations are most likely to be effective in targeting the benefits of the technology at poor farmers. Alternatively, providing improved broodstock to small-scale hatcheries, combined with appropriate technical and financial support, can also be effective in targeting small-scale farmers.

Recommendations for further research and modifications to on-going dissemination strategies to enhance the benefits of genetics-based technologies for the poor are made.

Background

Technological advances in aquaculture can impact upon poverty and livelihoods through three main routes: production by poor people; employment of poor people; and improved access to food by poor people through increased supply and/or reduced prices. This paper considers existing and potential impact of genetics-based technologies in aquaculture upon poverty and livelihoods, through a case study of the dissemination and impact of a high yielding, monosex male tilapia, in the Philippines. This technology was developed to address identified constraints to optimising tilapia production.

A Profile of Tilapia Culture in the Philippines

Aquaculture is in a phase of rapid expansion in the Philippines and is an important contributor to gross domestic product, to employment and to food consumption. Aquaculture production grew by an average 6.4% per annum between 1987 and 1996 and in 1996 it generated, at current prices, an estimated 1.4% of gross domestic product (BFAR, 1997). Tilapia culture in the Philippines, although a maturing sector of agriculture, is still highly labile. Although *Oreochromis mossambicus* has been present in the Philippines since the 1950s, tilapia culture began in earnest in the early 1970s with the widespread dissemination of *Oreochromis niloticus*. Production is commonly assumed to be in excess of 90,000 to 100,000t year^{-1} although FAO report 1998 production at only 72,000t year^{-1} (FAO, 2000). Monoculture of tilapia dominates inland aquaculture and is second only to milkfish (*Chanos chanos*) in terms of overall aquaculture production.

At least 50% of tilapia are produced in freshwater ponds, with 36% produced in cages and the remainder in brackishwater culture (Guerrero, 1996). Pond culture is concentrated in Central Luzon with cage culture dominating in Southern Luzon. Pond culture is commonly extensive or semi-intensive (some supplementary feeding of commercially available feeds) monoculture. There is relatively little integration of aquaculture with other farm activities and pond inputs, where present, are commonly from off-farm sources such as commercially purchased chicken manure. Tilapia grow-out is exclusively a private sector activity but there is a significant involvement of the public sector in the production of seed, with large hatcheries operated by the Bureau of Fisheries and Aquatic Resources (BFAR) and regional stations of the Department of Agriculture (DA). Nevertheless, the majority of tilapia seed is produced in private sector hatcheries, with a significant proportion being produced by a large number of small-scale hatcheries around Laguna de Bay, 50km south of Manila. Fingerling supply in 1996 was estimated to be in the region of 600 million per annum (Guerrero, 1996) with more recent estimates indicating a supply deficit with demand for fingerlings in excess of 1 billion per annum.

There are several social, economic, technical and environmental constraints to tilapia culture (such as high input costs, mortality, water supply and typhoon damage) but limited availability of quality seed has long been considered a major constraint by producers and researchers alike. In the commonly cultured *Oreochomis niloticus*, loss of genetic variation through founder effects and poor broodstock management, together with introgression with feral *Oreochomis mossambicus*, contributed to a decline in the performance of Philippine tilapia stocks compared to more recently introduced pure strains, including those based on wild-caught African fish, as indicated in growth trials conducted under Philippine culture conditions (Eknath *et al.*, 1993). Furthermore, the precocious sexual maturation and unwanted reproduction of tilapia have long been considered as major constraints to optimisation of marketable yields in tilapia aquaculture. Poor genetic quality of seed and the problems associated with early maturation were both considered as important researchable constraints that could be addressed through the application of genetic-based technologies. These technologies were focused on improving economic yields in response to demand from producers for higher yields of more uniform size. However, the extent of this demand, particularly among small-scale and poor producers and consumers was not well established.

The roles of small-scale farmers in tilapia production and aquaculture in small-scale farming systems are not well understood in the Philippine context.

Genetic Technologies in Philippine Aquaculture

The Philippines has been the centre for a number of development-related research projects on the genetic improvement of tilapia for aquaculture. The Philippines was one of the first countries in which the non-genetic technology of hormonal masculinisation of tilapia was attempted on a commercial scale (Guerrero and Guerrero, 1988) and the technology has been actively promoted to farmers since the mid-1970s. There are at least ten hatcheries in the Philippines producing sex-reversed tilapia (SRT) providing up to 15% of the fingerling supply. However, whilst demand for monosex fish is thought to be considerable, the technology faces important technical constraints with many hatcheries failing to produce SRT of sufficient quality. Sex ratios in SRT were often lower than 90% male (due largely to the failure to adhere to strict protocols) with sex ratio in excess of 95% male considered necessary to meet the major objectives of culturing all-male stocks (Mair and van Dam, 1996). Further expansion of SRT technology in the Philippines looks unlikely.

Selection programmes

There have been three major projects applying genetics-based technologies to the improvement of tilapia, two of these involving traditional methods of selective breeding. Bolivar *et al.* (1994) reported an average 18% increase in weight gain of tilapia compared to unselected controls after eight generations of within family selection for growth rate and further gains have been achieved in subsequent generations. However, this selected strain has not yet been made widely available to farmers and has thus had minimal impact upon production. Of greater significance is the Genetic Improvement of Farmed Tilapia (GIFT) Project coordinated by ICLARM. Using a combined selection methodology on a synthetic base population developed from newly introduced strains from Africa and domesticated Asian strains, this Project achieved genetic gains averaging 13% over five generations, providing an estimated cumulative increase of 85% in growth rate compared to the base population from which it was selected (Eknath and Acosta, 1998). The GIFT fish have been widely disseminated to farmers, initially through government agencies and since 1997 through the GIFT Foundation International (GFI). The GFI distributes GIFT strain broodstock through a network of large private sector hatcheries via formal licensing agreements.

The YY male technology and GMT

This technology has been developed as a direct alternative to hormonal sex reversal for the provision of monosex male tilapia as a solution to the problem of early sexual maturation and unwanted reproduction. The technology involves the combination of hormonal feminisation and progeny testing in a breeding programme to mass produce novel YY male genotypes which sire only male progeny known as genetically male tilapia (GMT) in crosses with normal females. The development of the technology is described by Mair *et al.* (1997). GMT commonly have sex ratios averaging > 95% male in controlled conditions although sex ratios in commercial production can be more variable, due largely to contamination events. Thus for GMT under culture, reproduction is not completely eliminated but is significantly reduced. On-station and on-farm trials have shown GMT to increase yields by 30-40% compared to normal, mixed-sex tilapia of existing Philippine strains and yields have also been shown to be superior to hormonally sex-reversed tilapia (Mair *et al.*, 1995). Farm trials in the Philippines demonstrated increases in the profitability of tilapia culture of over 100% through the culture of GMT compared to the alternative, mixed-sex tilapia (Mair *et al.*, unpublished data). The outputs of the research and development of the YY male technology are being disseminated through the provision of YY male and normal female broodstock to hatcheries which breed them using any one of a range of standard hatchery practices to produce GMT for grow-out. The key properties of the technology are:

- Hatcheries cannot produce their own YY males so the outputs of the technology cannot be adopted independently of the source of YY males.
- Hatcheries should be able to prevent contamination of broodstock to maintain quality of GMT (in terms of its sex ratio).
- GMT fry are close to 100% male so do not produce significant numbers of fingerlings under culture so growers need to purchase fingerlings for each production cycle.
- Growers stocking GMT should also be able to minimise contamination with females to prevent reproduction.

The dependence on the source of YY males requires that a sustainable development programme needs to be established to support dissemination of the outputs of the technology. This process limits to a degree the scale and speed of uptake but does permit a great degree of quality control over the process.

A dissemination organisation, known under the name Phil-Fishgen, has been established by the Freshwater Aquaculture Center of Central Luzon State University (the lead national agency for freshwater aquaculture research in the Philippines) to continue research and evaluation of the technology and to disseminate the products of the technology to producers, as illustrated in Fig. 15.1. Under this process private sector hatchery operators request accreditation as GMT producers. Accreditation is granted subject to evaluation of criteria related largely to the efficiency of the hatchery and its capacity to prevent contamination of broodfish, although geographic location is also a factor for some hatcheries. There are a total of 32 GMT accredited hatcheries in the Philippines with over 40,000 broodstock sets (1 YY male: 3 normal females) having been distributed since 1995. Broodstock presently in use are capable of producing in excess of 50 million GMT fry year^{-1} although accurate estimates of actual production are not available. In addition to supplying GMT-producing broodstock to accredited hatcheries, Phil-Fishgen also has its own GMT hatcheries with a target production of 1 million GMT fry month^{-1} and at the time of writing has distributed over 24 million GMT to over 1,400 farmers.

This dissemination process is a relatively passive one in that it is not deliberately targeted at any particular sector of production and the characteristics of the ultimate producer beneficiaries will depend upon the nature and location of the hatcheries that are accredited. To determine the impact of current dissemination strategies it is thus necessary to gain a greater understanding of small-scale tilapia farming and of the distribution pathways for tilapia seed.

Characterising Small-scale Tilapia Farmers in the Philippines

No previous published study has adequately characterised small-scale fish farmers in the Philippines. A fairly detailed study of tilapia farms was made under ICLARM's DEGITA project based primarily on a data set collected by the Bureau of Agricultural Statistics together with a number of smaller surveys. The results from this study, yet to be published in full, indicated that the average area of fish ponds on fish farms in the Philippines was 1.5ha (Eknath and Acosta, 1998). Furthermore, the study indicated that tilapia farmers principally culture fish as a cash crop (97% of farmers). Monoculture of tilapia in ponds yielded a relatively low average annual production of 3,559kg ha^{-1} compared with other countries although this had a relatively higher value estimated at US\$3,421 ha^{-1}. The overall picture was thus one in which tilapia farmers were generally farming on a small to medium scale and were well above the subsistence level of agriculture. There was concern, however, that data collected by the Bureau of Agricultural Statistics, with the assistance of the Department of Agriculture, did not adequately reflect the existence of small-scale farmers.

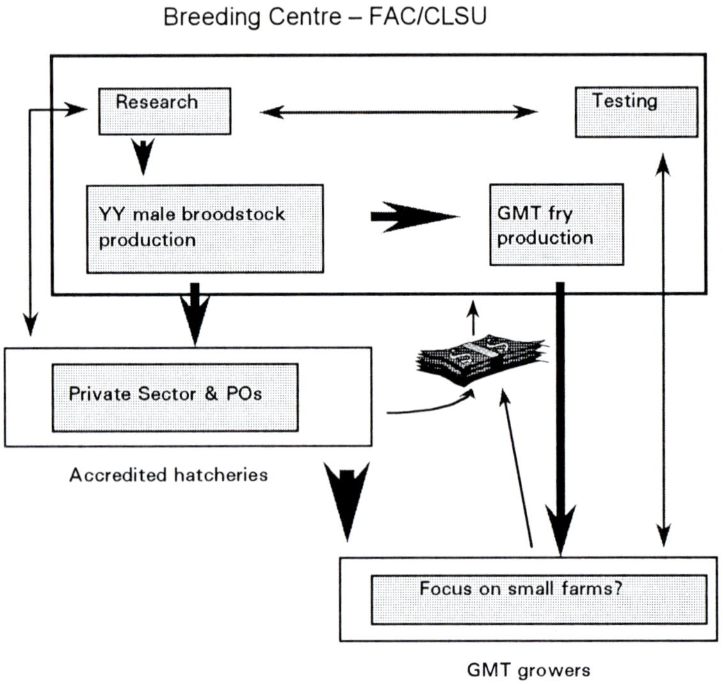

Breeding Centre – FAC/CLSU

Fig. 15.1. The mechanism for dissemination of the products of the YY male technology in the Philippines.

Methods

Three methods were used to gain a fuller understanding of the characteristics of small-scale farmers in the Philippines with a view to developing criteria to identify small-scale farmers who could be targeted as beneficiaries of genetics-based technologies and to evaluate their potential to adopt and gain from such technologies.

Participatory/rapid rural appraisal (PRA/RRA)

Participatory/rapid rural appraisal (PRA/RRA) exercises were conducted with groups of small- and medium-scale farmers, primarily, though not exclusively, in five targeted regions (Cagayan, Central Luzon, Southern Tagalog, Southern Mindanao and the Cordillera Autonomous Region – CAR). RRA utilised a suite of social science-based exercises in which researchers gathered socially sensitive data from farmers in a collaborative and participatory manner (Townsley, 1996). The type of exercises used included farm mapping, resource mapping, Venn diagrams, wealth ranking, activity scheduling, seasonal calendars, marketing maps, problem trees and historical transects. Participatory Rural Appraisal (PRA), in contrast, facilitated the planning of interventions by outside agencies with the active participation of local communities. In 'action research' of this nature, these research instruments were used not only to gather data passively, but also to plan field trials through the identification of potential participants.

Farmers surveyed in this manner were already hatchery and/or grow-out producers of tilapia including those operating ponds, cages and pens. The exercises were arranged with the help of a number of organisations, including regional BFAR stations and the Offices of the Provincial Agriculturalist (OPAg) in different provinces who assisted in the identification of participants representing the poorer and smaller-scale farmers in their areas. A total of 11 exercises were completed in five regions selected for the study from May 1998 to December 1999. Results were collated and summarised to identify characteristics of small-scale farmers involved in aquaculture and in particular to highlight the major constraints that they face. Limitations of this research approach were that only farmers involved in aquaculture were involved in the study and that there may have been some bias in the selection of participants in the exercises.

Questionnaire-based survey

To complement the qualitative data from the rural appraisal exercises, the research team completed a questionnaire-based survey of 108 small- and medium-scale producers in five targeted regions (Table 15.1). This survey sought to establish the characteristics of small-scale aquaculture and of small-scale aquaculture producers in these regions through questions that attempted to quantify their relative wealth and identify the major technical, social and economic factors related to their involvement in tilapia culture. The survey covered 29 hatcheries and 79 grow-out producers, and in the case of grow-out, it covered pond, cage and pen producers. Many respondents were identified during the course of PRA/RRA exercises and the completion of questionnaires with participants served to provide more quantitative data. Others were identified with the help of municipal staff of OPAg or BFAR staff who were requested to identify farmers they considered to be small-scale whilst others were selected on an *ad hoc* basis by project staff. In identifying respondents, no attempt was made to differentiate between medium- and small-scale producers prior to the development of criteria to distinguish them.

Table 15.1. Breakdown of respondents to the questionnaire-based survey of small- and medium-scale farmers in five regions of the Philippines.

Region	Hatchery	Grow-out	Total
Region 2: Cagayan	1	7	8
Region 3: Central Luzon	0	30	30
Region 4: Southern Tagalog	22	29	51
Region 11: Southern Mindanao	5	2	7
Cordillera Autonomous Region (CAR)	1	11	12
Total	29	79	108

Wealth ranking

This methodology had the overall objective to determine the proportion of small-scale fish farmers and their relative 'wealth' status in a total of nine communities (known as *barangays*) in Nueva Ecija (one of the wealthier provinces of the country). Three *barangays* were identified as representing the poorest communities in the province with three representing the median and three among the wealthiest. These communities were ranked according to statistics from the Provincial and Municipal representatives of the Department of Interior and Local Government (DILG). People involved in fish production in each community were interviewed using a basic questionnaire, which principally addressed criteria established for identifying small-scale farmers developed under the two aforementioned research activities (Box 15.1, Box 15.2). Wealth-ranking exercises were carried out in the same communities. Three or four respondents in each community carried out an exercise to rank the wealth of members of the community, as far as possible including all members, to determine the relative wealth status of the fish farmers within their own communities. The wealth ranking method used was based on a modification of that developed by Gregory (1999).

Box 15.1. Criteria developed for identifying small-scale hatcheries in Philippine tilapia culture. It is recommended that hatcheries meeting three of the five criteria be classified as small-scale.
- Output < 30,000 fingerlings month^{-1}
- < 1000m^2 of water surface area
- Inability to nurse fingerlings
- Insecure title over ponds, i.e. lease-holding or share-cropping all or some of their land or membership of an Agrarian Reform Community
- Vulnerability of hatchery to flooding or drought

Major Findings

Highlights from the surveys are presented separately for hatcheries and growers.

Box 15.2. Criteria for small-scale farmers (growers) in Philippine tilapia culture. It is recommended that hatcheries meeting three of the five criteria be classified as small-scale.

- For pond-based producers, a water surface area of less than 1,000m² and a total farm size of < 1ha, for pen-based producers < 1,000m² water surface area, and for cage-based producers a water surface area of < 200m²
- For all producers, no capital resources available for investment in aquaculture production, i.e. no access to formal credit
- < 30% of production sold as a cash crop
- For pond-based producers, insecure title over land, i.e. leaseholding or shareholding on all or some of land. For cage or pen-based producers, not owning the cage or pen
- For pond producers, unreliable or seasonal water supply; for cage or pen producers, vulnerability to water pollution or high salinity

---

Hatchery sector

Although attempts were made to identify small-scale hatcheries in each of the five regions in which research was conducted, the majority of respondents (19/29) came from the Laguna area in Southern Tagalog where historically there has been a concentration of small-scale hatcheries in the district of Kabaritan. This developed in the 1970s and 1980s to supply an expanding cage and pen culture production in the adjacent Laguna de Bay. Thus, the survey is biased to a great extent by the dominance of this group of hatcheries with a considerable degree of experience and maturity within the sector.

Among the 29 respondents to the survey of hatcheries, four were also tilapia growers. Most (62%) of the hatchery operators (or their families) owned the land whilst a small proportion (24%) leased all or part of the land. The survey data indicated that the land area of the hatcheries was indeed small, with 80% of respondents having a land area less than 2,500m². The water area for the hatcheries (excluding those also practising grow-out) was usually small, ranging from 70-7,000m² with a mean of 1,400m².

Seed production on the farms ranged from 6,250 to 300,000 fingerlings month^{-1} with a mean of 81,000. Data on numbers of broodstock, which would have provided an important indicator of production efficiency, were not recorded. Only 25% of hatcheries sold all their fingerlings directly to growers whilst 75% sold all or a proportion of their production to fingerling traders. Fingerling traders purchased fingerlings from many hatcheries, enabling them to meet orders from large-scale tilapia growers (normally using pens or large pond areas). However, this trading phenomenon was limited almost exclusively to the concentration of hatcheries in the Laguna area, with only two hatcheries outside this area selling any fingerlings through traders. The hatcheries sold fish either directly for cash or on credit in roughly equal proportions.

This group of hatcheries was not involved in the sale of monosex fish (either sex reversed or GMT), which indicates the existence of constraints to the adoption of these technologies by small-scale hatcheries, particularly for SRT, the technology that has been available to Philippine hatcheries for over 20 years. However, as an indication of demand for monosex fish, 35% of hatcheries had received enquiries for all-male fish.

Only five from 28 surveyed hatcheries produced their broodstock from their own production, with most being dependent on obtaining stock from other hatcheries or growers. Thus, the majority of hatcheries were not taking management responsibility for the genetic quality of their stocks although the apparent frequent exchange of broodstock between hatcheries should act to minimise inbreeding in the hatchery stocks. The fact that hatcheries are accustomed to buying broodfish from outside sources indicates that the necessity to purchase 'improved' broodfish would not represent a constraint to adoption, provided that prices were affordable and it was not required to transport large fish over long distances.

The level of awareness of the different technologies among the hatcheries was quite high, with over 75% of farmers being aware of GIFT fish and over 50% being aware of GMT (Fig. 15.2). Surprisingly, awareness of sex reversal, as the most mature technology, was less than one third. These data however are likely to be biased by the close affiliation of the Laguna farmers with a BFAR station located within the community. This BFAR station has been associated in the past with the adoption of GIFT and more recently with GMT. The relative importance of this station to the local farmers was apparent from the participatory exercises.

The major constraints identified by respondents of the survey were problems in marketing of their fingerlings and lack of extension support with lack of capital as the third-ranked constraint. However, a problem tree constructed during the RRA exercise in Kabaritan illustrates how the quality of their broodfish is considered central to their level of income. 'Genetic quality' did not, however, appear as an important element of this broodfish quality. The problem tree also illustrated that lack of capital to purchase improved broodstock would be a constraint to adopting genetic-based technologies if this involved increased purchase costs for broodfish. A simpler problem tree drawn by participants of a PRA/RRA exercise conducted with hatchery operators in Banga, South Cotabato, Mindanao also identified quality of broodfish as a major constraint.

Security of market would also be an issue in the adoption of genetically improved fish by these hatcheries as risk is increased if a more steady cash flow cannot be generated from using the more expensive improved broodstock.

The fingerling production data collected for individual small-scale hatcheries were low and would indicate that if they sell directly to growers, they can only supply small-scale growers (although not necessarily the very smallest). Thus, targeting of small-scale hatcheries that operate outside fingerling trading networks may represent one mechanism of targeting smaller scale producers.

Based on the participatory exercises and the survey data, we developed a set of criteria that could be adopted to identify 'small-scale' farmers who represent the poorer and least secure of farmers in the hatchery sector (Box 15.1).

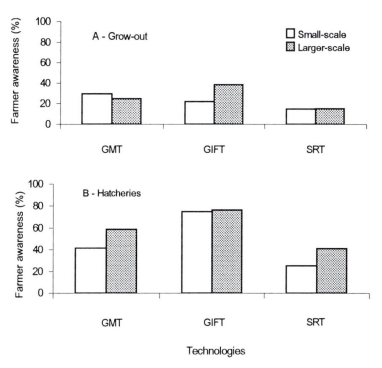

Fig.15.2. Levels of awareness (%) of different 'improved' tilapias among farmers surveyed. Data are presented separately for growers and seed producers and are divided up by scale of operation according to criteria used to identify small-scale farmers.

Grow-out sector

The participatory exercises and the survey data provided a generic picture of small- and medium-scale aquaculture production and producers in the Philippines as a whole. It also provided a snapshot of small- and medium-scale production in a number of different provinces and of pond and cage producers. In the questionnaire-based survey, 79 of 108 respondents (73.1%) were grow-out producers. The information collated indicated that medium- and small-scale aquaculture producers in the Philippines are comparatively well off and cannot be considered 'the poorest of the poor' in a Philippine context. In the questionnaire-based survey, respondents had farms with an average size of 1.7ha and with an average water surface area of 880m². The majority (60.3%) 'own' all the land area that they farm, although this includes farmers who have been awarded land under government agrarian reform programmes (two respondents) as well as farmers in mountainous areas where land is government-owned but where farmers have long-term leases. In both of these cases farmers do not have title to their land and as such cannot use their land as collateral to borrow money from formal credit sources. A few (17.7%) of respondents lease some or all of their farmland while 8.8% sharecrop some of their farmland. As such, these producers as a group have far greater security of tenure than the bulk of the rural population of the Philippines, a key indicator of socio-economic status, although many can still be regarded as 'poor'. According to Balisacan (1993), the Philippine rural poor are typically dependent on agriculture, are landless (relying instead on regular or irregular on-farm employment), have low levels of educational attainment and have weak access to modern technology, partly because of their poor access to credit. Some leaseholders, tenants and small-scale owner-cultivators are, however, included among the rural poor. In addition, respondents had water supplies for an average of 11 months of the year (including pond, cage and pen operators). Many 'poor' farmers in the Philippines, in contrast, do not have access to perennial water and can only grow one rice crop per year. Respondents also had better access to credit than the bulk of the rural population. A small proportion (24%) was able to save with co-operatives or government/private sector banks. Over 10% had loans from banks, co-operatives or informal savings associations. Only 22% borrowed from moneylenders.

In the RRA/PRA exercises, respondents used different terms to categorise different socio-economic strata, but usually placed themselves in the middle groups. In a more detailed follow up study in two communities in Nueva Ecija which was based on the preliminary results reported here, Cheftel (1998) concluded that fishpond owners represented 10-45% of all village households in the 'middle' strata and 23-55% of households in the 'rich' strata, confirming the conclusion drawn from our study.

The results from the participatory exercises and the surveys indicated that the tilapia farmers are weakly integrated into the cash economy or the market economy, at least as far as their aquaculture operations are concerned. Among the respondents in the questionnaire-based survey, the bulk (60%) obtained their fingerling supplies from the wild or from Government (DA/BFAR and FAC/ CLSU) hatcheries (Fig. 15.3).

With respect to the perceived need for monosex tilapia, 80% of respondents in the questionnaire-based survey reported reproduction among their tilapia stocks and of these respondents, 60% viewed such reproduction as negative while 40% viewed it as positive, presumably as they considered it as a free source of fingerlings and/or did not perceive the effects of the recruits on the growth of the stocked fish. Of those who viewed recruitment negatively, 40% said it led to over-population of the pond while 43% said it retarded growth. In terms of awareness of technologies, 32% of the respondents had some knowledge of GIFT, 26% of GMT and again surprisingly, only 14% of SRT (Fig. 15.3).

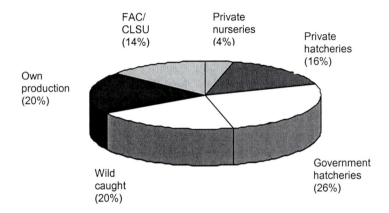

FAC/
CLSU
(14%)

Private
nurseries
(4%)

Private
hatcheries
(16%)

Own
production
(20%)

Wild
caught
(20%)

Government
hatcheries
(26%)

Fig. 15.3. The importance of the varied sources of fingerlings reported by the respondents to the questionnaire-based survey of small- and medium-scale tilapia growers.

Based on the participatory exercises and the survey data, we developed a set of criteria that could be adopted to identify 'small-scale' farmers who represent the poorer and least secure of farmers in the grow-out sector (Box 15.2).

Wealth ranking and the application of criteria for identification of small-scale farmers

In the follow up study of nine communities in Nueva Ecija (classed into three 'wealth status' classes), we applied the criteria described above to the classification of fish farmers in the nine different communities. This was conducted by way of verification of the criteria in that we expected to find higher proportions of 'small-scale' farmers meeting three or more of the criteria in the poorer communities.

Aquaculture production by farmers in these communities varied from very small-scale to large-scale operations. The low-income municipalities had higher proportions of smaller operations with 53.8% of the respondents having total farm area of less than 1ha.

Access to formal credit did not vary according to the wealth level of the municipality. The majority of the respondents from the three categorised municipalities have no access to formal credit. Normally farmers in the different communities use their own capital or borrowed money from informal moneylenders who were normally dealers of fishpond inputs or buyers of rice.

The results of the survey concerning the disposal of harvested fish were somewhat against expectation, showing that very few farmers sell less than 30% of their product. Only in the middle-income municipalities did the majority (66.7%) of the farmers sell less than 30% of their production (Table 15.2), indicating that the wealth level of municipality was not correlated with the amount of production sold. The explanation for this surprising finding may be that for small-scale farmers fish production was an important source of supplementary income from the farm and fish were sold for this reason. In the farmers in the mid-income communities, fish farming may have been seen as an unimportant secondary activity with the farmers taking fish for their own casual consumption. The farmers in the wealthiest communities were more entrepreneurial, commercial, and participated fully in the cash economy, selling all farm products.

Table 15.2. The number and proportion of farmers selling their tilapia production by quantity and category in three different 'wealth levels' of communities in Nueva Ecija province.

	Low income (n = 26)		Middle income (n = 27)		High income (n = 32)	
	(n)	(%)	(n)	(%)	(n)	(%)
Number of farmers not selling produce	2	15	18	67	2	6
Number of farmers selling produce	22	85	9	33	30	94
Percentage of produce sold						
< 30	2	8	0	0	1	3
30-60	4	15	1	4	8	25
> 60	16	62	7	26	21	66

Classifying farmers as small-scale if they meet at least three of the five criteria (Box 15.2), then 50% of farmers in the poor communities would be classed as small-scale whereas only 26 and 6% would meet the criteria for mid-income and high-income communities, respectively. It thus appears that these criteria may be of benefit in identifying small-scale farmers, i.e. those with less secure livelihoods, to which genetics-based technologies could be adapted and targeted. The criteria relating to access to credit and land tenure should, however, be further reviewed.

The wealth-ranking exercises were successfully conducted and enabled the 'wealth status' of the fish farmers to be characterised relative to the other members of the community. The proportion of fish farmers was low in all communities (3-10%) and there were no clear trends seen in any of the communities. Across all wealth levels of community, fish farmers were evenly distributed across the ranks from poorest to the richest members of their respective communities. There was thus no evidence to suggest that it is the poorer or richer members of communities that enter into aquaculture.

Passive Dissemination, Who is it Impacting?

The objective of this part of the study was to determine what type of producer is currently benefiting from the dissemination of genetically male tilapia through the previously described 'passive' mechanism adopted by Phil-Fishgen, with emphasis on the socio-economic status of beneficiaries. Data were collated for hatcheries and for grow-out operators.

Methods

Four approaches were taken to collect data to evaluate the nature and impact of the dissemination process of GMT:
- Summary data extracted from the application forms completed by hatcheries that apply for accreditation as GMT producers. These forms were initially designed for collecting data relevant to accreditation criteria and were not purpose-designed to extract important socio-economic information
- Summary data extracted from the application forms completed by tilapia growers when they ordered GMT directly from Phil-Fishgen at FAC, CLSU. These forms were also not purpose-designed to collect data and in many cases had not been properly completed
- A questionnaire-based survey of hatcheries producing fish using the different improvement technologies of GMT, GIFT and SRT. Data were collected from a similar number of hatcheries producing normal mixed-sex tilapia as a control
- A questionnaire-based survey of tilapia farmers growing fish produced using the available (and to some extent competing) technologies of GMT, GIFT and SRT. Data were collected from a similar number of growers producing traditional, unimproved, mixed-sex tilapia as a control.

It had been hoped to collect some data on the farmers growing GMT supplied by the accredited hatcheries. However, a certification system designed by Phil-Fishgen to generate feedback from the hatcheries had failed, with very few hatcheries returning the carbon copies from a book of certificates they should have been issued with their GMT.

For the surveys of adopters of the different technologies, hatcheries were selected from the list of accredited hatcheries (in the case of GMT and GIFT) and lists of hatcheries producing normal or SRT were obtained from BFAR. The data were collected in late 1998 when GMT had been disseminated by Phil-Fishgen for approximately three years. Dissemination of GIFT had been underway for a slightly longer period, with dissemination being coordinated through BFAR and a network of regional multiplier stations.

Hatchery Beneficiaries

Since accreditation had been offered by Phil-Fishgen, a total of 100 farmers had applied for accreditation of which 45 had been accredited. Despite limitations related to the design of the application form (for the purpose of data collection) and the number of incomplete forms, it was possible to draw a number of important points from the data. The average farm size of the hatcheries was 5.2ha (with those accredited having a lower average of 3.9ha) with average water area of 4.6ha (2ha for those actually accredited). This compares with the data collected from the survey of medium- and small-scale producers in which 80% of hatcheries farms were < 0.25ha and the mean water area was just 0.14ha. Thus, the applicants were clearly medium- to large-scale farms. Furthermore, significant proportions of the applicants were new entrants with an average for the applicants of only 3.4 years in operation. More than 87% of applicants owned their land with the majority being individual or family owned.

The average size of the orders for broodstock was 2,425 broodstock sets (1 male: 3 females), which had a value of P242,500 (US$5,166 using exchange rates prevailing at the time of writing), which would clearly limit accreditation to farmers with access to capital. Almost 90% of respondents used or intended to use their own capital to fund their accreditation.

The projected average monthly production for the accredited hatcheries was 500,000 fry month^{-1} (estimated fingerling production would be 300,000-350,000), much higher than the average production for small- and medium-scale hatcheries surveyed previously (85,000). This figure, however, represents the projected production for the hatcheries and not necessarily for GMT as many of the farms also continued to produce other strains of tilapia on their farm. The majority of applicants had extensive pond-based hatchery systems with some having more intensive tank-based systems.

The main difference between the overall data and the data from those hatcheries that were accredited is that the accredited hatcheries were on the whole smaller (2ha water areas compared to 3.9ha for all applicants). The fact that production estimates were similar indicates that the accreditation process selected hatcheries that are more efficient.

Although sufficient data were not available to test the criteria, it is unlikely that any of the accredited hatcheries matched the criteria (Box 15.2) outlined for identifying small-scale hatcheries (other than those accredited as part of a parallel targeted dissemination programme).

In the more general questionnaire-based survey of hatchery adopters of the different technologies (GMT, GIFT and SRT), the survey included 18 accredited GMT producing hatcheries, 20 GIFT hatcheries (only eight formally accredited according to procedures at the time) and 14 hatcheries producing

SRT. Ten hatcheries producing normal mixed-sex tilapia (MST) were included for comparative purposes. The majority of respondents came from the main tilapia growing regions of the country, Central Luzon and Southern Tagalog.

Fig. 15.4 illustrates the types of hatchery system operated by the adopters of the alternative technologies. This shows that hatchery systems being used to produce normal mixed-sex tilapia are also being used for the production of GMT and GIFT. However, a significantly higher proportion of hatcheries producing SRT have adopted *hapa*-based production systems. This reflects the requirement to collect young, sexually undifferentiated fry for sex reversal, which is difficult in pond-based systems. *Hapa*-based hatchery systems are more capital and labour intensive and this indicates that the adoption of SRT is not scale neutral as the technology cannot easily be adapted to the existing common hatchery systems. This is a major advantage of GMT over SRT in the production of monosex tilapia.

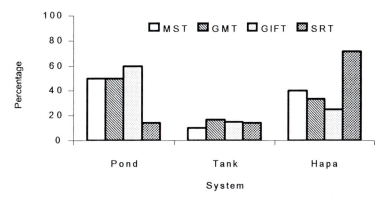

Fig. 15.4. The types of hatchery system used by hatcheries adopting the various alternative technologies for improved tilapia (GMT, GIFT or SRT) compared to normal unimproved mixed-sex tilapia (MST).

A large proportion (80-90%) of adopters had undergone some form of training in tilapia seed production and similar proportions were generally happy with the extension support they received (these figures were also high for the MST growers interviewed for comparison but this may have been an artefact of the way they were selected and the relatively smaller sample size). It is not clear whether this high level of training and extension support is due to the fact that they learned of the technologies whilst attending trainings or that their adoption of the technologies has resulted in the provision of this training and extension.

When asked for their view on the technical advantages of using improved tilapia, most (65-80%) cited a shorter growing period for grow-out farmers as the major benefit with uniform size being considered a further important benefit for the monosex GMT and SRT. Disadvantages related to adopting technologies included requirement of capital and difficulty in obtaining broodstock (or the treatment protocol/supplies in the case of SRT).

A separate survey (by telephone and farm visits) was conducted to obtain estimates of the selling price of fingerlings produced from the alternative technologies (GMT, GIFT and SRT) compared to normal MST. It was believed that the price of GIFT fry was fixed by the GIFT Foundation International, which coordinates the dissemination of this fish. Although the survey was limited (5 or 6 hatcheries per technology) the data were quite consistent. The mean prices for fingerlings of the common size categories are illustrated in Fig. 15.5. These clearly illustrate that fingerlings produced under all the different technologies were all more expensive than normal MST. It is further evident that GMT fingerlings were the cheapest of the improved fish, having a price premium of 12.5% over MST compared with approximately 42% for GIFT and SRT.

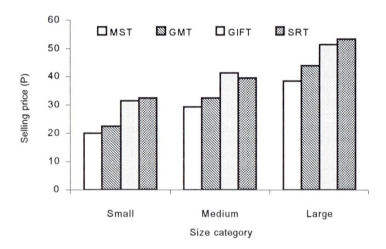

Fig. 15.5. Comparative selling prices for fingerlings produced using the different 'improvement' technologies for different size categories of fingerlings (US$1 = P50 approximately).

Grow-out Beneficiaries

These data were extracted from simple questionnaires, filled out by 594 farmers when they were requesting the purchase of GMT fingerlings directly from Phil-Fishgen. Again, despite limitations in the design of the form, some useful indications can be extracted from the data.

The vast majority of buyers cited tilapia culture as a secondary occupation. The average size of orders being placed for GMT was 13,680, reflecting the maximum purchase order size set by Phil-Fishgen (20,000 per order at the time of the survey) and also indicating that GMT was being used in relatively small culture units. The average farm size of the GMT buyers was 2.8ha and a total water area of 2ha. This compares with an average size of 1.7ha and with an average water surface area of 880m^2 for the small- and medium-scale growers surveyed under the general surveys reported above, indicating that the GMT buyers were somewhat representative of this sector. The vast majority of GMT buyers who answered the question owned their farm, with single (or family) ownership being the common form. The majority (83%) purchased GMT for culture in freshwater ponds with 7.7 and 5.5% respectively growing the GMT in cages or tanks. Only five respondents grew fish in polyculture with other species, reflecting the dominance of tilapia monoculture in Philippine aquaculture.

GMT buyers had an average experience of tilapia culture of only 2.9 years, indicating that the majority of buyers were relatively new entrants. The previous sources of fingerlings used by the farmers prior to their purchase of GMT were split almost equally among BFAR, private hatcheries and 'none', the latter again reflecting the fact that significant proportions of farmers were indeed new entrants to the tilapia culture. Only 4% had previously used recruits from their own pond production.

For a question relating to the intended disposal of the fish after grow-out, 23% of buyers purchased GMT entirely for subsistence and only 42% grew GMT entirely on a cash crop basis. This also reflects the usage found in the general survey among medium- to small-scale fish farmers. If we apply the criteria for classifying farmers as small-scale, 67% of farmers buying fingerlings directly from Phil-Fishgen could be classed as such.

One question asked of buyers was how they learned about GMT. The most common means (44%) was through visiting CLSU. Farmer-to-farmer contact through existing GMT growers or through technicians was also common (37%). Only 13% of buyers learned about GMT through the media or through formal seminars. This suggests that while promotion of GMT existed (merely informal media coverage at the initiative of the media) it was not effective and that most buyers learned of the technology through other people involved in tilapia production or by going out to search for fingerlings. There is clearly a role for targeted promotion in making small-scale farmers aware of the technology.

In the more general questionnaire-based survey of grow-out adopters of the different technologies (GMT, GIFT and SRT), the survey included a total of 210 growers (49 MST, 53 GMT, 57 GIFT and 51 SRT) from across six regions, again with the majority of respondents coming from the main tilapia-growing region of Central Luzon. The majority of the growers (75-90%) were operating pond-based culture systems with the highest proportion (92%) among the SRT growers. The majority (50-60%) of respondents operated semi-intensive culture systems (fertilisation to encourage natural productivity in the fish pond together with supplementary feeding), with slightly higher proportions of MST growers (19.4%) using extensive systems compared to those growing improved fish (6-14%). Slightly higher proportions of technology adopters used intensive systems (30-40%) compared to those using normal MST (30%).

A lower proportion of growers (compared to the hatchery respondents) had undergone training in tilapia culture (60-80%), with more GIFT growers having been trained than for the adopters of alternative technologies. Similar proportions of farmers appeared to be happy with their extension support (73-92%). This general level of satisfaction with extension support contradicts one of the main constraints of small/medium-scale growers surveyed under the general survey of small- to medium-scale tilapia producers who perceived very weak extension support. This anomaly is likely to be due to the fact that the farmers receiving these technologies were not small/medium-scale and/or that farmers were getting better support due to their adoption of these technologies.

Data on the experience of respondents in tilapia culture indicated that GMT, more than the alternative technologies, was being adopted by new entrants, with the average experience of the GMT growers being less than 1.5 years compared with 4 years for growers of MST and 2.5 and 3 years, respectively, for GIFT and SRT.

Shorter growing periods and larger harvest sizes were considered to be the main benefit from the improvement technologies, with the majority of farmers giving these responses. GMT growers had mean culture periods of 4.6 months compared to 4.8 months for GIFT and 5.1 months for SRT and MST.

Lack of capital and difficulty in obtaining fingerlings were listed as the two major constraints to tilapia growers, with the difficulty of obtaining fingerlings being greatest for GMT and SRT.

There was little difference in the marketing strategy of the adopters of the different technologies, with approximately 60% of respondents selling to middlemen (who sell the fish on to retailers), irrespective of the fish grown. There was also no difference in the selling price of the fish between the adopters of the different technologies and the control farmers selling normal unimproved fish.

When asked what factors would encourage farmers to increase their production of tilapia, all the adopters of the different technologies cited economic factors including access to cheaper credit, a higher market price for their fish and lower production costs. These factors were all more important for the technology adopters than for growers of normal MST, indicating their more entrepreneurial and commercial attitude to tilapia culture.

In the analysis of the impact of the technology, there was no clear evidence of the technology having contributed to employment at either the hatchery or grow-out level.

Alternative Mechanisms for Targeted Dissemination

Within the framework of the Phil-Fishgen organisation established to coordinate dissemination of GMT and GMT producing broodstock, three options were considered for initiating dissemination activities targeted at small-scale poor rural farmers:

- State sector (Department of Agriculture, BFAR and or State Colleges and Universities)
- Private sector hatcheries, specifically small-scale hatcheries
- NGOs and/or people's organisations (POs) including cooperatives and fisher folk associations.

These options were evaluated in a subjective way by attempting to set up field trials representative of the three strategies and evaluating their progress over a period of 18 to 24 months during 1997 and 1998.

Pathway 1: State Sector

In theory, a state research institution such as FAC-CLSU can disseminate a new agricultural technology to small-scale producers through state-controlled extension agencies. Such agencies should be widely dispersed throughout the country with a presence in remote provinces or communities in which small-scale producers are likely to be located. Where the government has a commitment to assisting small-scale producers to alleviate their poverty, as in the Philippines, such agencies should also have a mandate to help small-scale producers. In the case of the Philippines, governmental institutions working with aquaculture producers are primarily the DA and the BFAR. The DA has fisheries technicians in each provincial office, OPAg, which often supports a provincial hatchery. BFAR has hatcheries or research stations in each of the Philippines' 14 regions. The State Colleges and University system (in which CLSU plays a major role) has many Colleges of Fisheries (of which FAC is the lead agency), many of which have their own hatcheries.

To assess this pathway, field trials were established with two government-run hatcheries, one run by the Office of the Provincial Agriculturalist in Baler, Aurora and the other the regional research station and hatchery for Southern Tagalog (Region 4) run by BFAR in Laguna province. The trials indicated that government-run hatcheries had very variable production efficiency with production in the OPAg hatchery ranging from 3,000 to 19,000 fingerlings month^{-1} from 130 broodstock sets (1 YY male: 3 females) supplied to them, equivalent to 23–145 fry per set month^{-1}. In Laguna, the BFAR station manager reported average production of 50,000 fingerlings month^{-1} from October 1997, using 300 broodstock sets (166 fry per set month^{-1}), or approximately 800,000 in 16 months.

Furthermore, the two government-run hatcheries did not seem to promote strategies designed to deliver technologies such as GMT to small-scale producers. Of the total GMT fingerling production in Laguna, 10% of the production (80,000) was distributed free of charge to backyard pond producers in Laguna, in keeping with the objectives of targeting this sector. The BFAR station did provide training to producers, which was provided passively to farmers who became aware of it and could visit the station. The OPAg station in Baler meanwhile appeared to have no strategy for enabling small-scale farmers to access GMT production. Many of the problems with these government stations may relate to the lack of an incentive structure, for example, the capacity to plough back income from fingerling sales into the hatchery itself. The government hatcheries also sold other types of tilapia fingerlings and staff had little incentive to develop a customer base by providing reliable advice about the properties of different technologies, e.g. GIFT or GMT or to provide general management advice. Different pricing structures may also have alienated buyers, for example, in Baler OPAg charged 200 to 300% more for GMT fingerlings than for 'GIFT'. There was also some anecdotal evidence that government-run hatcheries did not maintain adequate control of stocks held at the station.

In addition to the trials described above, attempts were made by Phil-Fishgen and FAC-CLSU to establish a network of State Colleges and Universities (SCU) supplying GMT from their own small-scale hatcheries. This network had made limited progress, with attempts to establish GMT hatcheries meeting with little success due to difficulty in supplying technical support to locations distant to FAC-CLSU and inadequate quality control of broodstock transferred to the SCU hatcheries.

Pathway 2: Small- and Medium-scale Private Hatcheries

In theory, private sector hatcheries represent another potential pathway for the dissemination of GMT and related technologies to small-scale producers. It is assumed that larger-scale hatcheries sell their fingerlings primarily to meet the large orders of big producers. The larger hatcheries may also be located in the main aquaculture producing regions or close to areas where large-scale producers predominate. As such, they may not have a commitment to developing a clientele among small-scale producers. Small- and medium-scale hatcheries, on the other hand, may have a greater incentive. With lower production, they may be reliant on the small orders from medium- and small-scale producers. They may also be located in more marginal locations where small- and medium-scale growers are concentrated. Because they seek to make a profit and to maintain a stable clientele, such hatcheries also have an incentive to manage and maintain broodstock quality, and to provide good quality extension advice. Some hatchery operators may also act as informal extension agents or technology promoters.

To assess this pathway, two field trials were established with private sector hatcheries, one with a small-scale hatchery producer in Baler, Aurora province, and one with a medium-scale hatchery operator in Kabaritan, in Laguna province. By the end of the Project in March 1999, the Kabaritan hatchery had ceased to produce GMT fingerlings following a typhoon in October 1998 that flooded the hatchery, causing serious damage to facilities and loss of fish (highlighting the risk of investing significantly in any of the different technologies). The hatchery in Baler remained in operation and continued beyond the project research timeframe. Both trials provided evidence for the strength of this targeted dissemination strategy.

Both of these hatcheries proved more efficient than the government-run hatcheries in terms of production. The Kabaritan hatchery produced 300,000 fingerlings in 10 months from December 1997 to October 1998 from 80 broodstock sets (375 fry per set month^{-1}). The GMT was sold at a premium of 5 centavos per fingerling (+ 20%) compared to mixed-sex alternatives. No information was gathered on the support of this hatchery to the GMT buyers or of the socio-economic profile of these buyers. However, it is known that many hatcheries in the Laguna area sell fingerlings through fingerling traders. These traders are then able to supply some of the larger grow-out farms by combining output from a number of hatcheries. Thus, it is unlikely that hatcheries in this area would effectively target small-scale farmers.

In Baler, the small-scale hatchery also achieved relatively efficient output. From a small hatchery with a 65m^2 concrete tank and a 70m^2 earthen pond, containing a total of 100 broodstock sets, 20,000 to 30,000 GMT fingerlings month^{-1} were produced (200-300 fry per set month^{-1}). The hatchery owner developed a client base of local tilapia growers, being predominantly a direct supply to small-scale producers (86% of buyers met the criteria as small-scale producers – see Box 15.2), and also provided extension services to these clients. Whilst this was only one hatchery, it did provide a good example of the potential of small-scale hatchery producers, as informal extension agents, to promote GMT and to ensure that it is delivered to small-scale farmers on a sustainable basis.

As small-scale hatcheries produce only small numbers of fingerlings, those operating outside fingerling trading networks are limited to supplying relatively small-scale growers and thus represent a means for selectively targeting these growers as beneficiaries of GMT.

Pathway 3: NGOs and POs

NGOs also represent a potential, though largely untested, means of disseminating GMT to small-scale farmers. NGOs are in a relatively strong position to mobilise resources from a range of agencies to develop hatchery operations in remote locations, and to fund training and development of extension materials. Many also have a commitment to working with small-scale farmers and to represent their interests although all, to some extent, work to different agendas.

To test this hypothesis, a field trial was established with Mindoro Assistance for Human Advancement through Linkages (MAHAL, an acronym derived from the Tagalog word for 'Love'), an NGO based in Calapan in Oriental Mindoro. Following initial informal contact, a relationship between MAHAL and Phil-Fishgen was established, at MAHAL's request. Under this relationship MAHAL established its own hatchery and began to sell GMT fingerlings commercially as an income-generating activity and to provide GMT fingerlings on a subsidised basis to local small-scale farmers.

MAHAL's efforts to establish an NGO-run hatchery proved successful and the hatchery operations became financially self-sustaining. However, the success of the trial was due in large part to the extensive support provided by the Phil-Fishgen based researchers. The success could also have been due to links with other organisations, including the provincial and municipal government and national agencies. MAHAL was less successful in providing extension support to small-scale producers. Initially, it provided fish to pen producers having access to communal lagoon resources, selling GMT to the farmers at a 45% discount on normal prices. These pen culture systems failed, however, mainly due to excessive salinity of the water. Other pond-based farms also failed due to farmer inexperience, for example farmers stocked tilapia at excessively high stocking densities. However, through further interaction with Phil-Fishgen based researchers, including the holding of several participatory workshops for both farmers and extension agents, MAHAL developed the capacity to provide successful support and a flourishing network of small-scale private farmers culturing GMT (approximately 71% of MAHAL's farmer beneficiaries meet the criteria as small-scale farmers – see Box 15.2). Most small-scale farmer beneficiaries were new entrants to tilapia culture or had readopted it after previous failures, the introduction of improved fish having acted as a catalyst for them to diversify their agricultural practices. Some farmers sold their fish whilst others consumed their production from periodic harvests. The availability of fish also added to the social capital of these farmers in that they frequently shared the fish with their neighbours. All farmers interviewed planned to further expand their tilapia production. Whilst representing only an example, MAHAL's success confirms an important hypothesis of the research, that NGOs and POs can successfully establish and run hatcheries in support of small-scale farmers, in areas where shortages of quality fingerlings have historically impeded the development of aquaculture, especially among small-scale producers, and that these organisations can successfully target small-scale, poor farmers.

The Importance of Tilapia to Poor Consumers

Whilst it is apparent, with appropriately targeted dissemination, that genetics-based technologies in aquaculture have the potential to deliver benefits to small-scale and resource-poor farmers, such technologies also have the potential to deliver benefits to larger number of poor fish consumers who rely on tilapia for a significant portion of their animal protein needs. In surveys of members from 260 households identified in poor communities (as classified by DILG) in rural lowland (Nueva Ecija, Tarlac and Laguna) and upland (Nueva Viscaya) communities and also the poorest areas of two urban conurbations, Manila and

Baguio emerged, fish in general, and tilapia in particular, were found to be of considerable importance in the diet of both rural and urban poor. Data revealed that 85-100% of respondents ate tilapia regularly (1-3 times week^{-1} on average). In households in poor communities, many of which were food insecure, tilapia was the most common fish eaten due to its availability, preferred taste and lower price compared to other available fish. The widespread introduction of genetically improved tilapia by producers should permit more efficient utilisation of resources among tilapia farmers. Provided appropriate market conditions exist, this could ultimately bring about a reduction in the price of tilapia, which could have very significant impact upon the livelihoods of both urban and rural poor through improved food security and health.

Summary, Conclusions and Recommendations

In the Philippines, the YY male technology and its main product GMT, have been commercially disseminated by Phil-Fishgen since 1995. Whilst Phil-Fishgen is achieving its original objective of disseminating the products to tilapia producers in a financially sustainable way, crude monitoring suggests that the passive dissemination mechanisms adopted by Phil-Fishgen is failing to target small-scale, poor farmers and thus runs the risk of further marginalising these farmers. The original objective of the development of the technology was to enhance, which was successfully achieved. These increases are being passed onto the producers through on-going dissemination. However, under the current development paradigm, benefits of such technologies should be seen to impact positively upon the sustainable livelihoods of the poor.

Questionnaire-based surveys and participatory research methods were used to characterise small-scale, poor farmers in both the hatchery and grow-out sectors of tilapia production. This enabled the identification of criteria that can be used to target outputs of the technology in support of more sustainable livelihoods for these farmers.

Analysis of the current beneficiaries of the technology indicated that hatcheries selected by Phil-Fishgen for accreditation were predominantly represented by relatively large, wealthy, commercial enterprises. Many of these farmers were relatively new entrants and had available capital to invest in a new technology such as GMT. Few small-scale hatcheries have adopted the technology to date, primarily due to lack of information about the technology and the cost of securing the broodstock. The relatively small price differential placed on GMT compared to alternative improved fish indicates that GMT from these accredited hatcheries may be accessible to those small-scale farmers integrated into the tilapia fingerling market. A survey of direct beneficiaries of GMT, i.e. those purchasing GMT from Phil-Fishgen's own hatchery, showed that these were also relatively affluent small- to medium-scale farmers. Few were large-scale, due to the limitation placed on the maximum order size by Phil-Fishgen.

Comparisons of alternative improvement technologies in the hatchery sector showed that adoption of sex reversal was not scale or production system neutral, being adopted mainly by farmers operating more capital-intensive, *hapa*-based production systems. The majority of hatcheries producing GMT and GIFT used similar extensive and semi-intensive, mainly pond-based, systems also used by producers of unimproved fish. An important element of producing the genetically improved fish, particularly for GMT production, was the prevention of contamination of broodfish. None of the hatcheries adopting these technologies cited this as a constraint to adoption although it was not established whether the respondents fully understood how contamination events can occur or the implications of such events. It seems likely, however, that the genetics-based technologies are broadly scale-neutral, with the main factor determining uptake being the availability and cost of obtaining the improved broodfish.

Of the improved fish, GMT had the lowest price premium over the unimproved fish, with SRT having the largest. Whilst the cost of fingerlings was not listed as a constraint to adoption of technologies, the relatively lower price of GMT fingerlings should make them relatively more accessible to poor farmers or at least those accustomed to purchasing fingerlings. However, the limited availability of GMT was listed as a constraint to adoption. There were no major differences in the socio-economic profile of the adopters of the alternative technologies although GMT again tended to be preferred by relatively new entrants to tilapia culture and in general, the adopters of improved fish had a more entrepreneurial and commercial attitude to tilapia culture. Among tilapia farmers from different wealth-level communities, there was much greater awareness of GIFT than of the alternative improvement technologies, probably due to the high profile of the large Project which generated this technology and its close association with government fisheries agencies. Although awareness of GMT was low, it was distributed more evenly across the different wealth levels of communities. Clearly more extension of GMT technology is required to enhance awareness across all sectors of tilapia producers.

Early evaluation of different targeted dissemination pathways have indicated that NGOs and POs can be effective partners in delivering the benefits of these genetics-based technologies to poorer farmers provided that they receive adequate technical support. Also, accreditation of small-scale hatcheries not linked to seed trading networks may represent a viable mechanism for targeting small-scale tilapia growers as beneficiaries, although accreditation of such hatcheries may need to be subsidised due to lack of access to capital by these farmers.

Whilst more research is required, including livelihoods analysis, it is evident that genetics-based technologies such as GMT can enhance some aspects of the livelihoods of poor, small-scale farmers, both indirectly by acting as a catalyst for farmers to adopt aquaculture, thereby diversifying their agricultural practices and lessening their vulnerability, or directly through providing increased income or improving social capital.

When first established, Phil-Fishgen's primary objectives were to disseminate its products to the producers and generate income to ensure financial sustainability of its research, development and dissemination activities. It has been successful in achieving these objectives and is now positioned to target some of its products more specifically to benefit the livelihoods of poor farmers. With this approach, the early adopters have tended to be the better informed, more mobile middle and higher income sector of farmers. There are risks, with this sector being the early adopters, that they will take maximum gain from the technology and that ultimately the prices and profitability of culturing tilapia may be reduced due to increased supply, with later adopters (including small-scale, poorer farmers) gaining little or no benefit from the technology and thus becoming further marginalised from the mainstream of tilapia culture. However, the rate of adoption of culture of improved fish and the strength of the domestic market for tilapia are such that it is likely to be some time before the widespread adoption of improved fish will impact upon tilapia economics in this way.

It can therefore be concluded that, whilst genetics-based technologies have potential to benefit poor farmers and can be broadly scale-neutral (dependent on the delivery mechanisms and cost in particular), there is an inherent risk that small-scale farmers will be marginalised in the absence of appropriately targeted dissemination strategies. Such dissemination strategies must be well founded on a good understanding of the nature of small-scale tilapia culture by poor farmers and of the linkages between the different stakeholders in tilapia production. Conclusions drawn from this study should thus be treated with some caution as they are derived from research with relatively small sample sizes, carried out in only five of 14 regions in the Philippines and are biased towards the main tilapia producing regions of Luzon Island. It is perhaps less likely that benefits of GMT in the Philippines will accrue to small-scale hatcheries, particularly those linked to trading networks but there are large numbers of small-scale tilapia growers who stand to benefit from more efficient production. This benefit will likely be in the form of more diverse livelihood opportunities and improved economic and social assets. It is unlikely that introduction of genetics-based technologies such as GMT will have much impact, either positively or negatively, on employment of poor people. However, the widespread adoption of improved tilapia and a resulting enhancement in production efficiency in the Philippines may, in the longer term, have the greatest impact upon a large number of urban and rural poor who depend mainly on tilapia as an important and affordable source of animal protein in their diet.

A number of key recommendations are made concerning further research and dissemination activities related to GMT in the Philippines:

- As there are likely to be significant numbers of farmers practising subsistence level forms of tilapia culture in more rural and isolated areas of the Philippines, there is a need for research to enhance our understanding of the role of subsistence aquaculture in Philippine farming systems and the potential for such farmers to benefit from genetics-based technologies.

- The role of tilapia in the diet of poor people needs to be further quantified and the pathways by which the poor obtain their tilapia better understood. Economic models can be applied to predict the impact of improved fish and more efficient production upon the price and availability of tilapia to this sector of urban and rural societies.
- With regard to monitoring the impact of the dissemination of GMT, there is a need for an improved feedback mechanism to better understand the scale and nature of impacts and to profile the socio-economic status of beneficiaries, and to characterise the short- and medium-term influence of the technology upon their livelihoods.
- To better target the technology to the benefit of poor producers, Phil-Fishgen should adopt a more proactive dissemination strategy specifically targeted at small-scale, poor farmers. Improved dissemination strategies should focus on supported partnerships with NGOs and POs to target, possibly with subsidies, strategically located small- and medium-scale hatcheries. This policy should be backed up with greater extension support and improved access to affordable credit.

Acknowledgements

The majority of this work was funded by the UK Department for International Development (DFID) under its Fish Genetics Research Programme managed by the University of Wales Swansea (Project R.6937). Our gratitude is due to staff of Phil-Fishgen and of the Fish Genetics and Biotechnology Programme of the Freshwater Aquaculture Center of Central Luzon State University. We would also like to thank the UK Voluntary Service Overseas (VSO) for supporting this work and Mr. Alan Black who worked on the project as a VSO Volunteer.

References

Balisacan, A.M. (1993) Agricultural growth, landlessness, off-farm employment, and rural poverty in the Philippines. *Economic Development and Cultural Change* 41(3), 533-562.

BFAR (1997) Philippine fisheries profile 1996. Bureau of Fisheries and Aquatic Resources, Quezon City, Philippines. 48 pp.

Bolivar, R.B., Bartolome, Z.P. and Newkirk, G.F. (1994) Response to within-family selection for growth in Nile tilapia (*Oreochromis niloticus* L.). In: Chou, L.M., Munro, A.D., Lam, T., Chen, T.W., Cheong, L.K.K., Ding, J.K., Hooi, K.K., Khoo, H.W., Phang, V.P.E., Shim, K.F. and Tan, C.H. (eds) *The Third Asian Fisheries Forum.* Asian Fisheries Society, Manila, Philippines, pp. 548-551.

Cheftel, J. (1998) Genetically male tilapia: potential for small-scale fish farmers in the Philippines. MSc Thesis. Centre for Environmental Technology, Imperial College, London.

Eknath, A.E. and Acosta, B.O. (1998) Genetic improvement of farmed tilapia project final report (1988-1997). ICLARM. Manila, Philippines. 75 pp.

Eknath, A.E., Tayamen, M.M., Palada-deVera, M.S., Danting, J.C., Reyes, R.A., Dionisio, E.E., Capili, J.B., Bolivar, H.L., Abella, T.A., Circa, A.V., Bentsen, H.B., Gjerde, B., Gjedrem, T. and Pullin, R.S.V. (1993) Genetic improvement of farmed tilapias: the growth performance of eight strains of *Oreochromis niloticus* tested in different farm environments. *Aquaculture* 111, 171-188.

FAO Fisheries Department, Fishery Information, Data and Statistics Unit (2000) Fishstat Plus: universal software for fishery statistical time series. Version 2.3.

Gregory, R. (1999) Poverty ranking: bringing aid to the poorest. In: Gregory, R., Kamp, K. and Griffiths, D. (eds) *Aquaculture Extension in Bangladesh - Experiences from the Northwest Fisheries Extension Project (NFEP) 1989-1992*. MMS Dhaka, pp. 49-60.

Guerrero, R.D. III (1996) Aquaculture in the Philippines. *World Aquaculture* 27, 7-13.

Guerrero, R.D. III and Guerrero, L.V. (1988) Feasibility of commercial production of sex-reversed Nile tilapia fingerlings in the Philippines. In: Pullin, R.S.V., Bhukaswan, T., Tonguthai, K. and Maclean, J.L. (eds) *The Second International Symposium on Tilapia in Aquaculture*. Department of Fisheries, Thailand and International Center for Living Aquatic Resources Management, Bangkok, Thailand and Manila, Philippines, pp. 183-186.

Mair, G.C., Abucay, J.S., Beardmore, J.A. and Skibinski, D.O.F. (1995) Growth performance trials of genetically male tilapia (GMT) derived from YY-males in *Oreochromis niloticus* L.: on station comparisons with mixed-sex and sex reversed male populations. *Aquaculture* 137, 313-322.

Mair, G.C., Abucay, J.S., Skibinski, D.O.F., Abella, T.A. and Beardmore, J.A. (1997) Genetic manipulation of sex ratio for the large-scale production of all-male tilapia, *Oreochromis niloticus*. *Canadian Journal of Fisheries and Aquatic Science* 54, 396-404.

Mair, G.C. and van Dam, A.A. (1996) The effect of sex ratio at stocking on growth and recruitment in Nile tilapia (*Oreochromis niloticus*) ponds. In: Pullin, R.S.V., Lazard, J., Legendre, M., Amon Kothias, J.B. and Pauly, D. (eds) *The Third International Symposium on Tilapia in Aquaculture*. ICLARM Conference Proceedings 41, Manila, Philippines, pp. 100-107.

Townsley, P. (1996) Rapid Rural Appraisal, Participatory Rural Appraisal and Aquaculture. FAO Fisheries Technical Paper 358, Rome. 109 pp.

Chapter 16

Small-scale Fish Culture in Northwest Bangladesh: a Participatory Appraisal Focusing on the Role of Tilapia

B.K. Barman[1], D.C. Little,[2] and P. Edwards[3]

[1] *Northwest Fisheries Extension Project, Parbatipur, Dinajpur, Bangladesh*
[2] *Institute of Aquaculture, University of Stirling, Stirling FK9 4LA, UK*
[3] *Aquaculture and Aquatic Resources Management, School of Environment, Resources and Development, Asian Institute of Technology, PO Box 4, Klong Luang, Pathumthani 12120, Thailand*

Abstract

Tilapia are important for small-scale aquaculture in Northwest Bangladesh. Most farmers maintain tilapia in their ponds in addition to carp, which they stock regularly. Carp are sold to the market for cash income but tilapia are mostly used for household consumption. Their small size and ability to reproduce in the pond are both viewed positively by farmers. Few tilapia are found in markets which has led to the misconception that they are unimportant. A participatory study of four areas of the Northwest region of Bangladesh under two broad categories of aquaculture environment (more and less favourable) showed that of a total of 68 pond owners, 59 of them used their ponds for fish culture and 9 of them for wild fish production. Of the 59 farmers, 33 farmers had tilapia in their ponds, cultured mainly in polyculture with Indian major carp and Chinese carp. When the commonly cultured fish (7-8 species) were compared by matrix ranking with respect to yield, price, demand in the market, household consumption and seed availability, tilapia ranked second only to silver carp (*Hypophthalmichthys molitrix*). Matrix ranking of activities by gender and age indicated that there were differences in both participation and access to decision-making for fish culture activities within the household. Men dominated decision-making and participation in all types of aquaculture activity, whereas women's involvement in aquaculture was very low.

Introduction

In Bangladesh, inland fish production from capture fisheries no longer meets the consumption needs of the people. Undoubtedly, degradation and overexploitation of aquatic resources have been important factors, but increasing demand and a trend of exclusion of the poor from community water bodies are also important. The intensified use of such water bodies for agriculture and aquaculture by better-off sections of communities is believed to reduce access by, and benefits to, the poor (Custers, 1997; ITB, 1998). The importance of small indigenous species, mainly harvested from community water bodies, in the nutrition of the rural poor is clear (Roos *et al.*, this volume). Aquaculture has been promoted as an answer to these problems and the stocking of fish seed has become commonplace, especially in the last decade, as availability of hatchery-produced seed increased rapidly. Some critics, however, assert that fish culture has been mainly promoted to, and adopted by, the resource-rich (Lewis *et al.*, 1996).

The Northwest region is dominated by three of the 30 agro-ecological zones of Bangladesh, is relatively isolated and lacks perennial water (BARC, 1997; Karim, 1997). It has become the focus of several attempts to introduce and promote aquaculture (NFEP, 1996; Little *et al.*, 1999). Fish production has developed among poorer sections of the community in small borrow pits and ditches near the household, but how these resources are used and valued by people is poorly understood.

A total of nine species of carp and catfish seed is now available in the Northwest region (NFEP, 1995). Much of these are produced by hatcheries and nurseries in the South, imported into the region and distributed through a network of private sector fry traders (Lewis *et al.*, 1993). The quantities produced and traded have increased markedly over the last decade, and local producers have also developed (NFEP, 1995). Tilapia (*Oreochromis mossambicus*) was first introduced into Bangladesh in 1954 (Haque, 1996). Later introductions of Nile tilapia (*O. niloticus*) were made and most fish are probably hybrids of the two species. Such hybridisation has been a common feature of tilapia introductions in the last decades and has resulted in fish of questionable value (Macaranas *et al.*, 1986). There has been no formal promotion of tilapia seed production by the Government and a recent survey found little development of seed production in the commercial sector (Barman, 1997). An improved strain of Nile tilapia (GIFT) was imported in 1994 (Hussain *et al.*, 1996) but there is currently no plan for widespread dissemination to poor producers in the Northwest. This contrasts with the countries in Southeast Asia where tilapia seed production and dissemination have been actively promoted by Government and private sectors (Little and Hulata, 2000).

Little is known of the importance of tilapia to rural households, although national fish production statistics indicate tilapia production to be insignificant (Rahman, 1992). Market surveys in Northwest Bangladesh indicate that little tilapia enters into formal markets (NFEP, 1995, 1996). Some authors mention that tilapia are suitable for culture in small seasonal ponds or ditches using low cost technology (Shah, 1991; Rahman, 1992) but there is little evidence for the adoption of this or the impact compared to other species. A brief survey within a limited area (Mymensingh and Jaypurhat), however, has showed that tilapia are now popular with farmers and consumers, cultured both in monoculture and polyculture (Hossain, 1995). Many scientists and developmental professionals working in Bangladesh question these findings, however (personal observations).

The Northwest region is heterogeneous with regard to both agro-ecology and current status of fish culture. Understanding the importance of fish culture and the role of stocked hatchery-produced fish seed from a household perspective across the region, compared to the freely breeding tilapia, was the focus of the current study.

Materials and Methods

Research Framework

The study was conducted within 22 *paras* (communities) in four *thanas* (sub-districts) out of a total 58 *thanas* of the Northwest region of Bangladesh. Four zones of 'aquaculture environment' encompassing both agro-ecological and developmental criteria were identified under two broad categories on the basis of key informant interviews and prior knowledge of the study area as representative of the region. The more favourable category for aquaculture (MFA) included the two zones (a) areas with a high concentration of established culture ponds and (b) an area with a large fisheries development project which has promoted fish culture for 10 years. The less favourable category for aquaculture (LFA) included zones in (c) upland areas with sandy soils and few ponds and (d) riverine and flood-prone areas. The locations of the study areas are shown in Table 16.1 and Fig. 16.1.

Participatory Studies

A team of 6-8 members, balanced with respect to gender and discipline (aquaculture, social science, economics), as facilitators conducted participatory appraisals of small-scale aquaculture in a total of 22 communities in the selected areas. The facilitators, apart from the first author, were different in each location being drawn from governmental and non-government organisations working in the location. Village and community selection was based on key informant interviews with local *thana* fisheries officers. At the community level, older and better-educated people were selected as key informants. At each *thana* facilitators were divided into three smaller groups (2-3 members) which identified the location of representative *para* within villages sampled through a mapping exercise with a key informant. The status and roles of fish culture within each *para* were explored during community-based meetings.

Participatory study tools used included transect and local resource mapping to appraise physical resources.

Table 16.1. Areas of Northwest Bangladesh selected for the participatory studies. MFA, more favourable physical resource base and environment for aquaculture; LFA, less favourable physical resource base and environment for aquaculture.

Category of environment	Selected area		
		Thana	Community
MFA	Areas with a high concentration of culture ponds	Dinajpur Sadar	Ajiter Para, Burirthan Hindu Para, Chairman Para, Doctor Para, Majha Para and Sarker Para
	Areas with a large fisheries development project	Parbatipur	Dalali Para, Hindu Para, Kutir Danga, Paushim Para, Saha Para and Uttar Para
LFA	Upland dry areas with sandy soils	Atwari	Ban Para, Chhirashi Para, Majha Para and Sarker Para
	Riverine and flood-prone areas	Saghata	Arala Bari, Das Para, Fakir Para, Majider Vita, Sarker Para and Shil Para

Fig. 16.1. Map of Northwest Bangladesh showing the study areas (adapted from NFEP-2).

Social impacts were investigated using wealth matrix ranking and social mobility mapping. Except for wealth matrix ranking, all other tools were used by women and men as separate group exercises within the community. The wealth matrix ranking was carried out with three key informants individually for each community. The key informants grouped the households into different wealth ranking categories using certain criteria based on their knowledge and experience (Mikkelsen, 1995; Chambers, 1997, Table 16.2).

Table 16.2. Criteria used for wealth ranking of sampled households into different groups by key informants.

Category	I	II	III	IV	V
Relative wealth	poorest				wealthiest
Cultivable land holdings	Little to no land	Little land; sometimes share crop	Moderate amount of land	Sufficient amount of land	Enough to provide land to share cropper
Other resources	Little to no other resources (livestock, pond, machinery)	Limited amount of other resources	Have other resources sufficient to support household income	Have some other resources in large proportion	All kinds of other resources in large numbers or amounts
Main source of income	Agricultural or other labour and seasonal farmer	Agriculture and labour or run a small business	Agriculture and year round farming	Agriculture and also other source of income (service, business)	Large farmers and other large sources of income (business, service)
Own food production as a proportion of total consumption	Only fulfils a fraction of the annual food needs or other household needs	Sufficient for food and other household needs	Meets household needs for food and other necessities. Sometimes a surplus can be sold	More than sufficient to meet food and other needs of the household. Usually a small surplus is sold	Surplus to household needs; a large proportion is usually sold
Savings	No savings or incapable of accumulating assets; regularly use assets to fulfil regular household needs	No savings and some need to use assets to meet household needs	Small savings and some new assets acquired (house, machinery)	Sufficient savings to purchase assets (land, house, machinery)	Large amount of savings to invest to generate new income

The role of different species of fish with respect to their different characteristics was investigated using matrix ranking. This was obtained through participation of 11 groups of men (2-4 groups from each area) and four groups of women (one group from each area) in the communities studied. Participants (5-10 in each group) through discussion and analysis placed a maximum of 10 bean seeds in comparative rankings of individual species for each specific criterion. The criteria used for comparisons were: yield, taste, household consumption, price, market demand, availability of fry/fingerlings, suitability of culture in small ponds, suitability of culture in large ponds, harvesting problem and disease problem. The fish species were: rohu (*Labeo rohita*), catla (*Catla catla*), mrigal (*Cirrhinus mrigala*), silver carp (*Hypophthalmichthys molitrix*), silver barb (*Barbodes gonionotus*), common carp (*Cyprinus carpio*), tilapias (*Oreochromis niloticus, Oreochromis mossambicus* or their hybrids) and grass carp (*Ctenopharyngodon idella*).

Gender issues were explained through activity participation and decision-making profiling. The matrix on activity participation of men, women, boys and girls was obtained from nine groups of women (2-3 groups from each area) and six groups of men (1-2 groups from each area) of the studied communities. Participants (5-7 in each group) used a total of 10 bean seeds for each activity and through discussion placed for men, women, boys and girls. Like activity participation the participation in decision-making in fish culture activities were conducted by eight groups of participants (one group of women and one group of men in each study area) using similar methods of activity participation.

Semi-structured interviews were conducted with individual pond owning households (*n* = 68) sampled from each wealth category from two broad areas in equal numbers. The interviews took place over a period of one week to better understand individual pond management, resource use and attitudes to tilapias. Each household interviewed was asked to explain their attitudes to tilapias, including advantages and disadvantages compared to other species they culture.

The methods used in this study follow closely the standard PRA family of approaches and methods that enable local people to express, enhance, share and analyse their knowledge of life and conditions, leading to planning and action (Townsley, 1996; Chambers, 1994a, b, 1997). In the subsequent analysis, wealth ranking is used as a basis for categorising households.

Results

Pond ownership is common in Northwest Bangladesh, particularly among the wealthier section of the community. In the areas surveyed a mean of one third of all households owned ponds.

Most of the rich households had at least one pond (Groups V and IV, 78 and 73% respectively) whereas the far more numerous poorest section of the community were least likely to have a pond (Group I, 17%) (Table 16.3). Ponds took up a larger part of the land-holding of poorer people, reaching 16% in LFA (Table 16.4).

Table 16.3. Classification of all households in surveyed communities by wealth ranking category and pond ownership (figures in parentheses are number of households with ponds). MFA, more favourable physical resources base and environment for aquaculture; LFA, less favourable physical resources base and environment for aquaculture.

Category of environment		I poorest	II	III	IV	V wealthiest	Total
				Wealth ranking category			
MFA	Areas with high concentration of culture ponds	74 (15)	43 (19)	28 (18)	37 (31)	12 (10)	194 (93)
	%	20	44	64	84	83	48
	Areas with a large fisheries project	132 (21)	59 (20)	45 (26)	38 (29)	10 (7)	284 (103)
	%	16	34	58	73	70	36
LFA	Upland dry areas with sandy soils	97 (20)	62 (24)	52 (28)	47 (38)	8 (8)	266 (118)
	%	21	39	54	81	100	44
	Riverine and flood-prone areas	283 (43)	68 (23)	68 (22)	73 (44)	34 (25)	526 (157)
	%	15	34	32	60	74	30
	Total	586 (99)	232 (86)	193 (94)	195 (142)	64 (50)	1,270 (471)
	%	17	37	49	73	78	37

The categorisation of favourable and less favourable areas was indicated by large differences in the frequency of pond ownership ($P < 0.05$). In the MFA category, 41% of the households had ponds, the corresponding figure for the LFA was 35%. Ponds were more common in the upland dry areas (44%) than the area within a large fisheries project (36%). Ponds were most frequently owned in the area with well-developed aquaculture (48%). The differences in the frequencies of pond ownership for the subcategories of environment were highly significant ($P < 0.01$).

Of the total of 68 pond-owner households interviewed, 59 of them stocked their ponds for fish culture and 9 of them used their ponds only for wild fish production. Use of ponds for wild fish alone was most common in the upland dry areas (5 out of a total of 15 farmers). A majority of farmers who stocked their ponds for fish culture used tilapia (56%), mainly in a polyculture with Indian major carp and Chinese carp (Table 16.6a) but this disguised a major difference between riverine and flood-prone areas where most farmers managed tilapia (83%), and the other three areas where less than half of the farmers did (40-47%). The proportion of farmers managing tilapia exceeded 40% in every group and was not related to wealth, except among the richest group that were more likely to have the fish in their ponds (V, 73%). In wealth ranking categories, the proportion of tilapia farmers was the highest in the most better-off group (73%, Table 16.5).

Table 16.4. Land holdings of pond owning households by wealth ranking category in the study areas. MFA, more favourable physical resources and environment for aquaculture; LFA, less favourable physical resources and environment for aquaculture; 'f', number of farmers from different wealth ranking categories interviewed.

Wealth ranking category	'f'	Land holding by type (ha per household)				Pond as % total
		Cultivable	Homestead	Pond	Total	
MFA areas						
Poorest	6	0.52	0.07	0.07	0.65	10.70
Marginal	5	0.30	0.04	0.04	0.39	10.30
Less better-off	5	0.79	0.08	0.08	0.95	8.40
Better-off	11	1.22	0.14	0.12	1.48	8.10
Most better-off	7	4.92	0.54	0.43	5.88	7.30
LFA areas						
Poorest	7	0.18	0.08	0.05	0.31	16.00
Marginal	3	0.30	0.11	0.03	0.44	6.80
Less better-off	6	1.13	0.07	0.05	1.25	4.00
Better-off	10	1.19	0.16	0.07	2.12	3.30
Most better-off	8	3.60	0.19	0.12	3.91	3.10

Table 16.5. Classification of fish farming households based on wealth ranking category in the study areas (figures in the parentheses are number of farmers with tilapia). MFA, more favourable physical resources and environment for aquaculture; LFA, less favourable physical resources and environment for aquaculture.

Category of environment		Wealth ranking category					
		I poorest	II	III	IV	V wealthiest	Total
MFA	High concentration of culture ponds	2 (2)	2 (1)	2 (1)	6 (1)	3 (2)	15 (7)
	%	100	50	50	17	67	47
	Large fisheries project	4 (1)	2 (0)	2 (1)	4 (2)	4 (3)	16 (7)
	%	25	-	50	50	75	44
LFA	Upland dry areas with sandy soils	1 (0)	2 (1)	2 (1)	2 (0)	3 (2)	10 (4)
	%	-	50	50	-	67	40
	Riverine and flood-prone areas	3 (2)	1 (1)	2 (2)	7 (6)	5 (4)	18 (15)
	%	67	100	100	86	80	83
	Total	10 (5)	7 (3)	8 (5)	19 (9)	15 (11)	59 (33)
	%	50	43	63	47	73	56

Most farmers obtain tilapia fry from neighbours' ponds, with only a minority depending on traders. Trading of tilapia fry only occurred in the flood-prone area, where 15% of farmers purchased fry from traders (Table 16.6).

Table 16.6. Distribution of pond owners on types of fish culture, input uses, sources of tilapia seed and ponds in the study areas (figures in parentheses are percentages for respective areas). MFA, more favourable physical resources and environment for aquaculture; and LFA, less favourable resources and environment for aquaculture.

Description	Category of environment			
	MFA		LFA	
	High concentration of culture ponds	Large fisheries development project	Upland dry areas with sandy soils	Riverine and flood-prone areas
(a)Types of fish culture (*n* = 59)				
Tilapia with carp polyculture	7 (47)	6 (38)	3 (30)	13 (72)
Carp polyculture	8 (53)	9 (56)	6 (60)	3 (17)
Tilapia monoculture	-	1 (6)	1 (10)	2 (11)
(b) Input use for fish culture (*n* = 59)				
Application of lime	11 (73)	13 (81)	2 (20)	9 (50)
Cow dung	14 (93)	14 (88)	7 (70)	16 (89)
Inorganic fertiliser (Urea and TSP)	7 (47)	9 (56)	5 (50)	7 (39)
Use of supplementary feed (rice bran)	14 (93)	16 (100)	9 (90)	15 (83)
(c) Source of tilapia seed (*n* = 33)				
Neighbours	5 (71)	6 (86)	4 (100)	10 (67)
Fry traders	-	-	-	3 (20)
Fishermen	-	-	-	1 (7)
Other sources (office pond)	2 (29)	1 (14)	-	1 (7)
(d) Pond number and size (m^2)				
Number of pond owners	15	16	10	18
Number of ponds	20	20	17	22
Average size of ponds	988	543	437	668
Size range of ponds	121-2,429	283-607	61-972	202-1,700

A variety of inputs are now used for fish culture including lime, cow dung and supplementary feed (mainly rice bran) at a similar frequency by pond owners. In the favourable resource areas the proportion of farmers using inputs was higher than in less favourable areas. Pond use was more extensive in the upland dry areas, where they are generally managed as refuges for wild fish (6). Pond size tended to be smaller within the fisheries development project and upland dry areas. The average size of stocked ponds was 676m^2 with a range from 61 to 2,429m^2 (Table 16.6).

Farmers' opinions of tilapias revealed the high importance attached to them: 20 specific advantages were raised but only five disadvantages (Table 16.7). Their ability to produce seed in the pond was identified as a major advantage, as was their rapid growth rate. They have an important role in meeting household consumption needs in large amounts throughout the year and their easy harvest by angling was appreciated. Their suitability for culture with other species of fish in ponds was considered an advantage by the majority of tilapia farmers. Overcrowding of tilapia due to population increase in ponds was raised as the most important problem by a large number of tilapia farmers and most of the farmers who were not currently raising tilapia.

Tilapia were also ranked highly in terms of overall desirability for culture and consumption by farmers compared to other species, irrespective of pond ownership (Table 16.8). Overall, tilapia ranked on a par with the native Indian major carp (catla and mrigal), and second only to silver carp in level of performance. The characteristics of tilapia such as acceptable taste, ease of harvest by angling and suitability for culture in small ponds made them preferred for household consumption. A lack of disease was also an important advantage over many other species of cultured fish, especially the Indian major carp (rohu, catla and mrigal), silver barb and common carp which were otherwise equally popular. Silver carp scored highest; only its poor taste and low value reduced its attractiveness. Silver barb was reported to be very susceptible to disease. When the popularity of fish species was compared with respect to specific areas, silver carp was the most popular in all the areas surveyed. For other species, there were some minor variations. Tilapia were favoured more in the riverine and well-developed aquaculture areas than the other two areas. Grass carp was popular only in the upland dry areas with sandy soils.

There were great differences in the participation of household members in all kinds of aquaculture activities but few between the areas surveyed. Men were active in all types of activities relating to fish culture and also dominated decision-making. Women and girls had minimal participation in all activities and were not involved at all in pond excavation (except the upland dry areas) and sale of fish. In the upland dry areas, a small number of poor women participated in pond excavation with their male counterparts as a group in Hindu communities (Table 16.9). In contrast, women's role in decision-making was marginally higher, especially in terms of control over money and fish feeding. In general, children's roles in decision-making and activities was much lower than adults, and boys were more active than girls in both activities and decision-making, especially in terms of selling fish (Table 16.10).

Table 16.7. Characteristics of tilapia and their relative importance based on farmers' knowledge and experience (*n* = 33). +++ high response; ++ intermediate response from a minority of farmers; + minimal response from a few farmers.

Advantages of tilapia	
(a) Seed	
Breeds in the food fish pond and produces seed	+++
Can mature and breed within 3-month period	+
Produce seed frequently	+
Can collect fingerlings from neighbours' pond without any cash payment	+
(b) Culture management, growth, size and production	
Can culture with other species and no special care is required	+++
Rapid growth	+++
No disease problems	+++
High yield	++
Tilapia can be cultured with less investment than other species of fish	+
Less management required for production of tilapia than carp	+
(c) Harvesting and consumption	
Easy to harvest by angling	+++
Use in large amounts for household consumption	+++
Taste is similar to small indigenous fish Koi (*Anabas* sp.)	++
Responsive to harvest using supplementary feed as a bait	+
(d) Others	
High market demand	+
Good price	+
Not easy to poach	+
Use seed as gifts for neighbours to develop social relations	+
Can sell seed in addition to food fish	+
Share harvest with neighbours	+

Disadvantages of tilapia	
Rapid population increase of tilapia in pond hampers growth of other fish	+++
Difficult to harvest by using net if the water level is high	+
Sometimes high mortality of tilapia observed after their large population increase in pond	+
Do not grow to a large size	+
Lower market demand and low price for small size tilapia	+

The mobility maps collected from one group of men and one group of women in each of the four study areas showed that women were less mobile than men. Men travelled for a variety of reasons (classified as economic, recreational, or professional) but women mainly travelled for medical treatment of themselves or their children and for social reasons (relatives' house, religious festival, etc.). The poorest women also travelled to non-governmental organisations (NGOs) for loan collection and payment.

Table 16.8. Summary on matrix ranking of fish species. Fish species: rohu (*Labeo rohita*), catla (*Catla catla*), mrigal (*Cirrhinus mrigala*), silver carp (*Hypophthalmichthys molitrix*), silver barb (*Barbodes gonionotus*), common carp (*Cyprinus carpio*), tilapias (*O. niloticus*, *O. mossambicus* or their hybrids) and grass carp (*Ctenopharyngodon idella*).

	Fish species							
Characteristic	Rohu	Catla	Mrigal	Silver carp	Silver barb	Common carp	Tilapia	Grass carp
Yield	••••• •	••••• ••	•••••	••••• •••••	••••• •	••••• •	••••	•••••
Taste	••••• •••••	••••• •••	••••• ••	••••	••••• •••	••••• •••	•••• •	•••
Household consumption	••••	••••• •	•••••	••••• ••••	••••• ••	••••• •	•••• •••	••••• •
Price	••••• •••••	••••• •••	••••• •••	•••••	••••• ••	••••• •••	•••• •	•••••
Market demand	•••••	•••••	••••• ••	••••• ••••	••••• •	••••• •	•••• ••	••••• •
Availability of fry/fingerling	••••• •••	••••• ••	••••• •••	••••• ••••	•••••	••••• •	••••	•••••
Suitability of culture in small pond	••••	•••••	•••••	••••• ••••	••••• ••	•••••	•••• •••	••••
Suitability of culture in large pond	••••• ••	••••• ••••	••••• ••	••••• •••••	••••• •	••••• •	••••	••••• ••
Harvesting problem	••••• ••	••••	••••• ••	•••	•••••	••••• ••••	•••• •	••••
Disease problem	••••• ••	••••• •	••••• •••	••	••••• •••••	•••	Nil	Nil
Positive points	54	54	53	65	52	51	50	41
Negative points	14	10	15	5	15	12	6	4
Net points	40	44	38	60	37	39	44	37

Table 16.9. Summary of activity participation of household members in fish culture activities.

Fish culture activities	Men	Women	Boys	Girls
Pond excavation	••••• •••		••	
Pond preparation	••••• •••	•	•	
Manuring	•••••	•••	•	•
Selection of fish species	••••• •••	••		
Fry release	••••• ••	••	•	
Feeding fish	••••	•••	••	•
Harvesting fish	••••• •	•	••	•
Fish sale	••••• ••		•••	

Table 16.10. Summary of participation in decision-making by household members in fish culture activities.

Fish culture activities	Men	Women	Boys	Girls
Pond excavation	••••• ••••	•		
Pond preparation	••••• •••	•	•	
Manuring	••••• •••	•	•	
Selecting fish species to stock	••••• •	•••	•	
Fry release	••••• •••	•	•	
Feeding fish	•••••	••••	•	
Harvesting fish	••••• ••	••	•	
Fish sale	••••• •••		••	
Fish consumption	••••• ••	••	•	
Control over money	••••• •	••••		

Discussion

Pond ownership among better-off was much higher than that of the poorer households. About 80% of the wealthiest households had ponds compared to only 20% of the poorest. This finding was similar to that of Ahmed *et al.* (1993) who found pond-owning households in two *thanas* in central Bangladesh were wealthier than the rest of the households in the community. The importance of stocking ponds with fish seed in the region was indicated by the high proportion of households stocking fish ponds (mean 87%) compared with parts of wild-fish-rich Northeast Thailand (45%, Saengrut, 1998). This suggests a certain level of pressure on wild fish resources in the areas sampled, and that the practice is, at least to some extent, meeting needs. The lowest frequency of stocking ponds (70%) was found in the drier upland area with greatest availability of wild fish. A strong relationship between interest in stocking fish and abundance of wild fish was also found in Cambodia where Gregory and Guttman (1996) identified fish-deficit areas as having most potential for extension of fish culture. Saengrut (1998) found that wild fish were more abundant in trap ponds in drier areas away from the floodplain in Northeast Thailand, which he related to the migration habits of the common ricefield species.

The arbitrary distinction between less (LFA) and better-off (MFA) areas for aquaculture was not clearly reflected by differences in frequency of input use, except for lime that was more commonly used in MFA ($P < 0.01$). A similar level of frequency of other input use was apparent, although quantities and yields were not assessed and might be expected to be low following the observations of Ahmed *et al.* (1993). The use of nutrients was common, especially supplementary feed and cow manure (for both, mean = 90%) but even inorganic fertilisers were used (mean 48%). The frequency of lime, inorganic fertiliser and supplementary feed use appeared higher in the area with a large fisheries development project, suggesting that extension messages were having an impact. There was also evidence that the level and frequency of addition of inputs were higher in areas influenced by the development project. Mean levels of input use were also higher than those observed by Ahmed *et al.* (1993); double the numbers of farmers used rice bran, cow manure and inorganic fertiliser, and lime use was higher by a factor of 3 in our study. These data suggest that the concept of fertilisation of fish ponds was widely understood among pond owners but that available resources were used elsewhere. The identification of the 'easy' management of tilapia and compatibility with stocked carp as important advantages reflected that poorer farmers cannot, or do not, adopt more resource-intensive approaches to even small fish ponds. Ponds of the Northwest region, with the exception of areas with a high concentration of established culture ponds, were relatively small and varied little in respect of wealth ranking category. The small average pond size ($676m^2$) is still a significant investment given the total land holdings available for homestead and cultivation, and resources available for enhancing its productivity. Ahmed *et al.* (1993) found that small ponds and ditches accounted for only 5-6% of the operated land, while in the current study the level varied from as low as 3% for the wealthiest farmers in LFA to between 10-16% for the poorest in both categories. Our data suggested a greater relative importance of ponds as physical assets for the poorest people.

The popularity of tilapia found in the study reflects their inherent characteristics and suitability of the ponds and resource base for the species. It conflicts with the earlier study of Ahmed *et al.* (1993), in Gazipur District that found exotic species to be rare in farmers' ponds. In stocked systems, a mean of 56% of pond owners raised tilapia as a monoculture, or more normally as part of a polyculture with carp. The particular importance of tilapia in riverine, flood-prone areas suggests that the practice was an important part of the farmers' risk avoidance in these areas. Regular flooding of the pond may discourage them from stocking other species that required a cash investment. It also suggests that farmers regarded them in the same way as small, indigenous fish that do not need to be stocked or managed carefully and are low input (cash costs can be zero) compared to carp for which seed is always purchased. The tendency to favour tilapia to reduce cash costs was also found in Northeast Thailand among farmers with a subsistence bias (Little *et al.*, 1996). The relative popularity of tilapia among rural farmers raising and consuming fish, was also observed by Hossain and Little (1996) in two other areas of Bangladesh.

The particularly high frequency of tilapia culture among the wealthiest farmers (> 70%) is noteworthy, although as attitudes to fish species were not dis-aggregated by wealth, reasons for this cannot be concluded. In other areas of Asia where tilapia have been introduced into carp polycultures, tilapia become more dominant when the availability and level of inputs used increase (Little *et al.,* 1999). Phytoplankton-feeding tilapia are less sensitive to the variable dissolved oxygen regimes in very eutrophic ponds than carp, and can yield much higher under such conditions (Hassan *et al.,* 1996; Hossain, 1995). The evidence suggests that tilapia can meet both the needs of farmers for whom risk aversion and low cash costs are prioritised, as well as those with greater than average resources to invest. The reproductive characteristics of tilapia bring both major benefits and disadvantages. Obtaining both food and seed was identified as a major advantage but the impact of their breeding on their own growth and the performance of other fish bought contradictory responses. High densities of tilapia caused by free-breeding in the pond is known to reduce the growth of Indian major carp (Hossain, 1995) but partial harvest may be an important strategy to ensure regular food supply for the household and allow other stocked species to grow. Availability and low, or zero, cost of tilapia seed appears to be a major incentive for stocking, especially when the risk of loss of stocked fish is high. Most tilapia seed is currently obtained from neighbours' ponds. Tilapia are supplied from largely outside the trading networks that link carp hatcheries with farmers in rural areas, although there are indications that local transportation and sale of tilapia seed are becoming more important. Such mechanisms have allowed tilapia culture to develop and be sustained in rural areas but raises questions for the introduction and maintenance of new strains (Little *et al.,* 1999). Isolation from hatchery-supplied fish seed might also have impacts on the flow of information to rural farmers, although most stock tilapia in polyculture with carp, typically purchased from traders. This contact can be a valuable source of farmers' knowledge about fish culture (Lewis *et al.*, 1993). The positive impact of training and motivation of fry traders to enhance their role as informal extension agents has been demonstrated (NFEP, 1995).

The matrix ranking of fish species confirmed the relative importance of the exotic species, tilapia and silver carp, for producers and consumers of cultured fish species. Ahmed *et al.* (1997) observed that the increased trend to adopt tilapias by farmers after their promotion in Gazipur District was matched by increased consumption in the household. Apart from the relatively high yield and good taste, the possibility to harvest the fish by angling may be important for its importance as a regular food in the household. Although the gender analysis showed a very low participation by women and children in fish culture activities and decision-making compared to men, feeding and harvest of fish was undertaken and angling appears to be an acceptable activity for women and children. The gender and mobility analysis clearly indicated that the current minor role in, and exposure to, aquaculture by women requires attention. Women travel away from the household and pond less than men and participate in activities and decision-making to a significant degree. Although men dominate almost every category, women have a significant current role in species and pond input selection, daily pond management and control over money in the household. Given the complexity of decision-making by households towards ensuring food security and cash income and the species-related differences of cultured fish identified in this study, the importance of considering gender for education and extension is essential.

Acknowledgements

The study is part of the first author's dissertation research work for the degree of Doctor of Technical Science supported by the Department for International Development (DFID) and assisted by the Northwest Fisheries Extension Project (NFEP), Bangladesh. The authors wish to thank DFID and NFEP and also others who directly or indirectly helped with this study.

References

Ahmed, M., Rab, M.A. and Bimbao, M.P. (1993) Household socioeconomics, resource use and fish marketing in two thanas in Bangladesh. ICLARM Technical Report 40, ICLARM, Manila. 76 pp.

Ahmed, M., Rab, M.A. and Bimbao, M.P. (1997) Sustainable aquaculture in small waterbodies: experience from Bangladesh. In: Mathias, J.A., Charles, A.T. and Baotang, H. (eds) *Proceedings of a Workshop on Integrated Fish Farming Held in Wuxi, China*, pp. 392–402.

BARC (1997) Fertilizer recommendation guide. Soil Publication No 41. BARC, Farmgate, New Airport Road, Dhaka.

Barman, B.K. (1997) Analysis of the status of tilapia production system in Bangladesh. A special study submitted for the Doctor of Technical Science. Asian Institute of Technology, Bangkok, Thailand. pp. 34.

Chambers, R. (1994a) The origin and practice of participatory rural appraisal. *World Development* 22(7), 953-969.

Chambers, R. (1994b) Participatory rural appraisal: challenges, potentials and paradigm. *World Development* 22(10), 1437-1454.

Chambers, R. (1997) *Whose Reality Counts: Putting the First Last.* Intermediate Technology Publications, London. 297 pp.

Custers, P. (1997) Developmental feminism and peasant women's labour in Bangladesh. In: *Capital Accumulation and Women's Labour in Asian Economies.* Zed Books, London, and Vistaar Publications, New Delhi, pp. 201-227.

Gregory, R. and Guttman, H. (1996) Management of ricefield fisheries in South East Asia: capture or culture. *ILIEA Newsletter* 12(2), 20-21.

Haque, E. (1996) Fisheries statistics of Bangladesh. In: Mazid, M.A. (ed.) *Technologies and Management for Fisheries Development.* Bangladesh Fisheries Research Institute, Mymensingh, pp. 127-131.

Hassan, S., Edwards, P. and Little, D.C. (1996) Comparison of tilapia monoculture and carp polyculture in fertilized earthen ponds. *Journal of World Aquaculture Society* 28(3), 268-274.

Hossain, M.A. (1995) Investigation into the polyculture of two Indian major carps (*Labeo rohita* and *Cirrhinus mirgala*) and Nile tilapia (*Oreochromis niloticus*) in fertilized ponds. Doctoral Dissertation, Asian Institute of Technology, Bangkok.

Hossain, M.A. and Little, D.C. (1996) Farmer's attitudes to the production and consumption of commonly cultured fish species in two districts of Bangladesh. Paper presented at the World Aquaculture Symposium '96. Jan 29 –Feb 2, 1996, Bangkok.

Hussain, M.G. and Mazid, M.A. (1996) Progress of DEGITA activities in Bangladesh: National review on past performances and current status of tilapia industry. In: Mazid, M.A., Hossain, M.G. and Alam, M.J. (eds) *Current Status and Future Strategy for Dissemination of Genetically Improved Farmed Tilapia in Bangladesh.* Fisheries Research Institute, Mymensingh.

ITB (1998) Food, livelihoods and freshwater ecosystems: the significance of small indigenous fish species (SIS). Intermediate Technology Bangladesh and The Business Advisory Services Centre (BASC), Dhaka. 50 pp.

Karim, M. (1997) A review of aquaculture extension services in Bangladesh. RAP Publication 1997/35. Regional Office for Asia and the Pacific. FAO, Bangkok, Thailand. 54 pp.

Lewis, D.J., Gregory, R.G. and Wood, G.D. (1993) Indigenising extension: farmers, fish-seed traders and poverty focused aquaculture in Bangladesh. *Development and Policy Review* 11, 185-194.

Lewis, D.J., Wood, G.D. and Gregory, R. (1996) Trading the silver seed: local knowledge and market moralities in aquaculture development. University Press Limited, Dhaka. 199 pp.

Little, D.C. and Hulata, G. (2000) Broodstock management, hatchery and nursery technology. In: Beveridge, M.C. and MacAndrew, B. (eds) *Tilapias: Biology and Exploitation.* Kluwer, Dordrecht, pp. 267-326.

Little, D.C., Golder, M.I. and Barman, B.K. (1999) Rice field-based fish seed production - understanding and improving a poverty-focused approach to the promotion of aquaculture in Bangladesh. *AARM newsletter* 4(2), 7-10.

Little, D.C., Surintaraseree, P. and Innes-Taylor, N.L. (1996) Review: fish culture in rainfed rice fields of Northeast Thailand. *Aquaculture* 140, 295-321.

Macaranas, J.M., Taniguchi, N., Pante, M.-J.R., Capili, J.B. and Pullin, R.S.V. (1986) Electrophoretic evidence for extensive hybrid gene introgression into commercial *Oreochromis niloticus* (L.) stocks in the Philippines. *Aquaculture and Fisheries Management* 17, 249-258.

Mikkelsen, B. (1995) Methods for development work and research: a guide for practitioners. Sage Publications, New Delhi.

NFEP (1995) Annual report 1994/95. Northwest Fisheries Extension Project, Parbatipur, Dinajpur.

NFEP (1996) Annual report, 1995/96 on extension and training. Written and complied by C.P. Morrice. Northwest Fisheries Extension Project, Parbatipur, Dinajpur. 202 pp.

Rahman, A.K.A. (1992) Tilapia in Bangladesh. Country report presented at the fifth session of the Indo-Pacific Fishery Commission Working Party of Experts on Inland Fisheries, Bogor, Indonesia, 24-29 June 1991.

Saengrut, T. (1998) Role of wild fish in aquatic resource development in the lower Chi Valley of Thailand. MSc Thesis, Asian Institute of Technology, Bangkok. 100 pp.

Shah, M.S. (1991) Breeding biology of tilapia. In: Gupta, M.V. (ed.) *Trainer's Training Manual on Improved Fish Culture Management Practices for Fisheries Extension Officers.* Fisheries Research Institute, Mymensingh, pp. 103-109.

Townsley, P. (1996) Rapid rural appraisal, participatory rural appraisal and aquaculture. FAO Fisheries Technical Paper 358. FAO, Rome. 109 pp.

Chapter 17

Culture of Small Indigenous Fish Species in Seasonal Ponds in Bangladesh: the Potential for Production and Impact on Food and Nutrition Security

N. Roos[1], S.H. Thilsted[1] and Md.A. Wahab[2]

[1]*Research Department of Human Nutrition, The Royal Veterinary and Agricultural University, Rolighedsvej 30, 1958 Frederiksberg C, Denmark*
[2]*Department of Fisheries Management, Bangladesh Agricultural University, Mymensingh, Bangladesh*

Abstract

Small indigenous fish species (SIS) play an important role in the diet of the rural population in Bangladesh, not only as an animal protein source, but also as a source of a range of other essential nutrients such as vitamin A, calcium (Ca) and iron. Accessibility of SIS from capture fisheries is declining but development of low-input, semi-intensive, aquaculture technologies can potentially augment the production of this culturally accepted and affordable food for the rural poor. Some SIS, including mola (*Amblypharyngodon mola*) are extremely rich in vitamin A and since most SIS are eaten with bones, they are good Ca sources. A field trial on polyculture of SIS and carp was conducted in small, rural seasonal ponds in Northeastern Bangladesh to investigate the production potential and to survey the nutritional impact on the farmers' families. SIS contributed approximately 10% of the total fish production. The trial demonstrated that mola stocked in polyculture with carp in small seasonal ponds had potential to contribute significantly to the vitamin A supply in rural households.

Introduction

Small indigenous fish species (SIS) are an important component in the diet of the rural poor in Bangladesh. Fish is well recognised as a source of animal protein but SIS are also a rich source of a range of other essential nutrients such as vitamin A, calcium (Ca), iron (Fe) and zinc (Zn). SIS are culturally accepted and affordable food for the rural poor but declining capture fisheries in Bangladesh (Anonymous, 1995) have led to decreasing accessibility of SIS. To meet the present and future demand of SIS, development of low-input, semi-intensive, aquaculture technologies to augment production should be investigated. To ensure food and nutrition security, it is essential not only to increase the total food production but also to ensure availability for the poor (Kent, 1997). The vast numbers of small seasonal ponds in Bangladesh are often owned by poor and marginal farmers and are typically an underutilised resource for fish production. Seasonal ponds are a natural habitat for SIS and potentially suitable for their cultivation, alone or in combination with normally cultured carp. SIS are defined as species indigenous to the region which attain a maximum length of 25cm or less (Felt *et al.*, 1996) but many SIS of Bangladesh attain a length of less than 10cm.

SIS in aquaculture are suitable for home consumption as frequent (weekly or even daily) harvesting of small amounts of fish is possible since many of these fish species reproduce in the pond. By culturing SIS in polyculture with normally cultured carp, the household can potentially derive double benefit: poor people can increase nutrient intake by consuming the SIS and at the same time increase their income by selling the normally cultured carps and SIS. With regard to the beneficial effect on food and nutrient intake, culture of SIS can be compared to home gardening which is promoted as a strategy to combat vitamin and mineral deficiencies (Helen Keller International, 1994): small fish are rich in a range of minerals and vitamins and a steady supply can improve the nutritional quality of the diet.

Nutritional Value of Fish

Deficiency of vitamin A is widespread in Bangladesh, particularly in women and children (Ahmad and Hassan, 1983; Zeitlin *et al.*, 1992). Vitamin A is an essential nutrient for many physiological functions in the human body, including vision, physical growth, foetal development and immune response. Calcium is an important nutrient for foetal growth, growth in young children and milk production.

Analyses of a number of Bangladeshi fish species have shown great variations in vitamin A and Ca content (Thilsted *et al.*, 1997, unpublished data). Of 27 Bangladeshi fish species analysed, three species have been found to have extremely high vitamin A content. Table 17.1 shows the vitamin A content in two SIS, mola (*A. mola*) and puti (*Puntius* spp., mainly *P. sophore*), and two carp species normally stocked in fish ponds, silver carp (*Hypophthalmichthys molitrix*) and rui (*Labeo rohita*). The vitamin A is located in the eyes and liver and thus the cleaning procedure influences how much vitamin A is retained for consumption. Analysis of uncleaned mola and mola cleaned according to local practices to obtain raw edible parts showed that cleaning reduced the vitamin A content from 2,500 to 3,000 RE (retinol units) g^{-1}. Vitamin A is sensitive to sunlight and is destroyed by sun drying. Puti is one of the SIS which is low in vitamin A, even though both mola and puti have similar ecology. Analyses of normally cultured carps showed that they contain an insignificant amount of vitamin A.

Table 17.1. Vitamin A content in fish species. RE = retinol units.

Fish species	Vitamin A RE 100g^{-1}
Mola (*Amblypharyngodon mola*), not cleaned[1]	3,000
Mola (*A. mola*), raw edible parts[1]	2,500
Mola (*A. mola*), sundried[1]	100
Puti (*Puntius sophore*), raw edible parts[2]	40
Silver carp (*Hypophthalmichthys molitrix*), not cleaned, juvenile[2]	13
Silver carp (*H. molitrix*), raw edible parts[2]	20
Rui (*Labeo rohita*), raw edible parts[2]	30

[1] Roos, (unpublished data).
[2] Thilsted *et al.* (1997).

Since small fish are normally consumed with bones, they are also an important source of Ca. It has recently been shown in human studies that the absorption of Ca from mola is similar to that in skimmed milk and therefore mola is a very good source of Ca (Hansen *et al.*, 1998). SIS are also rich sources of Fe and Zn (unpublished data).

Some experimental trials on the culture of SIS have been carried out (Felt *et al.*, 1996) but these trials have not been extended to farmers' ponds. In 1997 and 1998 in rural Bangladesh, studies were undertaken to investigate the potential of production of SIS in combination with carp polyculture in seasonal ponds as well as the potential nutritional impact of production on the farming families.

Methods

A field trial on polyculture of SIS and normally cultured carp was conducted in Kishoreganj district, Northeastern Bangladesh from May 1997 to January 1998. The aim of the field trial was to investigate the production potential and resource management of SIS-normally cultured carp polyculture in small seasonal ponds, and to survey the nutritional impact on farmers' families. Fifty-nine small seasonal ponds (212-850m^2) belonging to poor or marginal farmers were stocked with silver carp, grass carp (*Ctenopharyngodon idella*), mrigal (*Cirrhinus mrigala*) (or common carp, *Cyprinus carpio*) and rui in the ratio 8:4:4:1 and at a density of 8,500 fingerlings ha^{-1}. Thirty-four of the ponds were treated with rotenone and adult mola were stocked with the normally cultured carps at a density of 25,000 mola ha^{-1} (mola model*)*. Brood fish were used for stocking since mola spawns and multiplies in ponds. The brood fish were collected from local ponds which had natural mola stocks. Twenty-five of the ponds were not treated with rotenone and the natural stocks of different SIS were left in the pond (mixed model). Rotenone was used in the mola ponds to control the trial conditions through removal of other SIS.

The culture period ranged from 181 to 240 days. The ponds were managed with low inputs of cheap, local feed (rice bran, duckweed and banana leaves) and were fertilised with cow dung, urea and TSP. Details on pond management are reported in Roos *et al.*, 1999.

The economics of the production models were compared based on the expenditures for fingerlings and mola brood stock and the value of the normally cultured carps and SIS harvested. Except for the use of rotenone in mola ponds, all ponds were managed according to the same guidelines regardless of the production and therefore the economics of the management practices have not been taken into consideration in this context.

In a follow-up trial in 1998, the 59 farmers who participated in the 1997 trial were offered extension services and technical assistance to continue normally cultured carp-SIS polyculture in their ponds but without credit or financial support.

Studies were conducted in three seasons to investigate the impact of fish production on fish consumption pattern: pre-production season (July 1997), production season (October 1997) and post-production season (February 1998). Fish consumption was surveyed by recall interviews. Fish consumption in the past 5 days starting from the day previous to the day of interview was recalled. The consumption survey covered the 59 fish-producing households and 25 control households without access to a pond.

Results and Discussion

Fish Production

There was no significant difference in the total fish production between the two production models. The total fish production was 2.8t ha^{-1} in both mola and mixed pond models over a period of 217 and 210 days, respectively. Fig. 17.1 shows the average production of each fish species based on a growth season of 7 months.

The carp production was 2.5t ha^{-1} 7 months^{-1} in both mola and mixed ponds, with trends for higher production of grass carp in the mola ponds and higher production of silver carp in the mixed ponds that were not statistically significant.

The average mola production was significantly higher in the mola ponds compared to the mixed ponds ($P < 0.001$), while production of other small fish species was significantly higher in the mixed ponds ($P < 0.001$). The highest mola production obtained was 0.8t ha^{-1} 7 months^{-1}. However, the average total production of all small fish species was not significantly different: 0.34t ha^{-1} 7 months^{-1} in the mola ponds and 0.44t ha^{-1} 7 months^{-1} in the mixed ponds. Small fish production in the mixed ponds was dominated by puti and darkina (*Esomus danricus*) but many other species were recorded. Mola contributed an average of 10.3% of the total fish production in the mola ponds.

Other small fish which had entered the pond after treatment with rotenone, most probably along with the duckweed collected from ricefields, contributed 2.9%. In the mixed ponds, mola contributed 2.7 and other small fish species 13.5% of the total fish production.

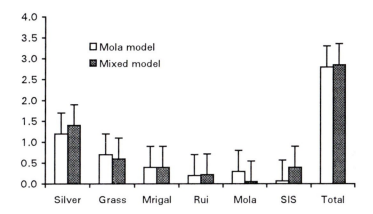

Fig. 17.1. Production of normally cultured carp, mola and other SIS in a 7-month growth season in mola ponds and mixed ponds.

A summary of the economics of the fish production in the two production models is presented in Table 17.2. SIS contributed about 15% of the net value of the fish production in both models. The investment in mola broodstock was compensated for by a higher market value of mola compared to most of the other SIS. The average economics of the two production models were similar and from a production aspect both models can be recommended for further innovation.

Table 17.2. The average economy for mola ponds (*n* = 34) and mixed ponds (*n* = 25) in the field trial.

Fish species	Expenditures (US$0.04 ha⁻¹ per season)		Net value of harvest (US$0.04 ha⁻¹ per season)	
	Mola ponds	Mixed ponds	Mola ponds	Mixed ponds
Silver carp	11	12	15	20
Other normally cultured carps	9	9	28	23
Total normally cultured carps	20	21	43	43
Mola	4	0	5	2
Other SIS	0	0	3	8
Total SIS	4	0	8	9

Rotenone should be avoided in future innovation of the technology as it is costly and has a limited effect, as wild fish often re-enter the pond. High biomass of wild fish can be reduced by repeated netting rather than by use of rotenone.

The majority of the farmers wanted to continued the normally cultured carp-SIS polyculture, and 44 of the original 59 farmers continued culturing the pond, indicating that the technology was acceptable to the farmers (Table 17.3).

Table 17.3. Numbers of farmers from the 1997 trial continuing fish farming with normally cultured carp-SIS polyculture in 1998.

1997: Production model	Farmers in 1997 field trial (number)	Farmers wanting to continue to culture the pond in 1998 according to the 1997 production models (number)	Farmers successfully culturing the pond in 1998 (number)	Farmers changing the production model in 1998 (number)
Mola	34	30	24	1
Mixed	25	24	20	0
Total	59	54	44	1

Nutritional Impact

There were no differences in the amount of fish consumed in households with mola ponds, mixed ponds and control households without access to a fish pond. The data from all surveyed households were therefore pooled. Fish bought in the market was the most important source, followed by captured wild fish (Table 17.4). Fish cultured in the pond was a minor source in terms of the amount of fish consumed as well as the frequency of consumption. The insignificant consumption of fish from household ponds explained why the control households without access to a pond consumed the same amount of fish as fish-producing households.

Farmers participating in the trial were restricted to harvesting the SIS only once monthly for practical reasons, which also restricted the use of SIS from the pond for home consumption since the fish must be consumed on the day of harvest. Excess SIS harvested were typically sold in the local market. Fish species reproducing in the pond are suitable for more frequent harvesting and can potentially lead to a higher home consumption than found in this survey. The potential contribution of the total mola and silver carp production from a small seasonal pond to the recommended intakes of vitamin A of a family of six persons is presented in Table 17.5. Puti is also included as it is one of the most commonly consumed SIS and is often found in seasonal ponds. Mola production from a small seasonal pond can contribute significantly to the recommended vitamin A intake of a family, whereas silver carp and other species can contribute only insignificantly. All fish species are nutritious food but mola has the added advantage that it is an exceptionally rich source of vitamin A. In general, all SIS are a good source of Ca.

Table 17.4. Sources of fish for consumption in three seasons in 84 households. The relative importance of sources is shown as % of total amount. The frequency of obtaining fish from the different sources is shown as number of batches of fish obtained in all surveyed households during the 5-day survey period.

	Season		
Source	July	October	February
Market			
Total amount (%)	68	57	69
Frequency	350	393	349
Capture fisheries			
Total amount (%)	30	37	16
Frequency	248	235	62
Own pond			
Total amount (%)	1	6	11
Frequency	3	41	38
Gift			
Total amount (%)	1	1	4
Frequency	14	8	11

Table 17.5. Example of potential contribution from fish production of silver carp, mola and puti in a small seasonal pond to the recommended intakes of vitamin A for a family of two adults and four children. The production is calculated from the average production rate during a 7-month growth season in a 0.04ha pond.

	Production potential in a 0.04 ha seasonal pond (kg)	Coverage of recommended intake of vitamin A per year in a family of two adults and four children (%)	Value (US$)
Silver carp	52	< 1	18
SIS:			
mola	11	20	5
puti	11	< 1	5

Conclusions

Some practical points must be considered in order to successfully culture mola and other SIS with normally cultured carp in small seasonal ponds:

- Good quality mola brood fish must be available for the farmers — guidelines for management of local brood fish ponds in the dry season should be developed.

- Well-developed guidelines for pond management are necessary — more knowledge regarding stocking combination and stocking density is needed.
- Simple methods for estimation of the biomass of SIS in a pond are needed.

There seems to be great potential for integrating mola and other SIS in the further development of rural aquaculture in Bangladesh. This new approach in aquaculture can play an important role in ensuring improved food and nutrition security for the rural poor and increased intake of vitamin A, Ca and other essential nutrients.

References

Ahmad, K. and Hassan, N. (1983) *Nutrition Survey of Rural Bangladesh 1981-82.* Institute of Nutrition and Food Sciences. University of Dhaka, Dhaka. 231 pp.

Anonymous (1995) *Potential Impact of Flood Control on the Biological Diversity and Nutritional Value of Subsistence Fisheries in Bangladesh.* Bangladesh Flood Action Plan, Environmental study, FAP 16. Irrigation Support Project for Asia and the Near East (ISPAN), (1-1)-(8-4). ISPAN Technical Support Center, Virginia.

Felt, R.A., Rajts, F. and Akhteruzzaman, D. (1996) *Small Indigenous Fish Species Culture in Bangladesh.* Technical Brief, May 1996. Integrated Food Assisted Development Project (IFADEP). Project ALA/92/05/02. Dhaka, Bangladesh. 41 pp.

Hansen, M., Thilsted, S.H., Sandström, B., Kongsbak, B.K., Larsen, T., Jensen, M. and Sørensen, S.S. (1998) Calcium absorption from small soft-boned fish. *Journal of Trace Elements in Medicine and Biology* 12, 148-154.

Helen Keller International/Asian Vegetable Research Center (1994) *Home Gardening in Bangladesh.* Helen Keller International, Dhaka. 9 pp.

Kent, G. (1997) Fisheries, food security, and the poor. *Food Policy* 22(5), 393-404.

Roos, N., Islam, Md.M., Thilsted, S.H., Md. Mursheduzzaman, M.A., Moshin, D.M. and Shamsuddin, A.B.M. (1999) Culture of mola (*Amblypharyngodon mola*) in polyculture with carps – experience from a field trial in Bangladesh. NAGA, *The ICLARM Quarterly* 22(2), 16-19.

Thilsted, S.H., Roos, N. and Hassan, N. (1997) The role of small indigenous fish species in food and nutrition security in Bangladesh. NAGA. July-December, pp. 82-84.

Zeitlin, M.F., Megawangi, R., Kramer, E.M. and Armstrong, H.C. (1992) Mothers' and children's intakes of vitamin A in rural Bangladesh. *American Journal of Clinical Nutrition* 56(1), 136-147.

Chapter 18

Gender Division of Labour in Integrated Agriculture/Aquaculture of Northeast Thailand

S. Setboonsarng

Aquaculture and Aquatic Resource Management, School of Environment, Resources and Development, Asian Institute of Technology, PO Box 4, Klong Luang, Pathumthani 12120, Thailand

Abstract

Integrated agriculture/aquaculture has been promoted in Northeast Thailand as an alternative to conventional agriculture but is faced with a labour shortage due to out-migration from the region. Since the financial crisis in July 1997, rural labour has again become abundant and there has been a renewed interest in promoting this form of agriculture as an answer to the current social and economic problems in Thailand. This chapter reviews the gender division of labour in the main activities of integrated agriculture/aquaculture ($n = 75$) and non-integrated farms ($n = 25$) in Buriram province, Northeast Thailand. There was a fairly well defined gender division of labour in all agricultural activities on both types of farms. The overall intensity of labour use was higher on integrated than on non-integrated farms. Hours of work increased both for women and men on integrated farms but increased more for women than men for all activities. The role of women in fish culture increased over time, especially in activities of feeding and pond maintenance. The status of women on integrated farms was enhanced by their increased access, participation and control over production resources. On the other hand, their workloads also increased significantly. The study revealed that women were more associated with on-farm recycling and environment enhancing activities. This implies that efforts to empower women through integrated farming will at the same time enhance the ecological sustainability of the system. Giving priority to the involvement of women in research, training and extension may be the key to the successful sustainable development of integrated agriculture/aquaculture in Northeast Thailand.

Introduction

In the predominantly risk-prone, resource-poor and rainfed ecosystems of Northeast Thailand, integrating pond aquaculture into existing farming practice has received widespread acceptance among the farmers of the region. Integrated agriculture/aquaculture offers farmers a more stable basis for an agricultural livelihood as diversifying production provides food security and minimises environmental and market related risks. Before the financial crisis struck in July 1997, several governmental and non-governmental organisations embarked on major extension projects but found that one of the main constraints to wider adoption was shortage of farm labour due to out-migration from the region. Since the financial crisis, the decline in employment opportunities in the service and industrial sectors has forced numbers of workers to return to Northeast Thailand. Consequently, there has been a renewed interest in promoting the integration of agriculture/aquaculture as an option to absorb labour and to enhance the livelihoods of the farmers in the Northeast region.

Integrated farming in this paper refers to a practice centred upon the farm pond where aquaculture is practised in combination with livestock rearing and vegetable and/or fruit cultivation. Pond water is used to water vegetables and fruit trees which are grown on the pond bank and is also used for raising livestock. The manure from livestock is used to fertilise the fish pond and cultivate the field.

The practice of integrated agriculture/aquaculture in Northeast Thailand is, for the most part, based on empirical wisdom of local farmers. Scientific understanding of the integrated system has been limited to micro-studies of the biological and technological aspects of the fish subsystem and its utilisation of livestock manure. There have been limited attempts to relate the micro-level biological understanding to the larger social and economic environment in which the integrated resource systems function. Despite the initial success of technological progress in pond fish production in terms of increasing fish availability in the rural households of Northeast Thailand, a number of questions on the socio-economic implications of integrating aquaculture into the existing farming system remain unanswered.

To contribute to the understanding of socio-economic aspects of integrated agriculture/aquaculture, this chapter is an attempt to understand one of the most important aspects of resource use in integrated farming systems, namely the implication on the gender division of labour in introducing a new element into the existing farming practice. This chapter uses gender analysis tools to explore the relationship of women and men in integrated farms and the inequalities in those relationships. Along with a detailed comparative analysis of gender division of labour in each subsystem of integrated and non-integrated farms, the paper aims to provide an understanding of how, if at all, aquaculture can contribute towards improving gender equity in the rural households of Northeast Thailand. Most importantly, it is an attempt to highlight the need to reconcile growth with equitable distribution and balance the partnership between men and women in aquaculture development.

Since 'gender and development' is a relatively new focus in the social science of development in recent years, the chapter begins with a brief review of the concept of gender and its evolution in development thinking for the benefit of natural scientists. The chapter then provides a short description of integrated agriculture/aquaculture practice in Northeast Thailand. This is followed by a description of the study, the findings and their relevance.

What is Gender?

Gender is a concept that deals with the roles and relationships between men and women which are determined by the social, political and economic contexts and not by biology. Gender is not a mass social movement that aims to divide men and women and cause conflicts that are not in existence. Rather it brings together those issues that have brought about unequal relations between men and women and draws the attention of development workers to address these with appropriate measures that will help to change rather than perpetuate them (ICIMOD, 1998).

Women in Development and Gender and Development

The movement 'women in development' (WID) which seeks to integrate women into the development process began in the 1970s after Ester Boserup, in her work on 'Women's Role in Economic Development', pointed out that women are major contributors to community productivity, particularly in agriculture, but that their contribution is not recognised in development efforts. She also pointed out that women are being displaced by modernisation and that their traditional productive functions, income and status are diminishing.

With this recognition, several studies were commissioned and, in general, showed that women had 'fared less well' from development efforts. A number of development projects were initiated to better integrate women into development activities. These projects tended to be 'welfare' oriented towards women and were designed based on the assumptions that gender relations will change naturally as women gain more economic power. However, the results of most projects showed that individual gain did not translate into overall betterment of women as a gender (March, 1996).

The 'gender and development' (GAD) approach emerged as an alternative to WID in the 1980s. Contrary to the WID approach, in which women were viewed as passive recipients of assistance, the GAD approach views women as active participants in development. It does not focus on women *per se* but on the social construct of gender and assignment of specific roles and responsibilities. The GAD approach critically examines and analyses the underlying assumptions of the current social, economic and political structures and seeks to bring about positive changes toward equitable participation in sustainable development (Moser and Levy, 1986).

Analytical Framework

An analytical framework sets out different categories of elements to be considered in analysis. As each framework selects a limited set or number of key issues for analysis out of a large number of factors that actually influence the situation, different frameworks have different assumptions, strengths and weaknesses. To attain a better understanding of the complex setting of gender issues in aquaculture, two formal frameworks, namely the Harvard Analytical Framework (Overholt *et al.*, 1985) and the Women's Empowerment Matrix (Longwe, 1991) were used in this chapter.

The Harvard Analytical Framework was one of the first frameworks designed for gender analysis. It was developed by researchers at the Harvard Institute of International Development for collecting information at the micro (household) level. The framework has four interrelated components: the activity profile; the access and control profile; the analysis of influencing factors and the project cycle analysis. By mapping out activities by gender, factors which shape gender relations and provide different opportunities and constraints for men and women can be traced out and improvement in overall productivity can be addressed (Carr *et al.*, 1996).

The Women's Empowerment Matrix Framework was developed by Sara Hlupekile Longwe in her work in Zambia. The framework aims to help one think through to what extent intervention is required to enable women to participate equally in achieving control over the factors of production on an equal basis with men in the development process. In this framework, increased equality and increased empowerment are evaluated by tracing women's welfare, access, consciousness, participation and control.

To gain insight into gender relations in farming practice in Northeast Thailand, this paper used the Harvard Analytical Framework to identify and organise information about gender division of labour in the form of activity profiles of farming activities for integrated and non-integrated farms. By mapping out the time allocation of men and women in different activities, analysis on access to, and control of, the use of resources were identified and listed. The paper then uses the Women's Empowerment Matrix model to trace women's equality and empowerment. The weakness of the Harvard Analytical Framework in not addressing power relationships is therefore addressed through the use of the Women's Equality and Empowerment Matrix Framework.

Gender Analysis in Aquaculture Development

Introducing aquaculture into the existing, predominantly rice-based farming system naturally brings about changing responsibilities and labour allocation of men and women in the household. Both men and women work to maintain the livelihood of their family in agricultural households. Their work tends to differ in nature and value. Several studies on gender in agriculture have shown that women usually concentrate on time-consuming and low productivity tasks whereas men concentrate on heavy labour tasks (Siwi *et al.,* 1989;

Chandrapanya *et al.*, 1990). A study in Chieng Mai, Thailand showed that land preparation and application of chemicals were done exclusively by men, while women's labour dominated crop activities such as planning, care, maintenance and harvesting. In animal production, collection of fodder, feeding the animal, providing water, collecting animal manure and putting the animal out to graze were done by women (Shinawatra *et al.*, 1986). While a considerable amount of work has been carried out on gender analysis in agriculture and fisheries, only a limited amount of work has been carried out on aquaculture in the context of the farming system.

Pritchard (1992), in his review of gender roles in aquaculture in 47 farm households in the province of Srisaket in Northeast Thailand, found that aquaculture was a male-dominated activity. In all activities from decision-making, fry purchasing, fish feeding to fish harvesting, the husband performed these tasks in over 80% of the households in the survey.

Gender and aquaculture issues were touched upon in the Asia-Pacific Regional Workshop on Gender Issues on the Participation of Women in Fisheries Development (Bueno, 1997). The workshop viewed problems of women in aquaculture to include being limited to mechanical, menial tasks; limited access to production resource, i.e. credit and land; and problems of lack of involvement of women in training and extension activities. Similar findings were discussed in the Seminar on Women in Fisheries in Indo-China Countries (Nandeesha and Hanglomong, 1997).

Gender analysis in aquaculture will provide a focused framework for making visible the roles of men and women in aquaculture practices in Northeast Thailand. It will enable researchers to identify who performs each particular task, who would be more responsive to different technologies to be introduced, and who would be negatively affected by the introduction of a particular technology.

In developing aquaculture technology to be introduced to rural households, aquaculture scientists should know at the onset of the technology development process who will be the 'user'. Scientists should know who influences the improvement of aquaculture, whose labour will be displaced, whose resources will be affected, and on whose initiative is the technology accepted by the farming household?

Knowing the user and beneficiaries embodies both equity and efficiency implications. It increases the efficiency of aquaculture research through better targeting and better specification as the patterns of activities and resource use are taken into account.

Integrated Agriculture/Aquaculture in Northeast Thailand

Northeast Thailand covers an area of 168,854km^2 and is home to approximately 20 million people. It is an area characterised by poor soils, erratic rainfall and degrading environmental conditions due to deforestation. Approximately one third of the land area is unsuitable for agriculture. From a total of 8.8 million ha of agricultural land, two thirds are under rice cultivation with the remaining third under field crops (Donner, 1978). Rice yields in Northeast Thailand are typically the lowest in the country. Important field crops in the area include sugarcane, cassava and *kenaf*. Irrigated land accounts for only 567,045ha or about 2% of agricultural land.

Traditionally, the farming system in Northeast Thailand was a rainfed, single crop of rice. In the 1970s and 1980s, massive land expansion took place and farmers were encouraged by the government to diversify into field crops, i.e. cassava and sugarcane, as well as livestock and vegetables. Availability of water always plays a critical role in determining the possibility of growing vegetables and raising livestock (Wigzell and Setboonsarng, 1995). The important livestock in the rural households of the region are buffalo, cattle and poultry. Traditionally, wild fish collected from rice fields and other water bodies were the main sources of animal protein for the rural population. With the decline in the wild fish populations due to population pressure and environmental degradation, aquaculture has been added to the existing farming system (Edwards *et al.*, 1996).

Natural resources constitute the basis of the rural livelihood systems of Northeast Thailand and hold the key to increased food production, and equitable and sustainable development. Thirty years of agriculture development following the path of diversification into export crops and intensification of production through the use of modern inputs has left farmers with increased environmental problems and increased debt. With the existing resource being only marginally suitable for agriculture and rapidly degrading, poverty has become a persistent problem of the region.

In recent years, an alternative agriculture practice has been introduced to the region. The combination of rice, fish, vegetables, fruit trees and livestock activities centred around the farm pond, known as 'integrated agriculture/ aquaculture farming system' or '*kaset-pasom-pasan*' in Thai has been spreading in the region. In integrated agriculture/aquaculture practice, integration of various sub-systems increases productivity and enhances environmental quality. With the success of pioneer farmers in diversifying production and marketing risks and providing food security for the farming household, integrated farming is seen as offering a viable alternative (Wigzell and Setboonsarng, 1995).

The success of several farmers was advocated by several non-governmental organisations (NGOs) and, starting about a decade ago, ponds were made available to some farmers through different development programmes funded by both the government and NGOs. Most ponds in Northeast Thailand were not built for fish but rather for water storage for small-scale irrigation and for household use. The main source of pond water is rainwater. At the beginning of the rainy season when the pond is filled with water and fish seed are made available through private seed sellers, most farmers stock fish in their pond. The purposes of growing fish are for home consumption, for feast and for sale. Vegetables, bananas and fruit trees are other important components of the integrated system. These crops are planted on the pond dike or nearby area, utilising pond water as source of irrigation. Vegetable and fruit cultivation are also for consumption and for sale. For many households, vegetables are the highest cash income earning activity.

Livestock is an important component of the farm household economy in Northeast Thailand. Almost every farm household keeps some ruminants and a small number of chickens. Raising pigs and poultry for sale at the household level has long been taken over by feedlot livestock production. However, due to a taste preference for native chickens, most farmers do keep a limited number of local breed chickens.

The Integrated Agriculture/Aquaculture Farming System Survey

There has been little understanding of the socio-economic characteristics of the integrated farming household in Northeast Thailand. To increase understanding of the role of small-scale aquaculture in integrated farming systems of Northeast Thailand, a survey was carried out in 1994 on 148 households in two provinces, Sakol Nakorn and Buriram, 37 non-integrated farms (without an aquaculture component) and 111 integrated agriculture/aquaculture farms with ponds of various ages, i.e. over 10 between 4-9 and 0-3 years. Since activities of integrated farms centre around the farm pond and the nature of activity differs over time, pond age is a good determinant of the different stages of farming system development. In Buriram province, preliminary information for household selection was gathered from local NGOs and the Rajapak Institute of Buriram. In Sakol Nakorn, the local officials of the Royal Initiated Agriculture Project assisted in providing a list of farmers for selection.

From preliminary data analysis, it was found that integrated farms in Sakol Nakorn province exhibited a limited level of integration while farms in Buriram province showed a high degree of integrated practice. This may have been because several NGOs were actively involved in integrated farm promotion in Buriram whereas none were present in Sakol Nakorn. Therefore, for this study, a sub-sample of 100 farms (25 non-integrated and 75 integrated) in Buriram only was used in the analysis.

This study investigated the gender division of labour of productive work, i.e. agricultural activities, to produce outputs for consumption and for sale. The study did not include two other categories of work women are involved in, i.e. reproductive work which refers to care and maintenance of the household and its members, bearing and caring of children, food preparation, etc. and community work which involves the collective organisation of social events such as services, ceremonies, community improvement activities, and participation in local groups and organisations. Since the original research objectives were to assess the socio-economic conditions of the households in general, specific measurement of gender variables, i.e. women's consciousness, participation and control, were not explicitly included.

Results and Discussion

Household Characteristics

The average size of the household in the study areas was relatively small at 4.2-4.3 people; with were no differences between non-integrated and integrated farms (Table 18.1). Comparison of the age of the household's head between the two groups showed that households practising integrated farming were older than households of non-integrated farms. However, as households practising integrated farming were originally sampled based on the age of their ponds, the group with ponds older than 10 years old may have caused the mean age of the integrated farmers to be higher. Concern that only older farmers were adopting integrated farming was not justified; rather the data showed that the population that remained in the agricultural sector in Northeast Thailand was an age group of over 40 years old. There were no differences in the number of children living away from home among the two groups showing that the higher labour requirement in integrated farming did not prevent the children from migrating to find work in the industrial and service sectors outside the region.

Table 18.1. Household characteristics.

	Non-integrated farms (*n* = 25)	Integrated farms (*n* = 75)
Total household members (number)	4.3	4.2
Age of father (years)	45.9	51.4
Age of mother (years)	41.7	48.6
Children living elsewhere (number)	1.0	1.0
Dependency ratio	1.0 : 0.6	1.0 : 0.3

Dependency ratios of both groups of farm households were relatively low. The slightly higher dependency ratio in the non-integrated farm group showed that 1 family member in the labour force was supporting 0.6 dependent persons compared to 0.3 dependent persons in the integrated farm group.

Education

Only primary education is compulsory in Thailand and the majority of farmers and their wives had attained this level of education. The survey showed that education levels of integrated farmers, both the father and the mother, were higher than those of non-integrated farmers (Table 18.2). Among the household heads of integrated farms, 18.5% of farmers attained secondary level and 6.3% attained a level above secondary school. On the other hand, 13.3% and 3.3% of household heads in the non-integrated farm group attained levels of secondary, and above secondary, respectively. The higher educational attainment of the integrated farm group was found amongst both men and women.

Table 18.2. Educational levels.

	Total households (%)	
Educational level	Non-integrated farm (n = 25)	Integrated farm (n = 75)
Education of father		
- None	0	0
- Primary	83.4	75.3
- Secondary	13.3	18.5
- Above secondary	3.3	6.3
Education of mother		
- None	5.3	6.1
- Primary	91.2	82.1
- Secondary	4.5	10.4
- Above secondary	0	1.4

When comparing the level of education between men and women, it was clear that the level of men was higher than that of women. There were no men in the category of no education while 5.3% and 6.1% of the women in non-integrated and integrated farms did not receive any education, respectively. There were also no women in the non-integrated farm category with education above secondary school level.

Land Holdings

Land holdings of integrated farms were larger than those of non-integrated farms with an average of 34.6rai (1rai = 1,600m^2) for integrated compared to 24.1rai for non-integrated farms. The percentage of land under rice cultivation was also higher in integrated farms despite the greater diversity (Table 18.3).

Table 18.3. Land holding and farming practice.

	Non-integrated farm	Integrated farm
Average land holding (rai)	24.1	34.6
Total subsystems[1] (number)	3.2	6.0
Land under rice (%)	66.6	71.5

[1]subsystems: rice, field crop, vegetable, banana, trees, livestock and aquaculture. 1 rai = 1,600m^2.

For non-integrated farms the average number of subsystems on the farm was far smaller at 3.2 subsystems. These systems were in general rice, field crops and livestock. On the other hand, the average number of subsystems on integrated farms was almost double. In addition to rice, field crops and livestock, water made available through the pond made it possible for farmers to grow vegetables, bananas, a few fruit trees and to practice aquaculture. Sources of income for integrated farms were also much more diverse than for non-integrated farms (Fig. 18.1).

Non-integrated farms   Integrated farms

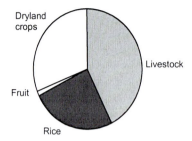
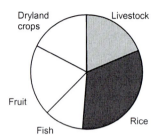

Fig. 18.1. Income profiles on non-integrated and integrated farms.

Labour Use in Rice Cultivation

Farms in Northeast Thailand are generally rice-based. Due to the nature of the rainfed ecosystem, a single crop of rice is grown in the region. Being the most important crop of the farming system, rice occupies over 60-70% of cultivated land and the labour use pattern in rice cultivation is of major importance. Labour use in rice cultivation reaches peaks during the rice transplanting season at the beginning of the rainy season and during the harvesting season.

The labour allocation pattern among different activities in rice cultivation showed a similar pattern in both non-integrated and integrated farms (Fig. 18.2). All the labour intensive activities, i.e. rice transplanting, manure application, weeding and harvesting, were dominated by women. In particular, rice transplanting was exclusively performed by women. Men dominated in more heavy labour activities, i.e. bund preparation, land tilling, and pest control using pesticides.

The study showed that farmers on integrated farms worked more intensively on their farms than those on non-integrated farms. Total labour inputs of males and females in rice cultivation for integrated farms were 84.4 and for non-integrated farms were 64 h rai^{-1}.

For gender-specific labour division in rice, women's hours were much higher than men's on both integrated and non-integrated farms. For non-integrated farms, women's work totalled 47.2 h rai^{-1} compared to 16.8 h for men. For integrated farms, women worked even longer hours, 58.1 h rai^{-1} compared to 26.3 h for men.

Due to the risk-prone nature of the region, rice cultivation was practised in an extensive manner. The survey showed that farmers did not apply chemical fertilisers in their rice field and only buffalo manure was used. Application of manure to the rice field was mainly carried out by women on both types of farms. With a slightly higher degree of intensity in production, the men of integrated farms did show more involvement in the activities of maintaining the rice field as they engaged in weeding, pesticide/herbicide application and other pest control activities.

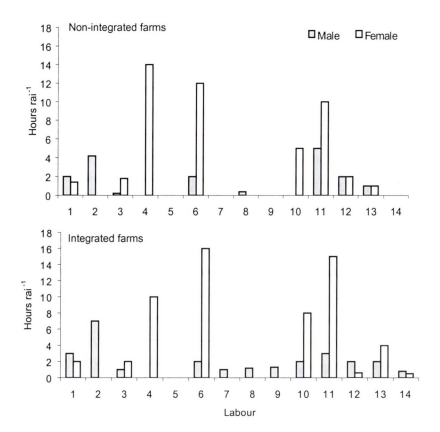

Fig. 18.2. Labour use pattern in rice cultivation on non-integrated and integrated farms. 1rai = 1,600m². Note: 1 = bund preparation; 2 = land till; 3 = seed bed preparation; 4 = transplanting; 5 = fertiliser application; 6 = manure application; 7 = pesticide application; 8 = other methods for pest control; 9 = herbicide application; 10 = weeding; 11 = harvesting; 12 = threshing; 13 = milling; 14 = marketing.

Labour Use in Fish Culture

By definition, non-integrated farms were farms without farm ponds and therefore without fish culture activity. Therefore, the analysis of labour use pattern in fish culture compared farms with experience in fish culture of less than three years and farms with experience over three years to assess the changing roles of males and females as the households gained more experience in fish culture.

Fish culture was clearly a male-dominated activity (Fig. 18.3). While participation of women in fish culture increased with time, even after 3 years of experience, fish culture remained a male-dominated activity. For farms with less than 3 years of experience in fish culture, fish fry purchasing and fish stocking were tasks performed exclusively by men. This may imply that decision-making in fish stocking was usually made with limited participation of women. For farms with more than 3 years experience, fry purchasing was still an exclusive male activity but women's participation in fish stocking increased. Increased participation of women over time was most clear in feeding fish using off-farm feed as it rose from 0.2 to 3.2 h rai^{-1}. In general, participation of women increased in all activities from conserving/repairing the dike, application of manure, purchasing off-farm feed, collecting and preparing feed, feeding, and harvesting, to marketing of fish. As fish culture requires only light labour inputs and the fish pond was usually situated not far from the house, women's participation increased sharply as women acquired the techniques. The limited participation of women in the early years of fish culture may be because training in fish culture, like training in agriculture, was so far given only to men.

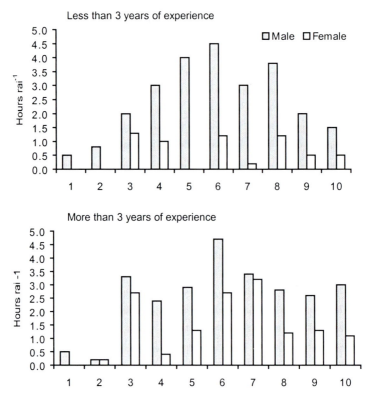

Fig.18.3. Labour use pattern in fish culture on farms with less than and more than 3 years experience in fish culture. 1 rai = 1,600m^2. (Note: 1 = purchasing fry; 2 = stocking; 3 = conservation/repair of pond; 4 = application of organic fertiliser; 5 = buying off-farm feed; 6 = collecting and preparing on-farm feed; 7 = feeding off-farm feed; 8 = feeding on-farm feed; 9 = harvesting; 10 = marketing).

While there was an increase in women's participation in fish culture activities, inputs of male labour did not decline over the years. There was a small replacement of female by male labour in the application of manure, buying off-farm feed, and feeding with on-farm feed. However, the total time allocated to fish culture increased over time from 32.8 to 39.9 h rai^{-1}. Total hours of male labour increased from 25.1 to 25.8 h while total hours of female labour increased from 7.7 to 14.1 h rai^{-1}. This implies some degree of intensification of the aquaculture system over time.

Labour use was most intensive to collect and prepare on-farm feed. For farms with less than 3 years experience, feed related activities constituted 72% of the time spent by men and 46% of the time spent by women. For farms with over 3 years experience, feed related activities constituted 62% of both male and female hours. At the same time, there was a sharp increase in hours spent feeding fish with off-farm feed on farms with more than 3 years experience, in this case pelleted feed. This may due to the promotion of hybrid catfish culture using pelleted feed by industrial feed mills, which was going on in the area. The replacement of labour by capital inputs was a common phenomenon in the labour-scarce economic system of Northeast Thailand. While the use of pelleted feed resulted in a lower degree of integration of fish with other subsystems, it did show that as farmers gained confidence in fish culture, they fed fish more and were willing to invest more in purchasing off-farm feed for their fish.

Fish marketing, which is generally known to be a women-dominated activity, was shown here to be a male-dominated activity, the reason being that on most of these farms, fish were marketed at the pond bank immediately after harvesting. This may be because live cultured fish were most preferred by consumers who were usually neighbours. Since fish harvesting was a difficult task for women, the men who carried out the harvesting activity usually performed the marketing activities as well. Whether the men gave the money to the women appeared to vary with household. A small portion of fish that could not be sold at the pond bank would be taken for sale to the market, usually by women. When fish were sold by women, they usually also kept the money and so control over benefits of fish cultured stayed with the women.

Labour Use in Banana Cultivation

As households diversified from rice to include other fruit trees, several kinds of trees were introduced. The most commonly cultivated fruits in the study area were banana, papaya, mango and tamarind. As banana was the most commonly grown, time allocation for banana growing was studied. Banana was usually grown on the pond dike or in some cases, on broadened bunds of the rice field. The overall pattern of labour use for banana cultivation did not differ significantly between non-integrated and integrated farms (Fig. 18.4). Total labour use by men was lower than that for women in both non-integrated and integrated farms. On non-integrated farms, the ratio of inputs of men to women was 43 to 61 h rai^{-1} whereas on integrated farms, it was 53 to 75.2 h rai^{-1}. The study showed a similar pattern of higher labour intensity on integrated than on non-integrated farms. The increase in labour on integrated farms was, again, higher for women than for men. The activities in which women's labour increased were land preparation, watering plants, weeding, fertiliser application, harvesting and marketing.

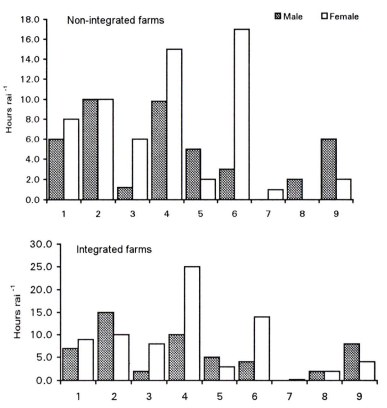

Fig.18.4. Labour use pattern in banana cultivation on non-integrated and integrated farms. 1rai = 1,600m². (Note: 1 = land preparation; 2 = planting; 3 = weeding; 4 = watering; 5 = fertiliser application; 6 = manure application; 7 = pesticide application; 8 = harvesting; 9 = marketing).

Labour Use in Vegetable Cultivation

Cultivation of vegetables was the most labour-intensive activity among the various subsystems of integrated farms. Labour use per rai for vegetable cultivation was more than double that of banana and rice cultivation. Total labour inputs by gender for vegetable cultivation on non-integrated farms was 93 h of male and 80.7 h of female labour. For integrated farms total hours of male labour were slightly lower at 92.4 but were significantly higher for females at 121.9, with a similar pattern of increase in female over male hours (Fig.18.5).

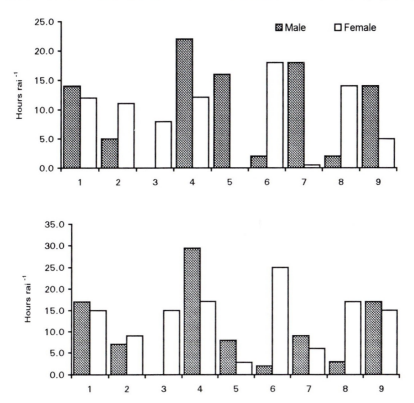

Fig.18.5. Labour use pattern in vegetables on non-integrated and integrated farms. 1 rai = 1,600m². (Note: 1 = land preparation; 2 = planting; 3 = weeding; 4 = watering; 5 = fertiliser application; 6 = manure application; 7 = pesticide application; 8 = harvesting; 9 = marketing).

Due to the short growing cycle, vegetables were a good cash earning activity for the households. Vegetables often competed with fish for water from farm ponds and were often grown on a significant scale only in areas where the water table was high or there were alternative sources of water besides the rainfed pond. The major vegetables grown in the study area were yard-long bean, tomato, chillies, cucumber and onion.

The study found the gender relationship pattern to be the same in all activities of vegetable production on both non-integrated and integrated farms. Activities dominated by men were land preparation, watering, fertiliser application, pesticide application and marketing. Activities dominated by women were planting, weeding, manure application and harvesting. However, the degree of dominance differed on the two types of farms. Domination of women in manure application, weeding and harvesting was greater on integrated farms. On the other hand, male domination in fertiliser and pesticide application was more pronounced on non-integrated farms.

There appeared to be a replacement of chemical fertilisers with animal manure on integrated farms. The total of 8 hours spent in fertiliser application by men on integrated farms was much lower compared to 16 h on non-integrated farms. However, the time spent by women in manure application increased from 18 h on non-integrated to 25 h on integrated farms.

Pesticide application by men was also reduced from 18 h on non-integrated to 9 h on integrated farms. Farmers explained during the field surveys that as most vegetables were grown near fish ponds or on pond dikes, they were more careful in pesticide application from fear of killing the fish and therefore reduced the amount of pesticide used.

Labour Use in Buffalo Raising

Livestock was one of the most important components of the farming systems of Northeast Thailand. Although most surveyed households kept some ruminants and a small number of chickens, this study focused only on the gender division of labour in buffalo raising due to the difficulties in data collection for chicken raising.

The overall intensity of labour use in buffalo raising exhibited a similar pattern to other subsystems in which labour inputs were higher on integrated than on non-integrated farms. Patterns of gender division of labour were similar on both types of farms (Fig. 18.6). Women's labour clearly dominated in feeding and providing water for buffaloes on both types of farms. Activities of stocking, collecting and preparing on-farm feed, cleaning buffaloes, cleaning and maintaining their stalls were shared by both genders. Men's labour input was prominent in marketing animals but men's labour inputs dominated only in pest and disease control using chemicals on integrated farms.

Female Labour in Non-IAA and IAA Farms

Female labour use in different farming activities on both types of farms is summarised (Fig. 18.7). Female labour use per rai was higher in all activities on the integrated farm and totalled 691 h rai^{-1} on integrated compared to 549 h rai^{-1} on non-integrated farms. In other words, the women of integrated farming households worked about 18 man-days longer than those from non-integrated farming households.

This suggests that women's participation may hold the key to the success of integrated farming practice. Women had more access on integrated farms to income generating activities such as vegetable and banana production and also fish culture. Women earned additional income by selling vegetables and fruits raised on small plots near the pond, as well as sale of fish.

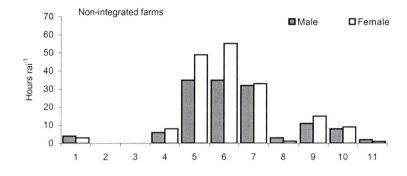

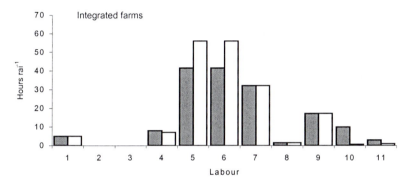

Fig.18.6. Labour use pattern in buffalo raising on non-integrated and integrated farms. 1rai = 1,600m². Note: 1 = stocking; 2 = buying and preparing off-farm feed; 3 = feeding off-farm feed; 4 = collecting and preparing on-farm feed; 5 = feeding on-farm feed; 6 = providing water; 7 = cleaning animals; 8 = cleaning stalls; 9 = maintaining stalls; 10 = pest and disease management (chemicals); 11 = marketing.

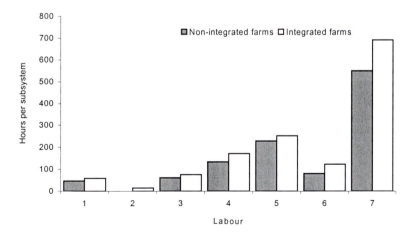

Fig. 18.7. Female labour on non-integrated and integrated farms. Note: 1 = rice; 2 = fish; 3 = banana; 4 = buffalo; 5 = cattle; 6 = vegetables; 7 = total.

Decision-Making in Farm Households

The pattern of decision-making differed to a large extent between non-integrated and integrated farms (Table 18.4). Decisions in non-integrated farm households were mainly made by the husband alone (56% of households), followed by joint decision-making between husband and wife (37.5% of households). In contrast, decision-making in integrated farm households was mainly made jointly by husband and wife (78.1% of households), followed by husbands alone (15.5% of households). Decision-making by the wife alone was not found in non-integrated farm households and was found in only one integrated farm household. The results showed that the greater participation of women in production activities on integrated farm households did translate into a slight increase in their control over resource allocation as reflected by the increase in joint decision-making.

Table 18.4. Decision-making in households.

Decision maker	Non-integrated farms (%)	Integrated farms (%)
Husband	56.3	15.5
Wife	0	0.1
Husband and wife	37.5	78.1
Husband and child	0	4.0
Whole family	6.2	2.2

Daily Food Consumption of Households

Welfare of the household in general, and in particular in relation to available food for household consumption, is generally of great concern to women. Frequency of daily consumption of food items on integrated and non-integrated households are compared in Fig. 18.8. The survey showed that except for sticky rice, all food items were consumed more frequently in integrated households, showing improved nutrition. Frequency of consumption of animal protein was also higher in integrated households. Fish were consumed daily in 46% of integrated but only in 29% of non-integrated farm households. Other forms of animal protein, namely chicken, duck and beef, were also reported to be consumed daily in 3-8% of integrated compared to 0% of non-integrated farm households.

Discussion

The study showed that women worked longer than men on both non-integrated and integrated farms in Northeast Thailand. Hours of work increased for both women and men but increased more for women than for men on integrated farms. Not only did women work longer than men in productive activities but, because of their triple role in reproductive and community as well as farming functions, their hours of work were extremely long. On the other hand, women also participated more in decision-making on integrated farms. Women's access to food, farm resources and control over resources also improved on integrated farms.

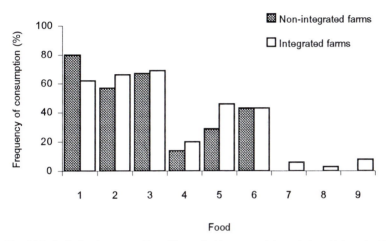

Fig. 18.8. Daily food consumption of households on non-integrated and integrated farms. Note: 1 = sticky rice; 2 = plain rice; 3 = vegetables; 4 = fruit; 5 = fresh fish; 6 = preserved fish; 7 = chicken; 8 = duck; 9 = beef.

The welfare impacts on women were inconclusive based on the statistics shown, as their working hours increased significantly in integrated households. In a number of interviews conducted with women on integrated farms during the field surveys, most expressed general satisfaction over the changes in their household since aquaculture had been integrated into the farming system. They were satisfied with the greater availability of food for the household and they gained higher self esteem in having more control over the use of farm resources.

The study also showed a fairly defined sexual division of labour in agriculture activities in Northeast Thailand. The overall intensity of labour use in rice production was higher on integrated than on non-integrated farms. Among activities in production of rice, women had more responsibility than men in rice transplanting, manure application, weeding and harvesting. Men were responsible for heavy labour activities including bund preparation, and land tilling. Men were also responsible for activities more hazardous to health such as pesticide application. The pattern of gender division of labour was also well defined in vegetable and banana production and in buffalo raising. A similar pattern of gender division of labour was found in Chiang Mai province in Northern Thailand (Shinawatra *et al.*, 1986).

Regarding the gender division of labour in fish culture, culturally, it was not surprising to find it a male-dominated activity. Fish culture has been recently introduced to Northeast Thailand and activities associated with fish were capture of wild fish which is traditionally a male-dominated activity. Furthermore, training in fish culture has so far only been given to men which may also be a factor for the limited participation of women in fish culture in the early years of aquaculture development. Women had not been properly trained in aquaculture although they were involved in agricultural activities.

Gender analysis in this chapter has made visible the roles of men and women in aquaculture practice in Northeast Thailand. The increased participation of women in fish culture activities, especially in feeding and pond maintenance, over the years suggests that technical knowledge on these activities should be provided to women. Training women in fish culture may be the key to successful aquaculture development of the region. In the case of IPM training in vegetable production in the Philippines, it was found that farmers' wives given training learned the technology faster than their husbands (WIRFS, 1986). To date aquaculture training sessions or extension programmes in Northeast Thailand have not been gender sensitive.

Conclusions

This study of the status of the gender division of labour in integrated agriculture/aquaculture and non-integrated households has provided considerable insight into the situation of women in integrated farming systems of Buriram province in Northeast Thailand. It has shown that women hold the key to the success of integrated farming practice in the region as they were major contributors to the practice. Their productive functions, income and status were enhanced through integrated agriculture/aquaculture practice. On the other hand, their workloads also increased significantly.

The study revealed the close relationship between women's roles and the ecology of the small-scale integrated farming system. Women were more associated with on-farm recycling and environmental enhancing activities such as improving soil fertility using organic manure, feeding animals using on-farm feed and weeding the field as opposed to using herbicide to kill weeds. Efforts to empower women through integrated farming would at the same time enhance the ecological sustainability of the system.

On the whole the study has shown that integrated agriculture/aquaculture practice can lead to sustainable agriculture and rural development in the rainfed agro-ecosystem of Northeast Thailand. There is an urgent need to integrate the needs of rural women in technology development and dissemination in the farming systems of Northeast Thailand to enhance the welfare of women, and thus the entire household, through better understanding by development workers and government officials.

Women's role in aquaculture became more prominent as the years of experience increased. Their role in aquaculture was observed to be complementary to their concern for food availability and well being of the household. It seems from the study that improved aquaculture management requires both male and female participation to be successful, without imposing particular constraints on the workload of the females. It can be concluded that on the basis of looking at the fish subsystem alone, aquaculture has brought about positive implications to gender equity of the households.

Gender equity can be further enhanced by involving women in research, training and extension in aquaculture which is still at the infancy stage in agriculture development in developing countries. In order to strengthen the development movement of women, institutional support to, and organisation of, women in some form of group at village level are essential. Gender analysis is only a tool that has to be followed by action plans to disseminate technology tailored to the needs of women engaged in agriculture, or more specifically, aquaculture production. When their access to resources, both farm resources and technical know-how, is assured, their status in both the family and society will improve.

The chapter has also shown that integrated agriculture/aquaculture could be a potential solution to the current social and economic problems of Northeast Thailand. The systems were shown to have a high capacity for absorbing rural labour; the systems also promoted on-farm recycling which enhanced the environment; and finally the systems were shown to provide the farmers and their families with a better supply of food. As women have been shown to be the major contributor to integrated farming, further efforts to promote the practice should focus on women.

Acknowledgements

The author would like to express sincere gratitude to Professor Peter Edwards for his continued support for this study; to DFID (former ODA), UK, for the financial support for this study; to the survey and data processing teams; and most importantly to the farmers of Northeast Thailand.

References

Boserup, E. (1970) *Women's Role in Development*. Earthscan, London.
Bueno, P.B. (1997) Gender issues on the participation of women in fisheries development. In: Nandeesha, M.C. and Hanglomong, H. (eds) *Proceedings of the Seminar on Women in Fisheries in Indo-China Countries*. Bati Fisheries Station-PADEK, Phnom Penh.
Carr, M., Chen, M. and Jhabvala, R. (eds) (1996) *Speaking Out: Women's Economic Empowerment in South Asia*. IT Publication, Women Ink, USA.
Chandrapanya, D., Kirithaveep, D., Viriyasiri, S., Sasiprapa, V. and Viratphong, S. (1990) Potentials of modernizing sericulture technology: a case study from Suphan Buri, Thailand. In: *Report of the Women in Rice Farming Systems Workshop, 4-8 June 1990, Puncak, Bogor, Indonesia*. IRRI, Los Banos.

Donner, W. (1978) *The Five Faces of Thailand: and Economic Geography.* Hamburg.

Edwards, P., Demaine, H., Innes-Taylor, N. and Turongruang, D. (1996) Sustainable aquaculture for small-scale farmers: need for a balanced model. *Outlook on Agriculture* 25(1), 19-26.

Edwards, P., Pullin, R.S.V. and Gartner, J.A. (1988) *Research and Education for the Development of Integrated Crop-Livestock-Fish Farming Systems in the Tropics.* ICLARM, Manila.

ICIMOD (1998) Gender and mountain development. Newsletter No. 29. Kathmandu, Nepal.

Longwe, S. (1991) Gender awareness: the missing element in the third world development project, In: Wallace, T. and March, C. (eds) *Changing Perceptions: Writings on Gender and Development.* Oxfam, Oxford.

March, C. (1996) *Concepts and Frameworks for Gender Analysis and Planning: a Tool Kit.* Oxfam, Oxford.

Moser, C.O.N. and Levy, K. (1986) A theory and methodology of gender planning: meeting women's practical and strategic gender needs. University College, London.

Nandeesha, M.C. and Hanglomong, H. (eds) (1997) Proceedings of the seminar on women in fisheries in Indo-China countries. Bati Fisheries Station-PADEK, Phnom Penh.

Overholt, C., Anderson, M.B., Cloud, K. and Austin, J.E. (eds) (1985) *Gender Roles in Development Projects: a Case Book.* Kumarian Press, Connecticut.

Pritchard, M. (1992) Gender for aquaculture development with rural households. AIT Aquaculture Outreach, unpublished.

Shinawatra, B., Diowwilai, C. and Bangliang, S. (1986) Evaluating the impact of technologies on women: a case study in Amphoe Phrao, Chiang Mai, Thailand. In: *IRRI Report of the Women in Rice Farming Systems Workshop 4-8 June 1990, Bogor.*

Siwi, S.S., Saenong, S., Supriadi, H. and Manwan, I. (1989) Interfacing technology and the needs of rural women in Indonesia. Paper presented at the International Conference on Appropriate Technology for Farm Women - Future Research Linkage with Development Systems in India, November/December, 1988.

Wigzell, S. and Setboonsarng, S. (1995) The diffusion of integrated farming in Northeast Thailand. *Thailand Environment Institute Quarterly Environment Journal* 3(2), 14-40.

WIRFS (1986) Women in rice farming system. Sta. Barbara Survey, Los Banos, Laguna, Philippines, Agricultural Economics Department, International Rice Research Institute, Los Banos.

Chapter 19

Farmer-managed Trials and Extension of Rural Aquaculture in the Mekong Delta, Vietnam

N.T. Phuong[1], D.N. Long[1], L. Varadi[2], Z. Jeney[2] and F. Pekar[2]
[1]*College of Agriculture, Cantho University, Cantho, Vietnam*
[2]*Fish Culture Research Institute, Szarvas, Hungary*

Abstract

The WES (West-East-South) Aquaculture Project of Can Tho University had the ultimate objective of assisting rural households in the Mekong Delta to improve and diversify production, nutrition and income sources through integration of aquaculture practices into their farming systems. The final stage of the project comprised different activities involving aquaculture extension for the development of aquaculture in the rural areas of the Delta. Several studies in this region have suggested that relatively small incremental changes to existing farming systems should be made, rather than attempting to introduce turn key, alien technologies which usually fail. Small-scale farmers were selected as the target group. The basic institutional infrastructure for extension in the region was available; but the efficiency of the national extension service was low due to limited funds, facilities and knowledge. The WES-Aquaculture Project applied a step by step approach during the implementation of the extension component, which included surveys of farming system and existing extension activities, training courses for both extension officers and farmers, and in the final stage farmer-managed on-farm trials. The procedure for farmer-managed trials was technical assistance through farm visits, training programmes, field school workshops and study tours, and also interest-free credit. The results from 25 farmer-managed trials demonstrated significant improvement of the productivity of various farming systems. The fish yield in rice-cum-fish systems increased from an average of 429 to 801kg ha^{-1} crop^{-1} (average increase 2.1 times) while the fish yield in pig-cum-fish systems increased from 3,860 to 7,592kg ha^{-1} crop^{-1} (average increase 2.6 times). The success of the farmer-managed trials indicated the potential for rural aquaculture in the Mekong Delta through appropriate extension activities.

Introduction

The Mekong Delta in the southern part of Vietnam covers 12% of the total area of the country and has great potential for increased agriculture and aquaculture production. Fifty per cent of the rice production of Vietnam comes from this area (Mekong Committee, 1992). The Mekong Delta is also considered as the area with the best potential in the whole country for aquaculture because of its favourable environmental conditions, extensive water surface area and abundant fish resources. Agricultural production in the Mekong Delta is rice-based with many linkages with other crops, livestock, fruit and fish production.

The development of the farming systems in the region is closely linked to the utilisation of natural resources, canal excavation, the process of settlement and land reclamation, the past war, and government policies (Nguyen, 1996). The population of the Mekong Delta was about 16 million in 1995, of which about 12 million obtained their livelihood from agricultural production, mainly rice farming. While the population density is 355 persons km^{-2} in areas with better infrastructure, in remote areas this figure is less than 100 persons km^{-2}. Agriculture production in the Delta is based on small private enterprises with an average farm size of about 1ha. As a result of years of war and international isolation, Vietnam is one of the poorest countries in Asia with an estimated *per caput* income of US$120-125 $year^{-1}$ (Nguyen, 1996). At the Sixth Party Congress in 1986 it was decided to change the economic policy of Vietnam as part of the process of '*doi moi*' (literally: radical change), while maintaining the political *status quo*. As a consequence, Vietnam is presently making a transition from a planned to a market economy. The open door policy in Vietnam has increased the overall recognition of the farming household economy as a basis for agricultural production.

An important component of the new agricultural policy in Vietnam has been the establishment of a nationwide extension service that extends from ministry to village level. Universities and research institutions also carry out extension activities often supported by international organisations. However, the efficiency of the work of the national extension service is low due to limited funds, facilities and knowledge. The farming systems approach has not been properly applied in research and extension activities; and there is inefficient co-ordination among agencies, institutions and projects involved in extension. The linkage between extension and agricultural credit is also poor. Therefore, in spite of the willingness of the farmers to adopt new methods and technologies, many poor farmers are unable to improve the productivity of their farming system and increase their income. As a result, the gap between poor and rich farmers is increasing. Poor farmers comprised about 18.5% in 1995 (Nguyen, 1996), and the widening of this gap is expected if efficient agricultural extension and rural credit systems remain unavailable to the rural poor.

To support sustainable aquaculture development in the Mekong Delta of Vietnam, a West-East-South Project (WES Project), 'Institutional Upgrading for the Sustainable Aquaculture Development in the Mekong Delta of Vietnam', was implemented between 1995 and 1999 at the College of Agriculture,

Fisheries Department of Can Tho University (CAFID) with the ultimate objective of supporting rural households in the Mekong Delta to improve and diversify production, nutrition and income sources through integration of aquaculture practices into their farming systems. The project was funded by The Netherlands Government and managed by the Fish Culture Research Institute (HAKI), Szarvas, Hungary. The immediate objectives of the project were to strengthen and upgrade the educational, adaptive research and extension capacity and capability in aquaculture in the southern part of Vietnam; to improve access to current knowledge and experience in aquaculture development; and to increase institutional cooperation between national partners and international development organisations involved in aquaculture. The extension component was only a relatively small part of the WES Project but special emphasis was given to the utilisation of improved knowledge and facilities of CAFID in the final stage of the project. The main extension activities and their results are summarised in this chapter.

Review of the Major Extension Activities of the WES Project

During the planning and implementation of the extension component of the WES Project, a conceptual framework was followed. The work was started with an appraisal of the situation in the target area of the Project, namely in Can Tho, Vinh Long, An Giang and Dong Thap provinces of the Mekong Delta (Le and Duong, 1997). A detailed baseline survey of the high diversity of farming systems in the region was then carried out by CAFID, HAKI and ICLARM staff on the eco-technological and socio-economic situation in the target area (WES Project Report, 1996/97). The survey covered 260 farms and provided information on their main characteristics, function, and needs of the various integrated farming systems. One of the main constraints identified was the limited knowledge of farmers of aquaculture. Most farms had excessive water exchange in the ponds by tidal action, farmers used low doses of lime, manure and inorganic fertilisers and they applied high stocking densities and stocking mixtures which were not appropriate in a given farming system.

It has also been shown that farmers were aware of the benefit of integration and were ready to integrate aquaculture into their farming activities. However, a major constraint to wider application of integration was the poor agricultural credit system, and also the limited knowledge of farmers on business management and marketing. A survey was also carried out on the structure and function of the agricultural extension services in the target region. The survey showed that extension activities in agriculture/aquaculture contributed to the enhancement of aquaculture and fisheries production to some extent; however, serious constraints have been also identified, mainly related to limited funds, facilities and knowledge of the extension organisations. Since the human factor is a very important element of extension activities, a survey entitled 'Technical

Manpower Demand for Aquaculture Development in the Mekong Delta of Vietnam' has been completed which revealed the problems related to the human resources for the development of aquaculture in the Mekong Delta, and suggested adjustments in educational strategies for the future (Le *et al.*, 1997).

National and international workshops on aquaculture extension have also been organised to exchange information among various players involved in extension programmes. Some main findings of the workshops are summarised below:

- Extension activities should focus on three important types of system as follows: fish seed production; rice-cum-fish; and garden-fish-livestock (VAC) systems;
- There is a need for more efficient extension services, higher quality training courses, and more written and audio-visual extension materials; and
- The relationships and collaboration among related institutes, universities, local government and farmers have to be improved.

The International Workshop on the Development of Aquaculture Extension in the Mekong Delta of Vietnam indicated the need for better collaboration among institutions and donors carrying out aquaculture extension programmes in the region. During this workshop an action plan was accepted as a framework for the implementation of the extension support programme of the Project (WES Project Report, 1997).

Training of extension officers and farmers has been an important activity of the extension component of the WES Project. Five training courses for provincial and district level extension officers were organised in Can Tho, Vinh Long, An Giang and Dong Thap provinces. A total of 184 extension officers received training in the programme, which focused on the following subjects:

- Extension methodology in agriculture and aquaculture
- Water quality management
- Integrated farming systems
- Hatchery management
- Fish nursing and rearing technologies.

Twelve training courses have been organised for farmers that were attended by about 500 farmers. The courses focused on some specific aspects of integrated aquaculture systems as follows:

- Animal (pig, duck, chicken) culture techniques
- Crop-fruit tree culture techniques
- Fish nursing and culture techniques
- Water quality management
- Fish health management
- Integrated farming systems.

Thirteen special courses for women farmers were attended by 560 women farmers. Although the courses focused on aquaculture development, two special aspects, namely educational psychology for children and health care for women and children have also been included. Four on-farm demonstrations, so-called 'Field School Workshops' or *Hoi Thao Dau Bo* in Vietnamese were also organised, when advanced farmers presented their results to the participants, and exchanged experiences among each other. A total of 200 farmers attended these workshops and study tours between 1997 and 1998.

The extension component of the Project also included limited technical assistance to regional agriculture/aquaculture extension services and the production of some written and audio-visual extension materials.

On-farm (Farmer-managed) Trials

The main activity in the extension component of the WES Project was the implementation of farmer-managed, on-farm trials. A central concept within the extension component of the WES Project document was the development of demonstration farms. Demonstration farms are generally considered as a top-down transfer of technology approach and have been widely criticised. The idea of the WES Project was, however, to provide assistance to those farms with the potential to become efficient aquaculture producers in the target area of the Project. Their activity would not be a 'show-like' demonstration but would test the viability of improved technology or management system through everyday operation.

The term 'on-farm trial' needs clarification, because the trials carried out at the farm sites and managed by farmers did not correspond to the classical form of such trials with a standard recommendation by the researchers. Due to the great diversity of the farming systems, and the significant differences in resources and management practices among the farms, the technology and management improvements proposed were built upon the best elements, identified individually at each farm, and contained only small changes in technology and management. No standard or relatively uniform technologies were tested on the various farms, because conditions differed substantially from each other. In fact, 25 different trials were carried out at 25 farm sites. The main aim of the trials was to assess the effect of small incremental changes applied to indigenous technology and management practice.

Based on a survey that included more than 200 farms in O Mon, Chau Thanh, Tam Binh and Long Ho districts in Can Tho and Vinh Long provinces (WES Project, 1996/97), 25 households were identified as collaborators in on-farm trials. Selection of farmers was based on:

- Resources (land, manpower, capital, etc.)
- Willingness to collaborate with the project

- Share know-how with other farmers
- Training attained
- Income status.

According to the standard of the Vietnamese Government there are three categories of farmers based on their income status as follows: poor (US$6.4-10.0 month^{-1}); medium (US$10.0-17.1 month^{-1}); and well off (US$17.1-24.3 month^{-1}). Among the 25 farms that were involved in the on-farm trials, 36% fell into the poor category, with 44% and 20% in the medium and well-off categories, respectively.

The technical advice given by CAFID staff to farmers was related to pond preparation, application of lime and manure, stocking structure and stocking rate, feeding, water management, disease prevention and treatment, financial record keeping and marketing.

Due to the diversity of the integrated farming systems, it was not possible to give clear well-defined descriptions of the various systems. However, three major categories were used for practical reasons as they showed substantial differences although various subsystems may have been identified within each, and there may have been overlap between them. The three main categories, which had been found to be the most important farming systems by previous workshops also, are listed below with basic information on the trials:

- Rice-cum-fish production system consisting of six trials with a total area of 4.2ha (area varied between 0.3-1.3ha per farm). Polycultures of different species (silver barb, common carp, tilapia, climbing perch, gray featherback, snakeskin gouramy) were applied with a stocking density of 2-3 fish m^{-2}.
- Pig-cum-fish production system consisting of 13 trials with a total area of 10.5ha (area varied between 0.025-0.22ha per farm). Polycultures in the ponds involved *Pangasius*, tilapia, common carp, kissing gouramy, giant gouramy, hybrid catfish and silver carp. The stocking densities in the ponds were 3-5 fish m^{-2}.
- Fish fingerling rearing system consisting of six trials with a total area of 0.94ha (area varied between 0.03-0.1ha per farm). The larvae of common carp, Indian major carp, snakeskin gouramy, kissing gouramy and hybrid catfish were reared to fingerling size.

It was also revealed by the surveys that even if farmers were ready to improve their farming system, learn new methods and follow advice, they had limited money to modify fish ponds or ricefields, to buy lime or other materials. Therefore, the Project established a credit scheme through which the farmers could get an interest-free loan for a 10-month period. This credit scheme was introduced after thorough discussion with provincial authorities and extension services. Taking into account the local conditions on the farm, the farming system, and also some criteria of the Project, a three-sided loan agreement was prepared and signed between the WES Project/CAFID, the Provincial Extension

Service Center and the farmer. According to the agreement, the farmer could use the money to buy stocking material and fish feed, but should pay it back to the Project without interest after 10 months when the fish were harvested. The WES Project provided 70-80% of the investment needed for the application of improved technology, while the farmer's contribution varied between 20-30%. The 25 households accepted the credit scheme, through which 60 million VND was paid to the farmers to support their on-farm trials.

As a result of the small incremental changes introduced into the technologies of two different types of integrated farming system, substantial increases in fish yields were achieved (Table 19.1). However, there were significant differences between the poor, medium and well-off farmer groups in the effectiveness of utilisation of project support. The most effective group was the poor farmers' group (average increase: 2.8 times), followed by the well-off (average increase: 2.0 times) and medium (average increase: 1.5 times) groups.

Table 19.1. Comparison of fish yields (kg ha^{-1} per crop, range in parentheses) before and after project support in 25 selected farms involved in on-farm trials. Yields in 1996/97 were before and in 1997/98 were after the project support.

| Farming system | Yield (kg ha^{-1} per crop) | | Times increase |
	1996/97	1997/98	
Rice-cum-fish	429	801	2.1
	(90-889)	(350-1,445)	(1.0-3.9)
Pig-cum-fish	3,859	7,592	2.6
	(1,400-2,755)	(5,856-22,577)	(1.5-4.6)

In the fish nursing system, there was no significant increase in the yield of the production. However, the average number of nursing cycles completed increased from 1.2 to 1.5 cycles year^{-1}. The survival rate of fish and the quality of the fingerlings improved considerably, which resulted in higher income for the farmer.

The farmers showed great enthusiasm in the application of improved technologies in their farming system and followed advice properly. All farmers could pay back the loan after the harvest of the fish, and 40% did not request further financial support as they had saved enough money to invest in their aquaculture production for the next growing season.

Based on the experience of the on-farm trials in 1997/98, a second series of trials was started in the growing season of 1998/99. Eighty-seven farms were included in these trials, out of which 36% fell into the 'poor' category, and 26 and 38% into the 'medium' and 'well-off' categories, respectively. The Project, however, provided only 37% of the average investment, while 63% was the farmer's contribution. After the positive experiences of the first trials in 1997-98, farmers were more willing to take bank loans, and banks also showed more readiness to provide credit for aquaculture. About 42% of the farmer's own contribution was a bank loan during the on-farm trials in 1998/99 on average.

Conclusions and Recommendations

The extension capacity of CAFID has been strengthened through the WES Project, and some direct support has been provided to local extension centres and to selected farmers. These preliminary results of efforts to improve the efficiency of extension work were promising. The on-farm trials demonstrated that the introduction of small, incremental changes into currently applied technologies in various farming systems could result in substantial increases in yield and productivity. Even if CAFID had the capacity to support extension activities in the region as a result of the WES Project, it would be important to find further external assistance to provide technical support in the future. Further measures are needed from local authorities and the government to establish an attractive and efficient credit scheme for fish farmers as it is one of the key factors for long-term, sustainable aquaculture development. Based on the results and experience of the on-farm trials, some recommendations may be summarised as follows:

- On-farm trials are required to determine the most beneficial water retention time, optimal stocking density and stocking structure for the various integrated farming systems.
- On-farm trials are also needed to optimise nutrient input to the culture systems during pond preparation and also during culture. Attention should be given to the development of simple, on-farm fish feed manufacturing methods to increase the nutritive value of locally available, agricultural by-products and other organic input materials.
- Besides the three main farming systems (rice-fish, garden-fish-livestock, fish nursing), attention should also be given to other farming models such as fruit-garden-aquaculture integration, and also to special technologies for flooded areas.
- No particular channel of information transfer was identified as the most efficient one through the implementation of the WES Project, i.e. WES-farmers; WES-provincial extension agency-farmer; or WES-provincial extension agency-district extension service-farmers). Extension should be carried out through various channels according to local conditions and specific needs. Close contact with, and involvement of, farmers' unions in the target area could contribute significantly to the efficiency of the extension activity.
- New methods and techniques can be introduced efficiently to farmers if adequate financial means are available. Otherwise, insufficient money may prevent them from developing their enterprise. Provincial and district agricultural banks still limit provision of credit to fish farmers because fish

culture is considered to be a risky enterprise. The involvement of the banks in aquaculture development should be strengthened.

- Farmers should be provided with more information on marketing and business management in training programmes. Certificate courses may also be organised, as farmers who complete the course and get a certificate will be favoured by banks when they apply for loans. Class training should be combined with field visits or field school workshops to stimulate the understanding as well as the interest of farmers in applying the acquired knowledge.

References

Le, N.X. and Duong, N.L. (1997) The status of the agricultural extension activities in Can Tho, Vinh Long, An Giang and Dong Thap provinces of the Mekong Delta of Vietnam. WES Newsletter, No. 5. College of Agriculture, Can Tho University, Can Tho.

Le, X.S., Nguyen, A.T., Korn, M., Demaine, H. and Jeney, Z. (1997) A survey on technical man-power demand for aquaculture development in the Mekong Delta of Vietnam. Project Report, West-East-South Project, College of Agriculture, Can Tho University, Can Tho.

Mekong Committee (1992) Fisheries in the Lower Mekong Basin. Review of the fishery sector in the Lower Mekong Basin. Main Report, 1992. Interim Mekong Committee, Bangkok.

Nguyen, V.S. (1996) History and future of the farming systems in the Mekong Delta. Paper presented at the Workshop of JIRCAS Project held on November 22, 1996 at Can Tho University, Can Tho. 27 pp.

WES Project Report. (1996/97) Eco-technological and socio-economic analysis of fish farming systems in the freshwater area of the Mekong Delta. Project Report, West-East-South Project. College of Agriculture, Can Tho University, Can Tho. 124 pp.

WES Project Report. (1997) Report of the International Workshop on the Development of Aquaculture Extension in the Mekong Delta of Vietnam. Organized by the WES Project (manuscript).

Chapter 20

Improving the Efficiency of Aquaculture Extension Activity in the Southeastern Provinces of Southern Vietnam

N.V. Tu and T.T. Giang
Faculty of Fisheries, University of Agriculture and Forestry, Ho Chi Minh City, Vietnam

Abstract

The Faculty of Fisheries of the University of Agriculture and Forestry (FoF-UAF) in collaboration with the Aqua Outreach programme of the Asian Institute of Technology (AIT-AOP) has implemented since 1994 farmer-managed on-farm trials in two different agro-ecological (rainfed and irrigated) areas of four provinces, Tay Ninh, Binh Duong, Binh Phuoc and Long An to develop fish culture in ponds of small-scale households in the southeast region of Southern Vietnam. Baseline surveys carried out prior to the trials have pointed out problems of fish culture of farmers in the two selected areas, e.g. pond preparation, cultured fish selection, stocking density, and water quality management. After 3 years implementation of the on-farm trials, appropriate recommendations of low-cost fish culture for small-scale farmers in the two agro-ecological areas have been produced. The average fish yield of 56 project farms in the three provinces, Tay Ninh, Binh Phuoc and Long An, has increased around 40, 25 and 70%, respectively. Since 1996 the project staff have collaborated with Agricultural Extension Centres (AECs) of the three provinces to train about 700 farmers on fish culture based on the proven recommendations. Moreover, three sets of leaflets on fish culture in ponds for different agro-ecological areas based on appropriate techniques have also been designed to support extension activities of the provincial AECs. To improve the efficiency of the AECs on aquacultural technique transfer, the FoF-UAF has organised several short training courses to upgrade knowledge and skills for AEC staff. In 1996, 64 extension staff of the three provinces were trained on integrated fish culture and extension methods and 41 extension workers have been trained on methods of fish culture on-farm trials and extension material production to date. The fish culture on-farm trials have been shown to be an efficient method to produce appropriate recommendations for small-scale farmers. However, some difficulties in implementation of the trials need to be overcome for further expansion.

Introduction

Fish is a traditional food of the Vietnamese people. In the past fish was supplied mainly from capture in natural water bodies. However, the distribution of aquatic resources is not even for the whole country. In Southern Vietnam, the average fish consumption of people who live in the upland areas, around 9kg per person year^{-1}, is lower than that of people in the Mekong River delta, 25kg per person year^{-1}. Recently, the wild fish supply has declined due to overfishing and environmental degradation. Aquaculture has been practiced for a long time in Vietnam in lowland areas but is poorly developed in upland ones.

A national extension service was established in 1993 in the country. Under this strategy aquaculture extension to transfer fish culture techniques to farmers in Southern Vietnam has had some preliminary success. However, the process of technical innovation and transfer to the farmers still has two main weaknesses. They are: (i) the weak cooperation between research and extension agencies oriented to solve problems of the farmers; and (ii) many recommendations produced from research institutions and transferred by extension networks have been found to be inappropriate to conditions of small-scale farmers who have poor resources to develop their production. How to develop sustainable aquaculture of households in regions where farmers have no or limited skills and experiences? How to assist the poor farmers to access appropriate techniques of aquaculture which are consistent with their objectives and resources?

Since 1994 one project between the Aqua Outreach Programme of the Asian Institute of Technology (AIT-AOP), a network of national institutions coordinated through AIT (AIT, 1994), and the Faculty of Fisheries, University of Agriculture and Forestry (FoF-UAF) has been implemented in the southeast provinces (SEPs) of Southern Vietnam (Tay Ninh, Binh Duong, Binh Phuoc and Long An) to improve the efficiency of aquaculture development activity of provincial Agriculture Extension Centres (AECs). The SEPs show a diverse potential for aquaculture with different systems of pond, cage and pen fish culture. Small-scale aquaculture plays an important role in nutrition and income improvement in rural areas. With the methodological approach of 'farming systems research and extension' (FSR & E) (Edwards and Demaine, 1998) through farmer-managed on-farm trials, the above weaknesses of aquaculture development activity have been reduced.

Methodologies

To better understand the efficiency and problems of extension activity to transfer techniques to fish culture farmers in the target region, a survey has been conducted to review the existing structure of, and methods applied by, the AECs of SEPs in Southern Vietnam. The efficiency of the aquaculture extension service was evaluated based on the percentage of technical adoption of the farmers who have received recommendations from the AECs through training. The efficiency was also analysed for the provinces with different aquaculture potentials.

From 1994 onwards, FoF-UAF adopted the approach of farmer-managed on-farm trials, which has been successfully practiced in Northeast Thailand (Edwards *et al.,* 1996), to produce appropriate recommendations for small-scale fish farmers of the SEPs. This approach also aimed to produce adequate extension materials to support the technology transfer activity of the AECs. The fish culture on-farm trials were implemented through a number of steps as outlined below.

Baseline Survey

One of the objectives of baseline survey was to identify potentials and problems of fish culture of the existing farming systems. In 1994, FoF-UAF staff carried out a baseline survey in the two target provinces of Tay Ninh and Song Be (now Binh Duong). In 1996, other baseline surveys were done for new sites when the on-farm trials were expanded to other parts of Tay Ninh, Binh Phuoc, and a rainfed area in Long An province which is in the upper part of the Mekong River delta.

Prioritization of Common Problems of Small-scale Fish Culture in the Target Areas

Based on the analysis of the baseline surveys, common problems of small-scale fish farmers were prioritised. Since then, promising recommendations which could solve the most common problems of fish culture were reviewed and proposed for testing during on-farm trials.

Project Farmer Selection and Promising Recommendation Testing

Promising recommendations were tested with project farmers who were willing to adopt new methods and technologies. The selected households for the trials were representative for fish culture systems in the target areas. After the farmers were selected, the project staff discussed with each farmer household all the elements that formed the farming system such as the amount and variety of available resources, the potential production of the farm, and the possible solutions for its problems. Project staff visited the farmers weekly or fortnightly to provide technical support and advice.

Appropriate Recommendation Dissemination

Tested recommendations, which were under the management of the farmers and the supervision of the project staff, were evaluated based on the percentage of adoption. The appropriate techniques then were disseminated to other farmers in the target areas through the activities of the AECs such as farmer training and extension material production.

The impacts of the on-farm trial programme to produce and disseminate appropriate techniques to the small-scale fish-culture farmers in the region were evaluated based on fish yield, economic efficiency and level of proven technique adoption.

Due to the lack of manpower to transfer aquacultural techniques to farmers, the FoF-UAF has organised several short training courses and study tours to upgrade knowledge and skills for AECs' staff of SEPs.

Results and Discussion

Organization and Activity of Provincial Agriculture Extension Centres

Human resource for aquaculture extension activity

An Agriculture Extension Centre is a government agency which is responsible for technique transfer in agronomy, animal husbandry, forestry and aquaculture at provincial level. The typical structure of one AEC comprises functional offices, extension stations and experimental stations as presented in Fig. 20.1. The annual plan of the AEC is approved by the Department of Agriculture and Rural Development at provincial level and the General Department of Agriculture Extension at central line. At lower levels, there are Extension Stations at districts and Extension Clubs at communes. The AEC also receives technical support from research and education organisations, and provides support to development projects of local non-government organisations (NGOs) such as the Farmer Union, the Women's Union and the Youth Association.

The common methods applied by the provincial AECs to transfer techniques to farmers are farmer training and setting up demonstration farms. The efficiency of any AEC, besides technical resources, depends on its institutional structure and human resource. However, the institutional structure and human resources are different from one AEC to another.

Among SEPs, the AEC of Song Be (former) was established very early, in 1993. In 1996, Song Be (former) province was divided into two new ones, namely Binh Duong and Binh Phuoc. The human resource of the Song Be AEC was also reorganised into two new centres. The AECs of Tay Ninh and Dong Nai were established in 1994. The institutional organisation and human resource base of each AEC are presented in Table 20.1.

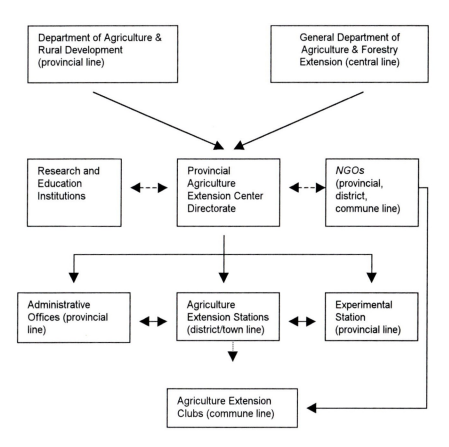

Fig. 20.1. Schema and relationships of a typical provincial Agriculture Extension Center. (◄►) functioning relationship; (◄►) cooperation relationship.

In the SEPs, the structure of AECs of Tay Ninh and Dong Nai are relatively complete with extension stations at district level, but those of Binh Duong and Binh Phuoc have no stations at the district. To increase the efficiency of technique transfer, some AECs have supported the establishment of extension clubs at commune level. These are groups of farmers who are willing to share their experiences and understandings on production practices. In Dong Nai the AEC has developed a network of co-operators who volunteer to transfer techniques to their neighbours at communes.

Table 20.1. Institutional organisation and human resources of the Agriculture Extension Centres of southeastern provinces.

Institutional structure/ Human resource	Tay Ninh	Binh Duong	Binh Phuoc	Dong Nai	Long An
Total	43	41	16	54	75
Directorate	3	2	2	4	3
Functional offices					
- Administration and Personnel	4	8	3	9	5
- Techniques	4	8	5	8	7
- Information and Communication	2	6	3	0	5
Staff					
- BSc in Fisheries	1	0	0	1	5
- BSc in Agronomy	18	8	3	20	33
- BSc in Animal Husbandry	7	9	5	7	11
- BSc in Forestry	0	0	0	2	0
- BSc in Economy	3	0	0	4	1
Agriculture Extension Station	9	0	0	8	14
Experimental Station	0	1	0	1	1

Although the AECs of Binh Duong and Dong Nai have established agronomy experiment stations, none of the AECs in the region has established aquaculture stations where adaptive research and trials are carried out to produce appropriate techniques to fish farmers.

One of the limitations of the aquaculture extension activity of the SEPs is the under-evaluation of aquatic resources which leads to a shortage of extensionists who are trained in fisheries to serve for technology transfer to the fish farmers (Table 20.1). Aquaculture extension workers, if available, are not well trained in extension methodology such as gathering data, being in contact with the farmers, and giving advice and assistance to solve farmer problems. The SEPs' administrations have not had resources to attract persons who have a background in aquaculture to serve for fisheries development. Working conditions of extension workers are poor, including low salary and lack of facilities and transportation means. In contrast to the SEPs, the AEC of Long An, one province in the upper part of the Mekong River delta, was established very early (in 1991) to serve the agricultural development plan of the province. The aquaculture potential is also appreciated by the provincial government which has invested in manpower recruitment, as illustrated by higher numbers of extension staff who were trained in fisheries.

The AECs have established good relationships with research institutions and universities, including the University of Agriculture and Forestry, to improve their activities through staff training and technical support. However, the efficiency of the support was low since these organisations have not considered the problems of small-scale farmers as a priority of their research and have not collaborated with the AECs to carry out on-farm trials to produce appropriate recommendations for them.

Types of Aquaculture Extension Activity

Common aquaculture extension activities applied by the AECs are farmer training and setting up demonstration farms which are similar for agronomy and animal husbandry. In contrast, there are no long-run extension programmes comprising planning, implementing, monitoring and evaluating stages to serve the aquaculture development strategy of the provinces.

Farmer training

Because of requests of farmers as well as the mandate of provincial governments, despite their lack of qualified staff, the AECs have to organise training to transfer techniques to the fish farmers. The procedure of organising farmer training is similar in all of the AECs. The need of training of the farmers is indicated to the AECs through local officials or NGO staff (mainly at district and commune levels). Based on the annual plan the AECs will allocate a number of training courses for each district. When the request is accepted, local officials are responsible for assembling farmers, preparing the meeting room and audio-visual materials, and the AECs are responsible for providing trainers, extension materials such as leaflets, and funding. The trainers can be extension staff and/or frequently, lecturers from universities and scientists from research institutions for specific topics. The amount of farmer training on aquaculture of the AECs during the last three years is presented in Table 20.2.

In general the number of training courses given by the AEC staff on aquaculture was lower than that on agronomy and animal husbandry. The number of training courses in aquaculture in Dong Nai province was increased yearly. In 1997, Binh Duong and Binh Phuoc could not organise any farmer training on aquaculture due to lack of staff with background in fisheries. The amount of training of Tay Ninh declined in 1997 but its AEC provided funding for training instructed by the AOP project staff. The amount of training of Long An also declined due to quality problems.

Training content depended upon the specific requests of farmers in certain aquaculture systems. Table 20.3 presents the training courses on different aquaculture systems of the AECs of Dong Nai and Long An provinces as an illustration. In Dong Nai province, the training on fish culture in cages and shrimp culture were highest, since the government had a policy to develop the two systems in 1995. However, due to an unsolved problem of cultured fish and shrimp diseases, the training on these systems was reduced quickly in 1997 and replaced by training on fish culture in ponds. Similar changes occurred in the AEC of Long An province.

The efficiency of farmer training of the AECs measured by the percentage of technical adoption of the farmers who attended the training is presented in Table 20.4. The adoption level was evaluated for basic techniques of fish culture in ponds such as pond drying, bottom-mud removal, predator elimination, liming and fertilisation.

Table 20.2. Number of farmer training courses and participants by Agriculture Extension Centres of southeastern provinces in Southern Vietnam, 1995-1997.

				Participants					
		1995			1996			1997	
Province/ topic	Courses	Total	Course average	Courses	Total	Course average	Courses	Total	Course average
Tay Ninh									
- Agronomy	79	3,956	50	82	4,115	50	55	2,347	43
- Husbandry	27	1,488	55	30	1,580	53	12	409	34
- Aquaculture	4	180	45	5	253	51	2	24	12
Binh Duong									
- Agronomy	31	1,488	48	40	2,500	63	45	2,925	65
- Husbandry	34	1,728	51	42	2,730	65	47	3,055	65
- Aquaculture	5	280	56	3	190	63	0	0	0
Binh Phuoc									
- Agronomy	-	-	-	-	-	-	20	1,000	50
- Husbandry	-	-	-	-	-	-	65	3,000	46
- Aquaculture	-	-	-	-	-	-	0	0	0
Dong Nai									
- Agronomy	95	5,510	58	138	8,280	60	92	5,508	60
- Husbandry	25	1,500	60	84	4,788	57	139	8,444	61
- Aquaculture	11	700	64	17	1,180	69	28	1,554	56
- Forestry	2	112	56	4	246	62	6	359	60
Long An									
- Agronomy	520	13,260	26	477	13,571	28	211	6,330	30
- Husbandry	-	7,012	25	152	4,295	28	178	5,730	32
- Aquaculture	-	1,736	23	109	3,011	28	53	1,590	30

The adoption level of the farmers of SEPs after the training was lower than that of Long An. In general, during the training the trainers always presented the whole package of techniques relevant to certain aquaculture systems. This caused difficulty to the farmers to identify appropriate recommendations, which could solve their problems. Because of the lack of manpower and knowledge, offering too many topics on different aquaculture systems (Table 20.3) also limited the quality of training due to low availability of appropriate recommendations. Moreover, the limited number of training courses on aquaculture due to lack of funding also resulted in high number of participants per course and low efficiency of the training. By contrast, the AEC staff of Long An province only trained the farmers in certain steps of the whole package of culture techniques and then upgraded their knowledge later. Moreover, the AEC of Long An have collaborated to train local NGO staff who later could train their members. This cooperation has increased the impact of training with less expense.

In general, there were some constraints of the farmer training on aquaculture as follows:

- The techniques given during training were mainly based on on-station research and were not consistent with the socio-economic conditions of the farmers in different agro-ecological areas.
- The method of training was very simple without training aids so it limited the understanding of the farmers.
- Due to lack of a survey before and after the training, there was no evaluation of impacts on the farmers.

Table 20.3. Amount and percentage of topics of farmer training carried out by the Agriculture Extension Centres of Dong Nai and Long An provinces.

| Topics | Number (%) | | |
	1995	1996	1997
Dong Nai			
- Fish culture in ponds	2 (18)	5 (29)	16 (57)
- Fish culture in cages	6 (55)	6 (35)	8 (29)
- Shrimp culture	3 (27)	5 (29)	0 (0)
- Disease prevention and treatment	0 (0)	1 (6)	4 (14)
Total	11 (100)	17 (100)	28 (100)
Long An			
- Fish culture in ponds	20 (27)	34 (31)	20 (38)
- Rice/fish culture	33 (45)	26 (24)	15 (28)
- Freshwater prawn culture	2 (3)	3 (3)	1 (2)
- Marine shrimp culture	19 (25)	40 (37)	16 (30)
- Fish nursing	0 (0)	6 (5)	1 (2)
Total	74 (100)	109 (100)	53 (100)

Table 20.4. Impact of farmer training on fish culture of Agriculture Extension Centres.

Province	Techniques applied by farmers before training (%)	Techniques adopted by farmers after training (%)	Surveyed households (number)
Tay Ninh	5-30	20-45	20
Binh Duong	0-29	0-57	21
Dong Nai	0-60	13-73	30
Long An	10-15	50-80	20

Setting up demonstration farms

Setting up demonstration farms is another extension method commonly used by the AECs. A demonstration farm is the model of a certain aquaculture system to be promoted. To set up one demonstration farm, the AECs provide a small amount of investment funding support such as seed and feed. The selected households were required to contribute the rest of the needed investment. The AEC staff also provide technical inputs for the households. The number of demonstration farms on aquaculture established by the AECs is presented in Table 20.5.

Table 20.5. Number of demonstration farms established by the provincial Agriculture Extension Centres, 1995-1997.

Topic	Number (%)		
	1995	1996	1997
Tay Ninh			
- Agronomy	85 (60)	85 (53)	Na
- Husbandry	49 (35)	60 (38)	Na
- Aquaculture	7 (5)	14 (9)	Na
Binh Duong			
- Agronomy	32 (34)	55 (47)	55 (44)
- Husbandry	40 (43)	62 (53)	70 (56)
- Aquaculture	21 (23)	0 (0)	0 (0)
Binh Phuoc			
- Agronomy	-	-	9 (11)
- Husbandry	-	-	75 (89)
- Aquaculture	-	-	0 (0)
Dong Nai			
- Agronomy	98 (63)	98 (48)	111 (35)
- Husbandry	41 (26)	75 (36)	162 (50)
- Aquaculture	15 (10)	28 (14)	46 (14)
- Forestry	2 (1)	4 (2)	3 (1)
Long An			
- Agronomy	1,011 (85)	1,169 (82)	94 (28)
- Husbandry	115 (10)	117 (13)	102 (49)
- Aquaculture	67 (5)	75 (5)	75 (23)

Na = Not available.

The amount of demonstration farms on aquaculture of Dong Nai and Tay Ninh increased whilst Binh Duong and Binh Phuoc did not set up aquaculture demonstration farms due to lack of staff having a background in fisheries. The AECs have tried to set up demonstration farms which could be representative for existing aquaculture systems in the provinces (Table 20.6).

Due to lack of manpower, the establishment of many demonstration farms has led to difficulties in technical support and monitoring and evaluation. To ensure the success of the demonstration farms, the AEC staff usually selected farmers who were better-off and experienced. Therefore, there are constraints to expand techniques through demonstration farms, even though they were economically efficient, due to the following reasons:

- The selection of better-off and experienced farmers leads to questions about their relevance for resource-poor, small-scale farmers.
- Demonstration farms may not work due to the large variation between farmers in terms of physical resources and socio-economic status.

Table 20.6. The aquaculture systems of demonstration farms established by the AECs, 1995-1997.

Topic	Number (%)		
	1995	1996	1997
Tay Ninh			
Cage fish culture	1 (14)	0 (0)	Na
Pond fish culture	4 (57)	12 (86)	Na
Freshwater shrimp culture	2 (29)	0 (0)	Na
Duck-fish integration	0 (0)	2 (14)	Na
Total	7 (100)	14 (100)	Na
Dong Nai			
Pond fish culture	7 (46)	17 (60)	50 (66)
VAC[1] systems	1 (7)	1 (4)	1 (2)
Rice/fish culture	1 (7)	0 (0)	2 (4)
Cage fish culture	3 (20)	9 (32)	9 (20)
Shrimp culture			
Marine shrimp culture	0 (0)	0 (0)	1 (2)
Freshwater shrimp culture	2 (13)	0 (0)	1 (2)
Specific animal culture (frog, soft turtle)	1 (7)	1 (4)	2 (4)
Total	15 (100)	28 (100)	46 (100)
Long An			
Pond fish culture	17 (25)	22 (29)	23 (31)
Integrated fish culture	2 (3)	10 (13)	20 (27)
Rice/fish culture	33 (49)	26 (35)	29 (38)
Fish seed nursing	0 (0)	5 (7)	1 (1)
Shrimp culture			
Marine shrimp culture	13 (20)	9 (12)	0 (0)
Freshwater shrimp culture	2 (3)	3 (4)	2 (3)
Total	67 (100)	75 (100)	75 (100)

[1] VAC: garden-fish pond-animal pen.
Na = Not available.

Small-scale Fish Culture On-farm Trials

Results of the fish culture on-farm trial programme

Due to the lack of manpower, funds, facilities, and knowledge, the dissemination of techniques developed by research organisations to farmers by extension agencies often follows a conventional, top-down method. As pointed out by Edwards and Demaine (1998), this mode of technology transfer is rarely effective because it usually fails to match the resource profiles of small-scale farms, which are diverse and complex, with relevant technology. A wider perspective on research to develop and disseminate technology appropriate for the widely varying resource contexts of poorer farmers is required.

The implementation of farmer-managed on-farm trials on fish culture of AIT-AOP in Southern Vietnam aimed to improve the efficiency of small-scale aquaculture and its extension in the SEPs. The main activities and their results are summarised in the following sections.

Baseline survey

The aim of baseline surveys was to identify potentials and problems of small-scale fish culture in the target areas. The surveys have been conducted for different agro-ecological zones: in irrigated lowlands of Tay Ninh, upland rainfed agriculture of Binh Phuoc and flat lowlands of Long An (Nguyen *et al.* 1994, 1995). The main characteristics of fish culture systems in ponds of the surveyed areas are presented in Table 20.7.

Table 20.7. Characteristics of fish culture in different agro-ecological zones of Tay Ninh, Binh Phuoc and Long An provinces.

Characteristics	Irrigated lowlands (Tay Ninh)	Rainfed uplands (Binh Phuoc)	Rainfed flatlands (Long An)
Fish culture experience	Low (< 3 years)	Low (< 3 years)	High (> 4 years)
Water supply	Irrigation system	Rainfall	Rainfall
Pond size	Small and shallow	Large and deep	Small and deep
Growing-out period	Year round	6 to 8 months	6 to 8 months
Purpose of fish culture	Subsistence and income improvement	Income improvement	Income improvement

Most farmers practised a polyculture with high variation of fish species and at a very high density, used low levels of poor nutritional inputs. The most common input was rice bran. The second most common input was pig manure in Tay Ninh and cassava leaf in Binh Phuoc. Crop by-products from vegetables were rarely used in Tay Ninh and Long An but were commonly used in Binh Phuoc. The concept of fertilising fish ponds to develop natural feed for fish was not widely understood and appreciated.

The main problems of households on fish culture development were lack of knowledge resulting in low survival rate, low yield, small size of harvested fish and poor water quality (polluted, turbid).

Farmer-managed on-farm trials

Fish culture on-farm trials were started in 1994 in Thuan An district of Song Be province (now Binh Duong) with a system of hybrid catfish (*Clarias gariepinus* x *C. macrocephalus*) culture in ponds and in Trang Bang district of Tay Ninh with the system of fish culture in an overhung latrine pond. In 1996, the trial in Binh Duong was discontinued due to urbanisation of the area. At the same time the trial was expanded into Chau Thanh, Dong Phu and Duc Hoa districts of Tay Ninh, Binh Phuoc and Long An provinces, respectively. The number of trial farms during the period is presented in Table 20.8.

Table 20.8. Number of project farmers in the target by year.

Year	Tay Ninh	Binh Duong	Binh Phuoc	Long An	Total
		Project farms			
1994/95	6	8	-	-	14
1995/96	6	9	-	-	15
1996/97	6	-	8	8	20
1997/98	36	-	10	10	56

Based on identified problems and resources available for various techniques, recommendations were proposed to the project farms. The acceptance and adoption of proposed recommendations varied depending on farmers' conditions (Tables 20.9 and 20.10).

Appropriate recommendations were identified as those which should improve fish yield and those adopted by a high percentage of project farmers. After 4 years of implementing the trials, the average fish yield increased (Table 20.11).

The yield of fish in the trial farms has increased yearly. However, the fish yield of the 1997/98 crop in Trang Bang district declined due to unstable water supply from the irrigation system.

Since 1996, the project staff have collaborated with the AECs of Long An, Tay Ninh and Binh Phuoc to train more than 700 farmers on fish culture based on the tested recommendations. From the fish culture trials in Tay Ninh, one extension booklet titled 'Fish Culture in Earthen Ponds' was designed and tested in 1996. Five hundred copies of the final version of the booklet were printed to support farmer training of the AECs of SEPs. However, the cost of the booklet was too high to be reprinted. Therefore, in 1997 three sets of leaflets on fish culture in ponds for different agro-ecological areas based on proven techniques were designed for the AECs' use.

A major constraint to the improvement of rural livelihoods through aquaculture is the inefficient transfer of research findings through service providers (Edwards and Demaine, 1998). In 1995, 64 extension staff of Long An, Tay Ninh and Song Be (now Binh Duong and Binh Phuoc) were trained on integrated fish culture and extension methodology. Since 1996, 41 extension staff have been trained on methods of fish culture on-farm trials and extension material production. In March 1998 one modular training programme on aquaculture systems was also offered to 22 local staff of the collaborating provinces. This training upgraded the knowledge and skills of staff to efficiently transfer techniques to fish farmers.

Table 20.9. Recommendations proposed for hybrid catfish culture and their acceptance in rainfed areas.

Problems	Causes	Recommendations	Acceptance	Reasons
Low survival rate	Stocking small fry	Nurse fry in hapas	-	Difficult to manage increases disease problems
		Purchase larger fry	+	Larger fry available
	Stocking diseased fry	Treat diseased fry with chemicals	-	Gives poor results
		Purchase healthy fry	+	Easy to adopt
	Escape of fish due to run-off	Use a net	-	High cost and limited life of net
		Improve the dike	+	Labour available
	Cannibalistic problem	Adjust density and give more feed	+	Easy to adopt
Low yield and small harvest fish	Very high stocking density	Adjust density	+	Easy to adopt
	Monoculture applied	Stock supplementary species	+	Fry available
	Poor feeding	Fertilise with animal manure	+	Manure available
		Increase on and off-farm feed	+	By-products and cheap feed available
Water turbidity	Clay soil	Develop algae with animal manure	+	Manure available

+: accepted; -: rejected.

Conclusions

The development of aquaculture has contributed to diversification of agricultural production, nutritional improvement and income generation in the rural areas of southeastern provinces of Southern Vietnam. Extension services play an increasingly important role in aquaculture development. Today universities and research institutions in Vietnam are required to be involved in rural development programmes, supported by international organisations. The involvement of the universities and research institutions in those programmes is mainly in human resource training and technical support. Their role in extension services is also appreciated by central and local government. The willingness of small-scale fish farmers to adopt appropriate techniques has been recognised in the region. The success of fish culture of the small-scale farmers may stimulate them to shift from subsistence to more commercial systems. This means that there will be increased needs of advanced techniques of fish culture in the future, e.g. from low-input and low yield systems to high-input and high yield ones (Edwards *et al.*, 1996). The approach of 'farming systems research and extension' implemented in the region has been effective in terms of production of appropriate recommendations for small-scale fish farmers and extension service improvement. This approach requires frequent visits of extensionists to target households during the process of testing recommendations and extension materials based on proven techniques. Therefore, the local government needs more incentives for implementation of the approach such as increasing resources for extension service and providing better working conditions.

Moreover, the farmers have also shown their own capacity to experiment with, and innovate in technology. Many of them have continued to develop their aquaculture systems with little to no technical support. The use of experienced farmers as promoters in extension services, e.g. farmer training and farm visits, also increases the efficiency of technology transfer.

Table 20.10. Recommendations proposed for fish polyculture and their acceptance in irrigated areas.

Problems	Causes	Recommendations	Acceptance	Reasons
Low survival rate	Predation	Eliminate predators	-	Ponds can be dried at the end of the season
	Stocking small fry	Nurse fry in hapas	-	Difficult to manage
		Purchase larger fry	+/-	Depends on fry availability
	Escape of fish due to run-off	Use a net	-	High cost and limited life of net
		Improve the dike	+	Labour available
	Incorrect stocking and feeding methods	Stock adequate seed and give more feed	+	Easy to adopt
Low yield and small harvest fish	Unsuitable species stocked	Stock supplementary species	+	Fry available
	Incorrect species ratio	Adjust species ratio	+	Easy to adopt
	High stocking density	Adjust density	+	Easy to adopt
	Poor feeding	Increase feed	+	By-products available
		Fertilise with animal manure	+	Manure available
Water pollution	Overloading with organic matter	Balance pond area and number of pigs	-	Depends on benefit of pig raising
		Manage water colour and exchange water	+	Water supply available

+: accepted; -: rejected.

Table 20.11. Fish yield of the project farms in different agro-ecological zones of target provinces, figures in parentheses are percentage change compared with the previous year.

	Extrapolated fish yield (t ha^{-1}year^{-1})			
	1994/95	1995/96	1996/97	1997/98
Thuan An district, Binh Duong province	3.94[1]	5.37 (36)	-	-
Trang Bang district, Tay Ninh province	3.95[1]	7.43 (97)	9.49 (27)	6.21 (-35)
Chau Thanh district, Tay Ninh province	-	-	3.66[1]	4.62 (26)
Phuoc Dong Phu district, Binh province	-	-	3.50[1]	4.98 (42)
Duc Hoa district, Long An province	-	-	2.70[1]	4.73 (75)

[1] The yield before on-farm trials.

Acknowledgements

Special thanks for financial support of SIDA and consultative support of the Aqua Outreach Programme (AOP) of the Field of Study of Aquaculture and Aquatic Resources Management (AARM) of the Asian Institute of Technology (AIT), and fruitful cooperation of the Agricultural Extension Centres of Tay Ninh, Binh Duong, Binh Phuoc, Dong Nai and Long An provinces.

References

AIT (1994) *Partners in Development, the Promotion of Sustainable Aquaculture.* Asian Institute of Technology, Bangkok.

Edwards, P. and Demaine, H. (1998) Completing the problem-solving cycle in aquaculture: the AIT Aqua Outreach. *Aquaculture Asia* 3(3), 10-17.

Edwards, P., Demaine, H., Innes-Taylor, N. and Turongruang, D. (1996) Sustainable aquaculture for small-scale farmers: need for a balance model. *Outlook on Agriculture* 25(1), 19-26.

Nguyen, V.T., Bui, T.H. and Demaine, H. (1994) Report on-farm survey in Song Be and Tay Ninh provinces, Vietnam, Working Paper No. SV-1. Aqua Outreach Program in Southern Vietnam, Ho Chi Minh City.

Nguyen, V.T., Huynh, T.N.A. and Demaine, H. (1995) Report on fish culture on-farm trials in Song Be and Tay Ninh provinces, Vietnam, Working Paper No. SV-2. Aqua Outreach Program in Southern Vietnam, Ho Chi Minh City.

Chapter 21

The Effectiveness of a Model Fisheries Village Approach to Aquaculture Extension in Northwest Bangladesh

M. Islam[1] and N. Mardall[2]

[1]*Extension Officer, Northwest Fisheries Extension Project, Parbatipur, Dinajpur District, Bangladesh*
[2]*Extension Advisor, Northwest Fisheries Extension Project, FMS, House 42, Road 28, Gulshan, Dhaka, Bangladesh*

Abstract

The Northwest Fisheries Extension Project (NFEP) operates in the Northwest region of Bangladesh. One of four main project outputs is the development of extension strategies appropriate to the needs of the poor. The Model Fisheries Village programme, or Model Village as it is simply known, is one such appropriate extension methodology first piloted by the NFEP in 1995. The community based approach now operates in each of 58 *thanas* (a local government administrative unit) within the command area on an annual basis with more than 180 pond resource rich villages operated to date. The programme aims to help stakeholders choose the most effective utilisation of available resources. The pond side training, an integral part of the programme, has seen benefits to the farmers steadily increase in terms of pond production after effective pond preparation, management and disease learning sessions. Production prior to the intervention was around 812kg ha^{-1} in 1994 increasing to 1,700kg ha^{-1} in 1998. There is also a significant socio-economic impact expressed in increases in family fish consumption and investment as well as less quantifiable benefits such as awareness and empowerment. The project has identified areas within the programme that will contribute to the future effectiveness and sustainability of the programme. These areas include working in villages and training the women who then operate the homestead ponds with their husbands. The project is also linking the Model Villages with a credit institution which disperses loans purely for aquaculture.

Introduction

The Northwest Fisheries Extension Project (NFEP) is a joint Bangladesh and British Government project. Extension methodologies appropriate to the needs of the poor is one of the Project's four outputs within the overall logframe. This model village programme is one such methodology. The programme aims to improve livelihoods through an effective utilisation of resources which fit in with the existing farming and household system. The programme has evolved as a means to reach a large number of stakeholders with a relatively low number of extension or field workers. Limited numbers of field workers is an issue for both government and non-governmental organisations (NGOs) and this approach addresses this. The broad coverage also ensures that, although we may have to work with some stakeholders who do not fall into the 'resource-poor' category, the programme is reaching a considerable number of low and middle-income producers, and more than if the project were to target only poor individual farmers.

The Project operates in eight districts divided into 58 units called *thanas*. Each *thana* is divided into 6-10 unions, each of which may contain 30 villages. It uses both government and project staff to implement the field programmes. Three government fisheries staff, a *thana* fisheries officer (TFO), an assistant fisheries officer (AFO) and a field assistant (FA) usually serve each *thana*. For each village the selection, training and follow up is conducted by the TFO and his two assistants and, for the duration of the NEFP, one of the project extension officers.

The Model Fisheries Village programme began with a single pilot village in April 1995. Twenty-seven farmers with 30 ponds between them were selected and participated in the programme. The programme is based around imparting a flexible production model which has been developed by the project from previous demonstration and trial programmes. The training includes management practices from stocking rates and species mixes, aquatic weed and wild fish removal, and inorganic and organic fertiliser application. Currently the programme operates with at least one model village per *thana*. This is over 58 new villages year^{-1}.

Programme Implementation

Selection of the villages is based on an approximate minimum number of 15 available ponds to ensure good coverage per village. Group interest and consensus are also prerequisites. The *thana* fisheries officers, assistant fisheries officers and field assistants undertake selection and training. A project extension officer helps to co-ordinate the selection and training of each village in his district. The village is also wealth ranked to measure the status of the stakeholders.

At a motivational meeting the farmers discuss and determine with the project and government staff how they wish to operate within the programme. At this meeting the participants are encouraged to form a group or club and to select a leader if one does not already exist. Experience and issues can be shared through this club. Training then takes place at three specific times, pond preparation prior to the rains, pond management once the culture season is underway, and finally disease and partial harvesting training as the weather cools and the likelihood of disease increases. The training sessions take place in the village at the 'pond side'. In this way the stakeholders are more relaxed and feel at ease to participate. The village is then visited on a fortnightly or monthly basis by the project and government staff in an advisory capacity. At the end of the growing season a rally is held in successful villages. Neighbouring village heads and influential persons are invited. Snacks are provided and neighbouring villagers exchange questions and views about how the programme has gone. It is usually organised around a pond harvest also. This always stimulates interest and draws people to the rally. All the stakeholders are given a record book for recording their inputs and production. In many cases, government staff assist the stakeholders in completing the books. This information is then used alongside a baseline which is conducted in each selected village.

The programme runs for two years. In this second year, one refresher training takes place for all the stakeholders. The project and government staff make informal visits and finally a rally is held again at the end of the season.

Findings

Information is gathered about inputs and production using the initial baseline survey and the pond record books. This baseline information summarised below is based on data taken from 1,708 ponds. Current production and practice are based on a sample of 350 pond owners. This sample of 350 pond owners was taken from 17 villages and is based on data taken from the baseline of 1996 and the production information from the pond record books in 1997. Both the baseline and the pond record book are used to collect information from every farmer involved in the Model Village programme.

Pond Inputs

Pond input practices are measured so production can be determined from certain management practices. The measurements are made through recording whether a pond owner carried out certain pond management practices and if there was an input then the quantity of the input is recorded.

The study has seen a rise in all recommended practices. The following practices are measured: stocking density, wild fish removal, liming, and cow dung and inorganic fertiliser application. The survey found that the level of both liming and inorganic fertiliser applications almost doubled. These along with the other practices are responsible for the increase in production. One issue regarding the practice is the stocking level which the farmers want to apply as opposed to that recommended: 34,975 fingerlings ha^{-1} (range 6,250-175,000) and 14,820, respectively.

A separate project baseline survey of the project's target group beneficiaries was conducted in 1995 to determine the current stocking levels in the north-west of Bangladesh (n = 93) although this baseline was not directly linked with the Model Village programme. The results showed an average stocking density of 50,000 fingerlings ha^{-1}. From this it can be seen that the stocking levels of the Model Village pond owners have decreased. It is assumed this is a result of training and experimentation by the pond owners.

Prior to project intervention, the addition of cow dung and rice bran was common to approximately 60% of pond owners but inorganic and lime use increased from 26% and 29%, respectively, to 63% and 62% following project intervention (Fig. 21.1).

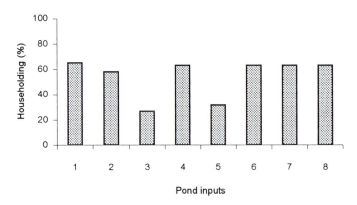

Fig. 21.1. Percentage of households using various pond inputs prior to (1, 3, 5, 7) and after (2, 4, 6, 8) project intervention. Cow dung (1, 2) inorganic fertiliser (3, 4) lime (5, 6) rice bran (7, 8).

Pond Production Information

Average pond size is 824m^2 (range 80 to 14,000m^2) and 48% are seasonal and 53% are perennial ponds. The average pond production in 1995 and 1996 was 1,250 and 1,300kg ha^{-1}, respectively (n = 1,378 and 1,182, respectively). In 1997 the average pond production rose to 1,700kg ha^{-1} (n = 335), a significant increase. This increase represents 105t of additional fish production annually. One of the selection criteria for a Model Village is that it has not received previous fisheries extension assistance. Therefore, the assumption that this increase is attributable to the Model Village Fisheries Programme is reasonable.

Production Costs

The mean input cost for an $824m^2$ pond was Tk1,277 (US\$1 = 49Taka). Taking a mean fish market value of $Tk40kg^{-1}$, this represents a cost-benefit ratio of 1 : 5.1.

Socio-economic Information

The Model Village programme baseline survey reported an average family size of 7.1 and a land area of $1.9ha^{-1}$ household. Twenty-four per cent of households had less than 0.6ha. According to a socio-economic survey conducted in 1995, the extension approach reached 78% of poor and marginal farmers ($n = 27$) in the pilot Model Village.

Households were categorised as poor, marginal or rich in the survey on the basis of no savings and a food deficit, no food deficit and no savings or a small amount of savings, and no food deficit and substantial savings, respectively. The quickest method during the baseline for ascertaining a level or degree of wealth in terms of resources is to ask the land holding. It is a much lengthier process to discover the savings and food deficit situation for the large numbers of farmers with which the project deals. It must be noted that the initial socio-economic survey conducted in 1995 was only for one village. To conduct the same survey annually for all Model Villages would not be logistically feasible. However, project experience shows that land holding is not the most accurate method of wealth determination due to the variation in land productivity. The Project covers eight districts and a large variation is seen.

Seventy-eight percent of pond owners were involved in agriculture as their main source of income. Of the 17 villages surveyed, information was collected on fish sales and fish consumption. On average 51kg were consumed and 113kg were sold, 31% and 69%, respectively. Using the same ratio of sale to consumption, the figures for previous years of 1995 and 1996 for consumption and sale would have been 32 and 71, and 33 and 74kg, respectively.

On a household member basis this represents a consumption increase over the fish culture season (approximately 6 months) from 4.5kg in 1995 to 7.1kg in 1997. However, it is recognised that the sale to consumption ratio would not necessarily remain constant but for the sake of comparison it has been assumed here. Also household size is ever changing with daughters moving to their husband's parents and so on.

Conclusions

Several issues have become apparent during the course of this study. Most important is the increase in inputs. Also the stocking density is approaching that which is recommended during training. Although the present mean of 34,975 fingerlings ha^{-1} is more than double that recommended, it is significantly reduced from the figure of 50,000 fingerlings ha^{-1} collected during a farmers' baseline survey prior to the introduction of the Model Village programme in 1995.

It is apparent from the data that the Model Village Fisheries programme has directly influenced an increase in fish production in selected villages. This increase has raised both fish consumption and income of the farmer. The progressive production increase is testimony to the motivation and uptake of skill and knowledge by the farmer after pond side training. Mention must also be made of government and project staff who are involved in the programme, as it is their work which is producing such results at the field level.

Future Issues

As NFEP would like to ensure that the effectiveness of the Model Fisheries Village programme is sustainable, it has evolved various ideas to help achieve this.

One such idea is the training of female stakeholders within selected villages so that for aquaculture a level of role sharing may take place which will ultimately contribute to the household. Selection of these Female-operated Model Villages, as they have become known, is similar to the original Model Village selection process. The *thana* fisheries officer and the project extension officer discuss the programme with various members of a village community but with the important difference that programme training is targeted only at women members of households. If the community is in agreement, and the village meets with the selection criteria of other villages, it is selected. At present these Female-operated Model Villages are being piloted with encouraging results.

Secondly the Project is linking the Model Villages with a formal agricultural credit institution. This will allow the stakeholders access to credit specifically for the purpose of aquaculture. This initiative gives confidence to both the villagers and the credit institution as the villagers have received a certain level of aquaculture training from both government and project staff.

Chapter 22

Participatory Development of Aquaculture Extension Materials and their Effectiveness in Transfer of Technology: the Case of the AIT Aqua Outreach Programme, Northeast Thailand

D. Turongruang and H. Demaine
Aquaculture and Aquatic Resources Management, School of Environment, Resources and Development, Asian Institute of Technology, PO Box 4, Klong Luang, Pathumthani 12120, Thailand

Abstract

The Aqua Outreach Programme (AOP) of the Asian Institute of Technology (AIT) in Northeast Thailand has been engaged in the development of technical recommendations and extension approaches for small-scale aquaculture for over 10 years. In the absence of a conventional extension service in the Thai Department of Fisheries, AOP experimented with a distance extension approach, centred on the participatory development of extension materials based upon successful results of on-farm trials. By making these materials appropriate to farmers' lifestyle, culture, language and learning experience, it was hoped that they would be an effective tool for information dissemination. Evidence of surveys on project impact substantiates that hope. Different groups of farmers receiving only materials, only training and both training and materials were surveyed. Over 70% followed recommendations to some degree, with higher levels amongst groups receiving the extension materials. Adoption of pond fertilisation was higher than nursing of fingerlings, although most farmers adapted the principle to their own resources. Typical yields were around 250kg rai^{-1} (1,600m^2) higher than before adoption and many farmers spread the knowledge gained to others. Given that over 6,000 sets of the materials have been distributed in Northeast Thailand, an incremental production value of US$0.5 million annually arising from the recommendations is estimated.

Introduction

The AIT Aqua Outreach Programme (AOP) in Northeast Thailand began in 1988 as an adaptive research project aimed at developing appropriate technical recommendations for aquaculture for small-scale, non-specialist farmers, through a farming systems research approach. By 1990, the Project began to feel confident that it had identified a package of low-cost recommendations which raised yields in earthen ponds by some 300% over the estimated baseline fish production of 60-80kg rai^{-1} (1,600m^2). The project then began to assess ways of disseminating information about this package more widely to a larger number of farmers taking into consideration the resource limitations of its main collaborating partner, the Thai Department of Fisheries (DoF). A SWOT analysis of the social and institutional context for extension revealed that DoF had very few extension officers operating from the provincial level, none of whom had been formally trained in extension techniques. On the other hand, it was dealing with a mobile and literate farm population, and had close links with a number of other Thai government agencies that were represented at the grass-roots (commune and village) level. AOP thus determined that a 'distance extension' approach relying on the dissemination of information through these non-specialist agencies through mass media might be the answer (Demaine and Turongruang, 1996).

Although it has been the dominant system for over two decades, there are growing doubts about the effectiveness of the conventional Training and Visit System (T and V) as a substantial basis for agricultural extension in developing countries (Farrington, 1994). The T and V system was born in the transfer of technology paradigm of extension and involves the passing on of technical messages through a dense network of extension agents working with contact farmers in the villages (Benor and Baxter, 1984). T and V systems have been established throughout Asia by a series of World Bank projects, but have proved to be both unsustainable in terms of current expenditure and unable to cope with the complexity of farming systems in resource-poor areas (World Bank, 1994). Such systems have also tended to be crop specific, but governments have been reluctant to allow other sectoral departments in fisheries, forestry and livestock to create parallel structures. Thus most fisheries departments find themselves with a minimal complement of extension workers, which mirrors the situation in the crop sector 20 years ago. Most of them are fisheries scientists with no formal training in extension (Melkote, 1991). The Thai DoF appears to have been no exception, until recently having a maximum of six extension agents per province (Komolmarl, 1992).

In the absence of conventional extension services, one alternative is the use of mass media in the form of leaflets, pamphlets, videos, television and radio. According to Van Den Ban and Hawkins (1988), 'newspapers, magazines, radio and television generally are the least expensive media to carry messages to large numbers of people'. While doubts have been cast about their effectiveness in influencing decision-making, as opposed to creating awareness of innovations and stimulating their interest in them, a distinction should be made here between audio-visual and printed materials. These can be complementary in that the former may be used to create awareness, while the latter supply 'a constant reference, as a person can keep a leaflet and refer to it at any time' (MacDonald and Hearle, 1984).

Unfortunately many such materials are designed by research scientists working with the staff of centrally located media units, who have little or no contact with farmers. The extension materials usually contain too much information presented in too scientific a format, written in too educated a language for farmers to understand. The Thai DoF Extension Division has used many types of media materials, but these have been made centrally with content deriving from the results of on-station research. The language used for the materials is central Thai, with many technical terms; pre-tests of the materials were carried out by circulating the materials amongst extensionists in the Division, with no involvement of fish farmers (Lertpradist, 1990). During 1987-1991, 42 different extension materials were produced, 31 in printed form, but only 4 to 5 had contents which were suitable for use by small-scale farmers, assuming they could cope with the language and the technical terms (Komolmarl, 1992). Not only were the materials of little use to small-scale farmers; farmers who were interested had to collect them from the provincial office.

The AIT AOP has sought to address these problems through a participatory approach to the production of such extension materials.

Design and Development of Extension Materials by AOP

The approach aimed to create greater efficiency in the use of extension materials to transfer technology so that farmers could use extension materials as points of reference in their farm practice. Audio-visual materials and printed materials were used complementarily. The former created awareness and explained where the printed materials could be obtained; the latter offered the details required to follow the recommended practices (Turongruang, 1995). In the design of the printed materials in particular, a great detail of attention was paid to making them appropriate to the farmers' lifestyle, culture, language and learning experience. A good deal of this information came from interaction with and observation of farmers during the process of on-farm trials, although topical surveys were also carried out to assess their habits in listening to radio and viewing television.

On the basis of these observations, the basic format of printed materials could be designed. Messages were thus clear, short and sharp, with the point of the message instantly obvious. They were written in straightforward language, using local dialect and with no technical terms. The structure of the message had to be clear and well arranged, while the material had to attract attention, be stimulating to read and, if possible, be entertaining as well. Even the size of print was considered; although most households included members with more recent schooling who could read fluently, most farmers had left school many years previously and were not used to reading. Abstract concepts such as tables were kept to a minimum. Care was also taken that pictures did not convey the wrong message.

To ensure that these principles were followed, the Project developed an iterative design process (Fig. 22.1). The development of printed materials began with a brainstorming exercise bringing together all Project staff who had been involved in the development of technical recommendations and in monitoring the on-farm trials. Extension messages were drafted based upon the information from the on-farm trials, with media professionals instructed by the field team. The media professionals could advise on technical design issues, but could not change the content or specific words. The 'media professionals' mentioned here were local people who were also trained in and took part in field research. Both understood the local Northeast Thai dialect and customs. Here the AOP approach differs considerably from the model of Mody (1991) in which the design professionals take the results of field research and are dominant in the production process.

Once promising extension messages have been developed and initial drafts drawn up, these are field-tested in the same way as the aquaculture technologies. Farmers who had not been involved in the technical trials were briefed at village meetings about the materials and were left to make their own decision about whether to try out the recommendations included in the leaflets. Those who did so were visited regularly by Project staff, with the emphasis on understanding and use of the messages. At the end of the trials, an analysis was carried out on whether anything was missing from the leaflets which caused problems in implementing the technical messages. The success of the materials was evaluated on the basis of the degree to which the farmers were able to implement the technology.

The promising materials were revised on the basis of the experience of these trials and a further round of evaluation by another representative group of farmers — this time without conducting the actual trials — was made. In this case, materials were left with the farmers for a few days for study and comment. Only when this further evaluation appeared positive were the materials moved to production and distribution. Great care was taken in the selection of farmers to take part in the evaluation process on the basis of age, level of literacy, level of education, main occupation and gender, ensuring that the group contained no government officials or people with members in the family with higher education. Care taken at this point would ensure that messages could be transferred from farmer to farmer and overcome some of the concerns about the atypicality of contact farmers in the T and V process.

The process of materials development is a lengthy one. On the basis of the experience of on-farm trials in year 1, extension materials are drafted in year 2 for testing with trialists who do not receive the same degree of technical monitoring from Project staff. Revised materials are then re-evaluated without

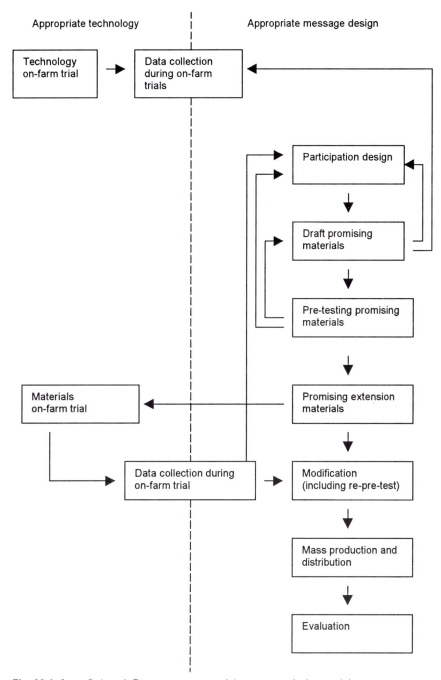

Fig. 22.1. Aqua Outreach Programme appropriate message design model.

trials by a further group of farmers before finalisation and distribution in a third year. This process may be unnecessarily long and efforts have been made to refine and shorten it since the initial round of experimentation. The Project has introduced a pre-test procedure which involves trialist farmers directly in the design of the initial materials and their evaluation without further on-farm testing with a separate group. Specifically Project staff:

- Test pictures alone to measure attention, first impressions and understanding
- Read the text associated with the pictures to farmers
- Interview them for their comments and suggestions.

This direct involvement of trialist farmers with extension and media staff has been tried successfully in the development of booklets and leaflets with the DOF Extension Division in Thailand and with DOF in Cambodia (Gregory *et al.*, 1996)

Effectiveness of the Extension Materials

AOP has conducted several studies to evaluate the use of media materials produced through this approach. The first of these specifically addressed the extension system, including the materials, by distributing the latter through non-specialist channels (schools, health centres, the agricultural bank, the general agricultural extension service) in two districts of Udorn Thani province not involved in the on-farm trials in 1992/93. The results of this 'extension experiment' by the project and by the DoF in three provinces of the region are presented elsewhere (Demaine *et al.*, 1994a; Demaine and Turongruang, 1996).

A more comprehensive evaluation of project materials has been underway since 1996 as the AOP has sought to assess its wider impact in Northeast Thailand. Over a period of 6 years since their initial testing, the AOP extension materials have been distributed widely to other organisations, not only the close collaborators in Phase II, the DoF and the Department of Vocational Education (DOVE), but also to NGOs, other development projects and individual farmers. Many of the materials have been distributed alongside training programmes, some of them conducted by government departments, some by NGOs, some independently, some in collaboration with Project staff. AOP has sought to find out how far such organisations have used the materials distributed, the degree to which farm households have taken up the technology and the benefit gained. Unfortunately the spread has been wide geographically and it has been difficult to obtain information about distribution from some of the organisations. NGOs in particular have changed personnel several times, leading to a loss in institutional memory.

For that reason, the AOP's evaluation has concentrated mainly on groups contacted by its main collaborators. Analysis of the contacts of these agencies reveals that there have been three different groups in receipt of AOP technical recommendations:

- Those who came to get extension materials from the Project itself from the DoF themselves and did not receive any training (49 cases)
- Those who received training and also got extension materials during the training (107 cases)
- Those who received training, but were not given extension materials (26).

This breakdown has enabled AOP to evaluate whether the materials are effective in their own right or whether they are better used supported by training, as suggested by some farmers in the initial extension experiment evaluation (Demaine *et al.*, 1994a).

Results

Source of Recommendations

Farmers who had not received training were asked details of their source of knowledge of Project extension materials. Of the 49 farmers in this category, over half had been told by the officers of the DoF, with almost 25% finding out from the radio. Most (65%) then went to seek them out themselves, while 20% got them by post after writing letters to the Project office. Others asked friends to collect the materials for them.

Almost all farmers obtaining the materials, both in the course of training and through other means, had received all three booklets concerning the three recommendations: *hapa* nursing (advanced nursing of fingerlings from 2-3cm to 6-8cm using a mixture of rice bran and livestock feed concentrate), making green water (fertilising ponds with buffalo manure supplemented with urea to produce phytoplankton) and the practice of polyculture (herbivorous and omnivorous species of fish low in the food chain).

The 133 farmers receiving training had received their information about the training from two main sources: DoF officers (30%) and from village and commune leaders (exactly 50% of all cases). A minor source of information (10%) was commune level agricultural extension officers. Most farmers entered the training with a view to receiving more information about fish culture, but a significant minority claimed that they had been told to do so by, or went as a substitute for, somebody else. Around 12% were expecting to receive free fish seed through the training.

Understanding of Leaflets

Almost every one of the 156 farmers receiving the extension materials had read them and almost everybody claimed to have understood what they had read. Of the four cases not understanding, two said they were not sure about fish feed for nursing, while the two others did not understand the steps in making green water and in nursing. Most farmers saw benefit in the materials: to increase their ability to rear fish by themselves; to increase knowledge in general; and to enable them to advise other farmers. Over half of those receiving materials still had the full set for reference and most had retained at least part of the set.

Use of Recommendations

Of these 182 farmers, 164 actually had fish ponds so that they were in the position to put the training/information to use. The distribution by group is set out in Table 22.1.

Table 22.1. Number of farmers surveyed who owned a pond by group (figures in parentheses are percentages).

Group	With pond	Without pond	Total
Only materials	44 (89.7)	5	49
Training/materials	98 (91.6)	9	107
Only training	22 (84.6)	4	26
Total	164 (90.1)	18	182

Of the 164 farmers with a pond, the adoption of project recommendations by group is set out in Table 22.2.

Thus over 70% followed recommendations to some degree, although the proportion was much higher amongst those who had received materials than amongst those only receiving training. In total, 63.5% of all farmers with a pond adopted the green water recommendation, while 45.8% followed the nursing recommendation.

Table 22.2. Number of farmers with a pond adopting recommendations (figures in parentheses are percentages of each group).

Group	Did not follow	Green water only	Hapa nursing only	Both	Total
Only materials	9 (20.4)	16 (36.4)	6 (13.6)	13 (29.5)	44
Training/materials	21 (21.4)	29 (29.6)	9 (9.2)	39 (39.8)	98
Only training	12 (54.5)	2 (9.1)	3 (13.6)	5 (22.7)	22
Total	42 (25.6)	47 (28.7)	18 (11.0)	57 (34.8)	164

Adoption of recommendations usually means the use of the technology for more than one year. The average length of time that farmers had been following recommendations when surveyed in late 1996 is presented in Table 22.3.

Nursing

Before receiving information from the Project, only six households (3.7%) had previously nursed fingerlings, with no major differences between the three groups. Of these, two were relatives of fisheries officers, while the remainder had obtained information elsewhere.

Table 22.3. Length of time farmers had adopted the recommendations.

Recommenda-tions/group	Only materials		Training and materials		Only training	
	Years following	Farmers (number)	Years following	Farmers (number)	Years following	Farmers (number)
Green water	2.7	15	2.5	27	0.7	3
Hapa nursing	2.0	6	2.2	9	1.9	2
Both		14		39		5
Total		35		75		10

Of those farmers with a pond who rejected the recommendation without trial (89 cases), 21.3% (mainly those who had received both training and materials) said that it was unnecessary since they already stocked large fingerlings and/or dried and prepared their ponds and three other farmers nursed fingerlings in cement or earth ponds without a *hapa*. Arguably this group had accepted the principle of nursing and their answers did not indicate outright rejection. However, 24.7% indicated that their pond was unsuitable for nursing; 13.5% answered that they only reared fish for consumption, apparently having enough to eat without improving their system; 10.1% claimed lack of labour; 6.7% lack of capital; and another 6.7% were engaged in more profitable business. Clearly these people totalling one third of all farmers with ponds regarded the technology as unsuitable for their system. A rather disproportionate proportion of these rejecters came from the group only receiving materials. Another four farmers from this group said that they did not understand the materials.

Among the group of farmers trying out the technology, the length of time they had adopted the innovation ranged from 1 to 5 years, with 49.3% of the 75 farmers adopting having tested it for only one year. Only 20 farmers had nursed fish every year after adopting, while as many as 46.7% had given up after a single year (Table 22.4). The table suggests that farmers receiving training were more likely to use the technology every year, or from time to time, than those who were only receiving materials.

Reasons for discontinuance which do not indicate a total rejection of the technology included the fact that the farmers had not restocked, having sufficient fish in the pond (22.6%), or that they had prepared the pond and released large fingerlings (15.1%). However, a further 15.1% indicated that the innovation was ineffective, 13.2% said they had insufficient labour and 9.4%

said that the technology was not suitable for their pond system (conceivably too shallow or easily flooded). These could be construed as rejection of the technology, as could the reply of 17% of farmers who said that they had other enterprises that offered a better return.

Table 22.4. Regularity of use of the nursing recommendation (figures in parentheses are percentages of each group).

	Materials only	Materials/ training	Training only	Total
Every year	3 (15.8)	14 (29.2)	3 (37.5)	20 (26.7)
Some years	4 (21.1)	14 (29.2)	2 (25.0)	20 (26.7)
Only first year	12 (63.0)	20 (41.7)	3 (37.5)	35 (49.3)
Total	19	48	8	75

Pond Fertilisation

Fertilisation was practised by significant numbers of farmers (14.6%) prior to their exposure to Project recommendations. Half of these had used only animal manure and a further quarter compost, consisting of animal manure and vegetable matter. Only four farmers had used chemical fertiliser. The introduction of inorganic fertilisation in the region is thus closely linked to the AOP and the fact that over 60% of pond owners adopted this recommendation (Table 22.2) is a particular indicator of the success of the Project approach and recommendations.

Of the 60 farmers who did not adopt the green water recommendation (Table 22.2), 18.3% were not actually culturing fish, but merely trapping wild fish. Arguably this group can be excluded from the analysis. Almost a quarter, however, said that their ponds were not suitable, probably indicating a too open system to make fertilisation effective. Another 15% said that they were too busy with other enterprises, 8.3% claimed lack of capital and 13.3% appeared to be satisfied with their existing production. More disturbing for the Project was that a similar number said they had not understood the recommendations, the bulk of these among those who had received only training.

Of the 104 farmers taking up the recommendation, the length of time they had followed it ranged from 1 to 6 years. Although 35.6% had only utilised it for one year, over 30% had adopted for more than 4 years. Farmers did not adhere to the green water recommendation every year, as indicated in Table 22.5. As many as 58% of farmers did not fertilise their water every year, although those who had received both training and materials tended to do so more consistently than their colleagues.

As many as 80.7% of those farmers adopting green water as a single recommendation made changes to the recommendation to suit their systems, indicating that it is not appropriate to be too prescriptive about the technical package (Table 22.6). Rather it is the principle of the recommendation that matters. In fact 56.7% of those making their pond water green followed the first round Project recommendations of mixing urea with buffalo manure, but quite a number changed the amounts of inputs specified. Others changed the recommendation according to their resources: 14.4% used other animal manure with urea either simply or with some supplementary feed; 22.1% used only animal manure; and 6.7% animal manure with supplementary feed. Among farmers who tried to make water green but did not follow the urea-buffalo manure recommendations, half claimed they did not have sufficient capital to buy urea, while almost all others practised integrated farming with small livestock (pigs or poultry) reared over the pond. This was most common among those who had received both materials and training. Overall those receiving training tended to follow the package more exactly, while more of those receiving materials tended to use animal manure alone (Table 22.7).

Table 22.5. Regularity of making green water in farmers' ponds (figures in parentheses are percentages of each group).

Group	Did every year	Not every year	Total
Only materials	5	10 (66.7)	15
Training/materials	14	13 (48.1)	27
Only training		3 (100.0)	3
Total	19	26 (57.8)	45

Table 22.6. Proportion of farmers making changes to fertilisation recommendation (figures in parentheses are group percentages).

Group	Changed recommendation	No change	Total
Only materials	24 (82.7)	5	29
Training/materials	54 (79.4)	14	68
Only training	6 (85.7)	1	7
Total	84 (80.7)	20	104

Table 22.7. Method of making water green as adopted by farmers (number of cases).

	Only materials	Training/ materials	Only training	Total
Urea + buffalo manure	12	43	4	59
Urea + buffalo manure + crop wastes	4	1		5
Buffalo manure only	10	11	2	23
Urea and other animal manure	2	7	1	10
Manure + others		2		2
Manure + crop waste	1	4		5
Total	29	68	7	104

Changes in Levels of Production

Farmers were asked about their previous fish production compared to their current production after adopting recommendations. While these are rough estimates, the results of the three groups of farmers in relation to their adoption of the two main recommendations showed significant increases, especially where both recommendations were adopted (Table 22.8).

Table 22.8. Estimated levels of fish production before and after using recommendations, in kg rai^{-1} (1rai = 1,600m^2).

Group/ recommendation	Green water only		Hapa nursing only		Both recommendations	
	Before	After	Before	After	Before	After
Only materials	182.7	294.3	202.0	281.5	306.9	538.3
Training/materials	106.7	255.4	207.5	258.3	332.5	550.9
Only training	17.5	15.0	130.0	196.1	58.3	251.2

Asked about the change in production in another way, 55.4% of 101 farmers offering an opinion said that their production had increased by a large amount and 17.8% by a reasonable amount. A further 16.8% said they had achieved a small increase. Those farmers receiving materials tended to have better results than those receiving training only. Just over half the farmers expressed great satisfaction with this production increase, with another 31% claiming general satisfaction; only 17% said they were not satisfied. Of 122 farmers following recommendations, the vast majority (94.3%) felt that the recommendations were suitable for them; of the remainder, six farmers said that their fry had died after nursing.

Farmer-to-farmer Spread of Information

Farmers were asked whether they had spread information on the recommendations to others. Of the total number of 182 farmers (those receiving training, but not having a pond, could have told other people about their experience just as much as those with a pond), slightly under half had passed on the message. Farmers receiving only training were very much less likely to pass on information (Table 22.9). The availability of written material is clearly a factor in farmer-to-farmer spread of information.

Of those farmers giving information, the average number of fellow farmers contacted is presented in Table 22.10, although the averages hide the fact that one farmer receiving training claimed to have contacted over 100 others. A total of 624 farmers had been contacted, over three times the number surveyed and on average over seven farmers per sample farm.

Table 22.9. Farmer-to-farmer spread of information (figures in parentheses are group percentages).

Group/spread	Farmers who spread	Farmers who did not spread	Total
Only materials	28 (57.1)	21	49
Training/materials	59 (55.1)	48	107
Only training	3 (11.5)	23	26
Total	90 (49.5)	92	182

Table 22.10. Number of farmers contacted.

Group	Number of farmers contacted	Average number contacted
Only materials	142	5.1
Training/materials	362	6.5
Only training	120	60.0
Total	624	

Most of the farmers contacting others explained their experience verbally (55%), while smaller numbers (almost 30%) gave their friends the extension materials to read. In a few cases, a village meeting was called to explain the technology or the information was broadcast on the village loudspeaker system.

Conclusions

This study confirms our earlier findings (Demaine *et al.*, 1994b) that low-cost technical recommendations, well-tested in field trials, can be taken up by farmers without regular extension advice such as is received under a typical T and V framework. More important, carefully designed extension materials, developed in dialogue between farmers and government officers, can be an effective means of extension of technical messages, are more effective than training on its own, and just as effective as when training is given at the same time as the materials are distributed. Perhaps we should qualify this statement to 'more effective than poor quality training'. It is well known that a good deal of formal training carried out by government agencies is neither well-targeted nor carefully adapted in mode and content to its farmer audience. Government agencies tend to emphasise numbers rather than careful selection of the audience; many trainees are just there 'for the ride/beer/*per diem*'. By contrast, farmers who made the conscious effort to seek out the extension media materials themselves may be presumed to be more committed to learning in the first place.

The analysis may also allow some modest speculation about the overall impact of the AOP's recommendations in Northeast Thailand. Between 50 and

70% of persons receiving extension materials tried them out and perhaps half continued to use the recommendations regularly or from time to time when the need arose. Between 1991 and 1995, some 6,000 sets of extension materials were distributed to various groups in the region. Assuming a modest 40% of these materials led to uptake of recommendations, with an average increase in fish production of about 200kg (in line with the figures in Table 22.8, but also supported by the evidence from the on-farm trials carried out under the AOP), then the incremental fish production is 480tonnes annually, valued at Baht 19.2 million (US$500,000). Since 1996, the Thai DoF alone has distributed a further 6,000 sets of such materials in Northeast Thailand, a figure only limited by budget constraints. Assuming a similar impact, this represents a significant annual increment to production in the region from aquaculture. According to DoF statistics, annual fish production from aquaculture in Northeast Thailand in 1994 was just 16,094t (Department of Fisheries, 1996).

What the evidence presented does not indicate is the distribution of benefit. In fact, farmers' 'guesstimates' of their fish production prior to the adoption of project recommendations presented in Table 22.8 appear to be high. This might suggest that the households adopting recommendations as a result of receiving materials, either directly or through training, could be the rather better-off farmers, chosen for training from farmers usually in contact with government and more mobile.

The AOP rapid survey did not address this problem, but a more recent survey both widened the sample and included more evidence of socio-economic status. This survey is of 201 families receiving training or materials from the Department of Fisheries in 1996/97 in the three provinces of Udorn Thani, Sakol Nakorn and Nakorn Phanom. Preliminary results show that 74% of households had adopted some part of the technical recommendations. This included 20 out of 28 families who had received information from other farmers. There was little difference in uptake between the groups receiving training only, training and materials and materials only. However, analysis of the socio-economic status of farmers adopting project recommendations in comparison with non-adopters, did show the former to be slightly better-off. There were no differences in land holding, with both groups having holdings significantly below the minimum size to make a living from agriculture, but the livestock holdings of adopters were larger and ownership of household assets slightly higher. Comparison of income levels proved to be difficult because lower incomes from agriculture were offset by off-farm earnings (Demaine *et al.,* in preparation).

To address this major problem of targeting by government agencies in their training activities, recent support (training and materials distribution) has been provided to marginal groups in the border areas of Ubol Ratchathani province. Preliminary results show excellent take-up of recommendations and positive results are emerging in fish production, which indicates that use of a mass media approach to aquaculture extension may be relevant for poor, small-scale farming households.

References

Benor, D. and Baxter, M. (1984) *Training and Visit Extension*. World Bank, Washington, DC.

Demaine, H. and Turongruang, D. (1996) Distance extension for rural aquaculture in Northeast Thailand. Paper presented at the World Aquaculture Society, Bangkok.

Demaine, H., Turongruang, D., Innes-Taylor, N. and Phromthong, P. (1994a) The AIT Outreach extension experiment 1991-92. Working Paper No.14 (revised edition), AIT Aquaculture Outreach Programme, Udorn Thani.

Demaine, H., Turongruang, D., Phromthong, P. and Mingkano, P. (1994b) Monitoring survey of extension experiment farmers 1993. Working Paper New Series T-2, AIT Aquaculture Outreach Programme, Udorn Thani.

Demaine, H., Turongruang, D., Phromthong, P. and Mingkano, P. (in preparation) Impact of project recommendations. Results of the Survey of Farmers Receiving Advice from Training or Materials Distribution 1995-97. Working Paper New Series T-10, AIT Aqua Outreach, Udorn Thani.

Department of Fisheries (1996) *Fisheries Statistics of Thailand 1994*. Fisheries Economics Division, Bangkok.

Farrington (1994) *Public Sector Agricultural Extension: is There Life After Structural Adjustment?* Natural Resource Perspectives, 2, Overseas Development Institute, London.

Gregory, R., Turongruang, D., Vathana, H. and Kekputhearith, T. (1996) Participatory development of an information leaflet for new entrant fish farmers in Cambodia. AIT Aqua Outreach, Cambodia, Working Paper C-4, Phnom Penh.

Komolmarl, S. (1992) Role of the Department of Fisheries in inland aquaculture development with emphasis on Northeast Thailand. MSc Dissertation, Asian Institute of Technology.

Lertpradist, P. (1990) A study of the application of a selective evaluation methodology in an extension setting. MSc Dissertation, University of Newfoundland.

MacDonald, I. and Hearle, D. (1984) *Communication Skills for Rural Development*. Evans Brothers, London.

Melkote, S.R. (1991) *Communication for Development in the Third World: Theory and Practice*. Sage Publications, India.

Mody, B. (1991) *Designing Messages for Development Communication: an Audience Participation Based Approach*. Sage Publications, India.

Turongruang, D. (1995) Examination of media design in relation to appropriate technology, with particular reference to the AIT Aquaculture Outreach Project. MA Dissertation, Agricultural Extension and Rural Development Department, Reading University.

Van den Ban, A.W. and Hawkins, H.S. (1988) *Agricultural Extension*. Longman Scientific and Technical/ Chee Leong Press, Malaysia.

World Bank (1994) Agricultural extension. Lessons and Practices No. 6, Operations Evaluation Division, Washington, DC.

Chapter 23

Issues in Rural Aquaculture

P. Edwards, D.C. Little and H. Demaine

This final chapter outlines key issues in rural aquaculture and discusses the extent to which the chapters in this volume have added to our general understanding. The main aims of rural aquaculture are to improve the livelihoods of the rural poor and to promote food security through improved food supply, employment and income. Increasing emphasis is being given in development programmes of national governments, non-government organisations and international donors to the goal of eradicating poverty which is regarded as an ethical, social, political and economic imperative. The concept of sustainable livelihoods, first proposed by the World Conference on Environment and Development, is also being increasingly followed by development practitioners.

Rural aquaculture defies simple definition although the beneficiaries are clear: the poor. In this respect, the use of the term is a reaction to the fact that, in the past, poor people have been largely by-passed in aquaculture development. It has emerged in the same way as rural development, as an approach to address the problems of rural poverty. Even when projects and programmes have claimed to be oriented towards poor people, frequently they have failed to address that target because the right approaches have not been used. All too often the poor have not directly benefited and may even have been adversely affected.

Technically, rural aquaculture has been defined previously mainly with reference to small-scale households using low-cost extensive and semi-intensive husbandry for household consumption and/or income. However, it may be better to broaden the definition to also include more intensive systems of production. The poor may be able to benefit also from operating small-scale but higher input and higher cost systems of production. Nor is rural aquaculture the only way in which aquaculture can help the poor. Rural poor people may benefit also through employment on, or through providing services to, the farms of better-off farmers and from provision of low-cost fish, irrespective of the scale of farming operation.

The 22 contributions to the book and the discussion below are arranged under six main topics as follows:

- Environmental context
- Integration with agriculture
- Specialised and intensified systems
- Seed
- Social aspects
- Development models.

The contributed chapters reflect a growing trend for holistic, systems-based approaches to rural aquaculture rather than the traditional and narrow, scientific focus on aquaculture technology. Most chapters cover two or more topics with varying degrees of emphasis (Table 23.1) which reflect the diverse nature of rural aquaculture and the need for a multidisciplinary approach to address its complexity if it is to fulfil its potential contribution to improving the livelihoods of the poor.

Issues

The following major issues are addressed in the discussion with respect to rural aquaculture:

- Environmental context
  - what is the relationship between aquaculture and wild fish resources?
  - how does water availability impact on potential for aquaculture?
- Integration with agriculture
  - do appropriate technologies exist?
- Seed
  - what is the role of the private sector?
  - are large centralised government or small, decentralised hatcheries more appropriate?
- Social issues
  - does aquaculture contribute to the alleviation of poverty?
  - are the poor early adopters of aquaculture?
  - do fish species influence the suitability of aquaculture for the poor?
  - do fish provide nutritional benefits for the poor
  - does aquaculture improve the situation of women?
- Development models
  - should individual poor farming households be targeted or communities with significant number of poor households?
  - should sectoral extension services or distance extension be developed?
  - should research and development projects work within the existing institutional framework or outside it?
  - is a top-down, transfer of technology research and extension or a bottom-up, 'farmer-first' approach more appropriate?

Table 23.1. Classification of chapters by major (●) topics under which they are ordered in the book and subsidiary topics (○). Country code: Bangladesh, 1; Cambodia, 2; India, 3; Indonesia, 4; Laos PDR, 5; Philippines, 6; Pakistan, 7; Sri Lanka, 8; Thailand, 8; Vietnam, 9; Vietnam, 10; Regional, 11.

Topic	Chapter number																					
	1	2	3	4	5	6	7	8	9	10	11	12	13	14	15	16	17	18	19	20	21	22
Environmental context	●																					
Rice field fisheries	●	●																				
Irrigation systems			●																			
Environmental impact	○	○	○	○		○	○	○	○	○	○					○			○	○		○
Integration with agriculture																						
Rice-fish culture	○		○	●	●			○											○	○		
Pond culture			○	○		●	●									○	○		○	○	○	○
Fish species			○													○	○			○		○
Specialised and intensified systems								●														
Macrobrachium									●													
Inorganic fertilisation										●												
Cage culture											●											
Seed																						
Supply	○											●		○	○							
Quality													●	●	●							
Social aspects																						
Benefits from tilapias															○	●						
Food security															○		●				○	
Gender																		●				
Risk			○																			
Economic aspects									○							○						
Development models																						
Approaches	○	○		○	○	○	○		○				○	○	○	○			○	○	○	○
Surveys	○	○		○	○	○	○						○		○				○	○	○	○
On-farm trials				○	○							○				○			○	○		
Institutional aspects	○	○		○	○						○		○			○	○					
Extension																			●	●	●	●
Training					○								○						○	○		○
Country code	2	2	3,7,8	1	10	10	6	1	9	1	4	5	3	11	6	1	1	9	10	10	1	9

Environmental Context

Environment in the context of rural aquaculture refers here to the resource base in which aquaculture may or may not be an appropriate social, economic or technical option, rather than the classic issue of adverse impact on the environment. In contrast to intensive culture of fish and shrimp, rural aquaculture may have positive impact and lead to increased sustainability: construction of a pond provides a focal point for agricultural diversification by providing a source of water; and farmers reduce pesticide usage when stocking fish in ricefields.

It is perhaps ironic that the first two chapters in a collection of chapters drawn together under the title 'Rural Aquaculture' actually question the focus of the book. Gregory and Guttman in Chapters 1 and 2 question the conventional view that aquaculture should be promoted uncritically by scientific and development communities to fill the gap between a perceived declining resource of wild fish and increasing demand from expanding human populations. They ask whether the focus should be on rural aquaculture at all, but rather on fish or even more generally, on aquatic animals and plants, and their role in rural livelihoods. Official estimates of the consumption of aquatic animals, much caught and foraged in and around ricefields, are usually grossly underestimated.

In the 'rice-fish societies' (the 'rice-fish cultures' of the second Gregory and Guttman chapter) covered in this review, most specialists would probably argue that fisheries is an appropriate entry point for rural development, given the importance of fish in the diet of local populations. In fact, all three editors are currently engaged in a project, funded by SIDA, actually entitled 'Rural Development through Aquatic Resources Management'. The most important question under debate is what is the source of that fish?

Gregory and Guttman argue that the starting point for assessment of the need for development project intervention in a particular area should be fish and that projects should start neutrally with no particular bias towards wild or stocked aquatic organisms. As the ricefield fish catch continues to be important in some contexts, especially floodplains in Southeast Asia, aquaculturists should be circumspect in promoting the enterprise. These authors call for the development of methods which serve to identify the contexts in which aquaculture is an option and propose that fisheries scientists develop approaches for research and development of both sectors.

The importance of ricefield-based fisheries to rural livelihoods has 'fallen between two stools' as it has been neglected by both fisheries specialists who have concentrated on river and large impoundment fisheries and aquaculturists who have focused on ponds and cages. Furthermore, the lack of holistic thinking, planning and implementation by planners, agronomists and irrigation engineers aiming to optimise rice yields in a mistaken belief that food security equates only with 'rice security' has had major negative impacts on sustaining nutritious aquatic foods for poor people. However, doubts are also cast on the feasibility of restoring/maintaining the wild fishery. Several studies have emerged recently which have improved understanding of the ricefield fisheries, but these have not led yet to readily identifiable ways and means to improve or maintain the resource.

On the whole, however, there is growing agreement that efforts should be made on both fronts. In relation to the ricefield fishery, it is important to marshal the evidence of the value of aquatic resources in comparison with alternative land uses, particularly in areas that may be threatened by development for, and intensification of, rice cultivation. There should be appraisal at the project design stage, with careful economic analysis including the true value of the existing system that considers the 'hidden' harvest of fish and other aquatic animals and plants that often disproportionately benefit the poor. This would at least underline to policy makers the potential negative impacts on fish of irrigation, flood protection and draining schemes undertaken to intensify rice production.

One key management step to sustain ricefield fisheries appears to be to maintain critical habitats for fish that adjoin the paddy fields, i.e. streams and perennial swamps. Maintaining supplies of naturally spawned and recruited wild seed of fish such as climbing perch and snakehead is very important as these species fetch a high price and may contribute significantly to household livelihood. This may even be the case in areas where aquaculture has begun to develop. Although often seen by aquaculture specialists as predators, such fish may contribute up to 10% of the total harvested catch, even with good pond preparation. There is growing evidence that what are now being termed 'self-recruiting species', which include not only the high value species mentioned but a wide range of aquatic organisms, are of considerable but unrecognised significance in aquaculture. A recent DFID-funded project is setting out to examine the situation in five countries in the region: Bangladesh, Cambodia, India (West Bengal), northern Vietnam and Thailand.

More intensive use of water at the household and community level and changes in access by poorer members of communities suggest that a broader view than the aquatic foods themselves is required, perhaps at the level of the water resource itself. In some cases, of course, irrigation development will go ahead but to mitigate the worst effects of its development, it is important to obtain a better understanding of the ways in which different types of irrigation systems affect fish and how irrigation systems themselves offer opportunities to integrate fish production, These issues are addressed by Murray *et al.,* in Chapter 3, who distinguish between farmer-managed, often small and seasonal irrigation systems and those constructed and operated by outside professionals and institutions. Identifying irrigation system designs and management that have neutral or positive impacts on the quality and productivity of aquatic stocks is a key need that is currently being addressed by Lorenzen and colleagues in Lao PDR.

The nature of the irrigation development will pose very different questions to sustaining or increasing the production of aquatic animals. Whereas maintaining the productivity and diversity of current systems is a key issue in very fish-dependent cultures in Southeast Asia, opportunities elsewhere may require stimulation of demand along with production.

In the Punjab, a historically dry area where pre-irrigation there was little dependence on aquatic foods, canal and ground water irrigation has raised novel opportunities for food production that have largely been taken up by resource-rich entrepreneurs. A similar situation has occurred in India where the major increases in cultured freshwater fish have occurred in irrigated areas such as Andhra Pradesh distant from the major markets for fish in eastern India. An analysis of this issue suggests that technologies and approaches appropriate for more resource-poor farmers have not been developed. The realisation that all agricultural development will require improved approaches to water use, often based on local and traditional approaches, which offers opportunities for introductions of fish culture. Recent fieldwork suggests that latent demand often exists even when consumption of fish is not traditional or widespread.

A further challenge is to ensure the sustained production of fish from community reservoirs and tanks that are common in water-scarce, seasonal environments. The dependence of poor rural people on tilapia stocks in perennial 'tanks' in Sri Lanka has recently been highlighted and opportunities for value addition through incorporating culture (fattening) of caught wild fish developed. In this case 'stand-alone', full-cycle aquaculture was found to be inappropriate to the needs of producers themselves or the numerous poor people who trade and consume the fish throughout the Dry Zone. Fish culture will often be a secondary or 'niche' activity within a broader water resource development framework and opportunities need to be developed locally with this firmly in mind.

Integration with Agriculture

Generic technologies for aquaculture integrated with aquaculture do exist as aquaculture is an ancient and traditional practice in some countries and regions in Asia, e.g. China and northern Vietnam, respectively. As aquaculture probably evolved in response to human population pressure causing declines in wild fish supply, it is relatively new farming practice in many Asian countries, e.g. Bangladesh, with considerable unrealised potential to contribute to improved welfare of the poor. The four chapters in this section cover ricefield and pond integration in lowland and upland areas where aquaculture is both a traditional and newly introduced practice.

Much integrated aquaculture is located in peri-urban areas where it depends on re-use of nutrients contained in manure from feedlot livestock enterprises as well as agro-industrial wastes and domestic wastewater. In contrast, integration in rural areas where there is relative nutrient scarcity is likely to be characterised as much by integration of space and labour as by integration through nutrient flows.

The promotion of rice-fish culture in Bangladesh, described by Gupta *et al.* in Chapter 4 is an example of the introduction of aquaculture as new farming practice at relatively low risk within the current resource base. An ancient practice in China, rice-fish culture has been promoted in recent decades in both rainfed and irrigated rice production systems in many parts of Asia. Ricefields may be used to produce food fish, usually for domestic consumption because of the small size of harvested fish but also as nurseries to produce fingerlings to stock other grow-out systems. Its much touted role as a low-cost entry point for poorer farmers to raise fish may have been overemphasised, however, as the Bangladesh study revealed that farmers with better access to resources (land, labour, knowledge) were more likely to adopt as has been observed also in other studies. The promotion of rice-fish culture also plays a role in integrated pest management (IPM), with better weed control in rice-fish plots than in control plots and farmers dramatically reduced the use of insecticides when stocking fish.

Papers by Luu *et al.* and Pekar *et al.* in Chapters 5 and 6, respectively, describe studies carried out on traditional farmer practice of so-called VAC integrated pond systems which integrate garden (*vuon*), pond (*ao*) and livestock quarters (*chuong*), hence the Vietnamese acronym. In recent decades a growing number of farming households have sought to increase fish production, or even to integrate aquaculture within their farms. Polyculture of carps is carried out in ponds in the north, while in the south a variety of fish (gouramis, carps, catfish and tilapias) and freshwater prawns are stocked in ponds, pond-ricefield complexes and irrigation canals. A common feature is the nutrient-limited productivity of most of the systems despite widespread rearing of small numbers of pigs in sties and scavenging poultry within the farm. In the north, production has been intensified chiefly through collection of wild grass to feed grass carp, but to collect grass, women have to travel increasing distances and spend an increasing amount of time as fish culture gains popularity. Both studies have led to significant increases in fish yields through introduction of better aquaculture practices, such as pond preparation, stocking larger fingerlings at lower density and introduction of multiple stocking and harvesting strategies. However, a limit to further intensification of grass carp dominated ponds was revealed in the north. Other studies indicate that the amount of organic manure is limited as it is used strategically by farmers to maintain the productivity of rice, other field crops and vegetables.

Clearly there is a need to introduce off-farm inputs to intensify production further because of this yield ceiling in traditional, semi-intensive, integrated pond culture. Trials also conducted in the Luu *et al.* study with inorganic fertilisers, the cheapest off-farm source of nitrogen and phosphorus, indicated this as a promising option in ponds stocked with a high proportion of the phytophagous Nile tilapia, an observation confirmed elsewhere with other field-based research (e.g. see Chapter 9). Another but more costly option is for farmers to use off-farm feed for either fish (see next section) or for feedlot livestock, with the manure used to fertilise the pond. However, promotion of feedlot livestock/fish integration for small-scale farmers has failed, following withdrawal of project support in several cases because of various constraints such as capital and operating costs, input supply and marketing, and susceptibility of feedlot livestock to disease.

The role of multipurpose ponds stocked with fish as a component of watershed management in fragile, upland areas through stimulating greater bio-resource recycling has been researched by Prein *et al.* in the Philippines (Chapter 7). Other recent work indicates that seasonal tanks in upland watersheds in southern India and Sri Lanka fulfil key livelihood roles in addition to their primary role in irrigation. In many cases fish production may be integrated, but research indicates that a poor understanding of current access to, and benefits from, such resources frequently results in failure. The critical nature of social issues in sustaining benefits to the poor from aquatic resource management is considered below in the section on social issues.

Specialised and Intensified Systems

Economic pressures are leading to a growing level of specialisation in rural aquaculture. At its simplest, specialisation into particular species of food fish and fish seed production has marked an important step in the evolution of aquaculture in the countries where it has grown most rapidly and is generating most impacts. The role of a range of entrepreneurs in developing and exploiting new species, production technologies and markets has yet to be fully analysed but they are clearly critical in supporting commercialisation of rural aquaculture. The relationships between promoters of specialised approaches, frequently alliances of government and commercial interests, are often obscure but typically there has been little primary concern with poverty alleviation or impacts on the environment. However, there is increasing evidence that poor people can benefit, particularly through employment opportunities. The rapid development of *gher* farming of the giant freshwater prawn (*Macrobrachium rosenbergii*) in converted ricefields in Bangladesh has raised issues as to both its environmental impact, and its role in changing land use and outcomes for the poor as reported by Chapman and Abedin in Chapter 8. The export-oriented system is highly profitable but is dependent on supplies of snail meat as the main feed input. Whilst collection of snails currently provides employment for the poor, its sustainability and potential impacts on the wider ecosystem are less certain. Moreover, the production system is high risk and depends on access to loans, which excludes poorer producers. The CARE GOLDA project has attempted to promote dike cropping but this has not proved attractive to profit-orientated *Macrobrachium* farmers. It is likely that the expertise developed by *Macrobrachium* farmers will endure and be applied as opportunities arise for other species and approaches. Transferable skills are a feature of entrepreneurial farmers who are typically involved in these types of specialised systems.

Encouraging rice farmers to specialise and intensify production within a limited resource base is a different issue, but one that has met with some success in Northeast Thailand (as reported in the study by Pant *et al.* in Chapter 9). Earlier attempts to increase production of carps and tilapias using on-farm resources and limited amounts of off-farm inorganic fertilisers built up confidence in some farmers to the degree that they began to seek means of further intensification. Although yields far exceeded control levels, impacts on farm incomes were relatively modest. Encouraged by consistently high yields of monosex Nile tilapia achieved in on-station trials and indicative returns, farmers were encouraged to use much higher levels of inorganic fertilisers. Average fish yields were more than three times previous level figures and returns were highly favourable. Ready availability of both nitrogen and phosphorus fertilisers and high local demand for fish, both common in other areas of Asia, were important to the success of the approach. This represented a 'jump' in intensification rather than a gradual process but both the culture system, and inputs and markets were familiar to the farmers used to lower input/output systems.

Elsewhere radically new forms of specialised aquaculture have been promoted and adopted. The promotion of cage culture in Bangladesh (outlined by McAndrew *et al.* in Chapter 10) and Indonesia (Munzir and Heidhues in Chapter 11), both attempt to focus on benefits to the poor, but in very different developmental contexts. These two examples demonstrate a fundamental difference in approach which may be summed up in terms of whether people and situations be identified that are appropriate for a given technology or whether technology should be adapted to meet the needs of people?

Cage culture has become a major culture system in Indonesia over the last decade whereas previous attempts to introduce the system to Bangladesh had failed. In contrast, whereas semi-intensive culture of carp in ponds has developed rapidly in Bangladesh, it is less important in Indonesia. These differences probably arise because of the popularity of common carp and, increasingly tilapias, in Indonesia which perform particularly well in cages, provided complete feeds are available. In Bangladesh the typical polyculture of native and exotic carp thrives in fertilised ponds that are a common feature of the floodplain landscape. Pond-based systems have developed in the absence of formulated feeds but the introduction of intensive, cage-based systems is likely to stimulate their development.

Specialisation in which large quantities of fish are produced consistently is also likely to change demand for fish. In Bangladesh, improved strains of tilapias are being favoured by cage operators and there are indications that this in time may begin to change perceptions of the fish from its current image as a low value trash fish (but see Barman *et al.* in Chapter 16). Undoubtedly specialisation increases risk to poor people and the CARE project in Bangladesh has attempted to reduce both entry costs and overall risks by reducing the size of cages, and by using local materials to fabricate them and feed the fish. A major risk is that once cage culture is determined to be profitable, control of inputs, production and/or marketing will become dominated by the resource-rich as has happened in cage culture in large reservoirs in Indonesia.

Uncontrolled uptake of cage culture in common property water resources risks deterioration of the environment. This has already occurred at several sites in Indonesia and the Philippines. Successful development of intensive cage culture in water bodies of limited carrying capacity requires careful regulation. There are currently few working models in developing countries but issues such as the role of producer associations, and the legal framework for access to water bodies and use of cages require urgent attention by policy makers if cage culture is to be sustainable and benefit the poor.

Seed

Carp-based aquaculture, which continues to dominate inland aquaculture in Asia, in the past tended to be limited to areas close to wild seed supplies. This may explain the tendency for fish seed production to be concentrated close to the rivers where hatchlings were harvested. The development and adoption of modern hatchery technologies, and additional species, has begun to change the nature of fish seed supply but the distribution of private sector hatchery and nursery operations often remains 'clustered'. Recent history has shown that although government and international promoters have often been involved in the early stages of hatchery development, the activity has typically subsequently been taken up by the private sector. Although some governments and agencies have been active in facilitating uptake by entrepreneurial households, administrative or resource constraints have often hampered this process until the knowledge has 'leaked out'. Even relatively modest efforts can 'kick start' private hatchery development provided that the demand to purchase fish seed, and thus the financial incentive, are strong.

The chronicling of the development of a single village in Orissa, India that has developed into a fish seed production centre by Radheyshyam in Chapter 13 suggests that considerable economic benefits can accrue to a single community and, possibly, the importance of farmer-to-farmer knowledge transfer. In some communities in northern Vietnam where nursing and trading seed are central components of livelihoods, efforts are made to retain the knowledge within the community. In contrast, in Thailand open training opportunities and widespread media coverage offered by promoters in the last two decades, together with different cultural attitudes, have ensured relatively widespread access to information and uptake. Both positive and negative impacts of these producer 'clusters' on the quality of seed have been identified by Little *et al.* in Chapter 14. Clearly, there are economic advantages for the clustering of any specialist enterprise and these must be understood if the supply of quality seed is to be improved.

Decentralisation of seed production, at least the latter stages of advanced fingerling production, may be critical for promoting aquaculture in areas where poor infrastructure and/or knowledge hampers distribution of seed. This contradicts much of the efforts over the last few decades in the region that focused on construction of large, centralised government hatcheries with little regard to meeting the needs of potential fish farmers. Litdamlong *et al.* in Chapter 12 report on the case of stimulating fish seed supply networks in southern Laos through promoting household-level spawning and nursing of fish seed. A key advantage of this approach is that it has allowed targeting of some of the communities' poorest households as *hapa*-based production can occur in public water bodies so that lack of land or a pond does not exclude participation.

For the Lao institution that introduced and promoted the approach, it has also been fundamental to efforts to increase capacity for rural development generally (see section on development models below). It also demonstrates the value of transferring generic technology, with local adaptation, as the general principles had been introduced from a related project in neighbouring Northeast Thailand (see Chapter 22). Local supply of seed has been shown to be a critical catalyst in interesting ordinary rice farmers in Laos to try including fish culture within their farming system. Recent evidence suggests that locally produced seed also outperforms fish transported from distant locations.

The production and dissemination of 'improved' quality seed has become a major objective of many promoters in the light of widespread perceptions of poor and declining quality of fish seed. The GMT (genetically male tilapia) technology has been one approach among several that have aimed to improve performance of Nile tilapia. Mair *et al.* in Chapter 15 explored the impact of the technology over a four-year period and found that mid- and upper-income people were the main beneficiaries of the technology. Although poorer people were not necessarily excluded from the largely 'passive' approach to dissemination of the technology, the authors make the case that targeting of assistance to the poor would improve the benefits they receive. Moreover, impacts on consumers and intermediaries were likely to be positive.

However, a major constraint to production of improved tilapias has been the difficulty in scaling-up production to have a significant impact on current systems that use mixed-sex fish, and then to maintain consistent levels of quality for the farmer. Breeding with feral tilapias makes the maintenance of 'improved' germplasm difficult under most conditions and it has proved particularly difficult to consistently produce monosex stocks that perform predictably in farmers' systems. Generally, the needs of farmers and consumers have also been misunderstood and 'quality' has been equated simplistically with individual fish growth. Very often farmers prefer mixed-sex tilapias, even when monosex are available, to reduce their costs and provide extra revenue through sales of seed locally. In many situations small tilapias are retained for consumption in the household and Barman *et al.* in Chapter 16 raise the issue that efforts to produce larger fish may prejudice household food security.

Social Issues

There is a growing debate on whether aquaculture can contribute to the alleviation of rural poverty. Although projects have only recently begun to assess the contribution of aquaculture to the welfare of poor people through use of the livelihoods approach, there appears to be evidence from a number of papers in this volume that, at the very least, poor people are not usually early adopters of new technology in aquaculture. In discussion of their paper in Chapter 4, Gupta *et al.* even stated that it does not make much sense to promote 'a richer man's technology for a poor man's problem'. Chapman and Abedin's Chapter 8 reviewing the experience of the GOLDA project for promotion of

Macrobrachium rosenbergii in ricefields in Bangladesh indicates that the new phase of that project is moving away from *Macrobrachium* because it is the domain of richer farmers. Conversely, it is argued by Gregory and Guttman in Chapters 1 and 2 that poor people may benefit from improvement in the management of small-scale aquatic resources which, as open access or common property resources, may be the only remaining livelihood option available to them. A similar argument is used for recommending cage culture as a possible option for the landless, provided that they can gain access to water bodies.

Aquaculture certainly contributed towards alleviating poverty in poor rural societies in the past in the few areas of the world in which it is traditional practice, e.g. China, Indonesia and Vietnam. Recent experience in Northern Vietnam indicates that poor farming households do adopt aquaculture since they perceive the value of fish as food and increasingly as an attractive option to diversify farming practice to generate income. It is also important that they already have a ricefield or pond in which to stock fish, they are familiar with a technology that fits in with the resources available to them, and seed is available at reasonable cost. As potential new entrants in areas with little tradition of aquaculture, however, poor farmers are likely to view aquaculture as a risky investment and it is unlikely they will adopt it without institutional support. Gupta *et al.* in Chapter 4 point out that in the early stages of technology adoption, small farmers require support in terms of training, input supply and credit. A modest subsidy to 'kick-start' the process may be justified if sustainable systems exist which can continue the process. Access to information and credit was also critical to poor people attempting cage culture of grass carp in Northern Vietnam. Cage culture has been adopted throughout the river systems of the north as part of complex livelihood strategies of both landless and landed people. Cage operators have sought to continue raising fish despite high risks of loss from disease, learning how to minimise losses and viewing the method as a flexible way to accumulate savings, as opposed to a regular source of cash or food.

This debate is raging widely in aquaculture, particularly where key donors in the sector in the region are placing more emphasis in their development strategies on poverty alleviation. Many of the papers in this volume refer to projects in which poverty alleviation is implicit; indeed the definition of rural aquaculture as discussed above assumes a poverty focus. Some countries have even established programmes with the explicit objective of addressing the needs of poor people. In Vietnam for instance, the Ministry of Fisheries has drawn up a strategic framework known as SAPA (Sustainable Aquaculture for Poverty Alleviation) as a contribution to the wider government policy for Hunger Eradication and Poverty Reduction. Unfortunately, the evidence that benefits reach the poorest groups is not so clear from the papers in this volume. Most talk of small-scale farmers and low-cost technology and contribution to food security, but there is little analysis of the socio-economic status of the target groups involved in the research and extension activities. Turongruang and Demaine in Chapter 22 offer evidence that the farmers taking up the recommendations of the AIT Aqua Outreach programme in Northeast Thailand are not better off than the regional average.

The attitudes of small-scale farming households towards relatively recently introduced tilapias in Bangladesh, where aquaculture is dominated by farming Indian major carp and Chinese carp were surveyed by Barman *et al.* in Chapter 16. Surveys revealed that tilapias are important for poorer households in Northeast Bangladesh, belying the misconception that they are unimportant in culture because few are sold in markets. The study revealed that farmers culture both tilapia and carp but the former are mainly used for household consumption, thus contributing directly to family nutrition whereas the latter are sold to the market to generate income. Recent evidence also supports the findings of Barman *et al.* of the emerging role of silver carp in diets of poor people. Despite an image of a poor quality fish, its low price and ready availability have stimulated huge demand. Interestingly, as institutions continue to concentrate on the silver carp's ability to grow fast to a large size, poor consumers prefer small fish, and farmers and seed traders benefit more from fish raised quickly and at high density. This paper underscores again the need for holistic, participatory studies to more accurately inform future strategies to improve livelihoods and that scientists engaged in technical development remain engaged in this process.

It is well known that fish contribute significantly to food security in many countries in Asia, particularly in rice-fish societies with large numbers of poor people. Fish provide at least 40% of dietary animal protein in nine Asian countries, highly digestible energy, fat and fatty acids, and water-soluble vitamins and minerals. Roos *et al.* in Chapter 17 describe the importance of fish in the diets of the rural population in Bangladesh and report how small indigenous species (SIS) are a particularly important source of vitamins and minerals for the poor. In an on-farm trial they demonstrated that one species of SIS (mola) contributed about 10% of the total fish production in polyculture with carp likely to be sold on the market and therefore it has the potential to make a direct and significant contribution to household welfare. Studies are now urgently required that relate the nutritional impact of both wild and cultured fish on an intra-household basis and within the context of the diet as a whole. A better understanding is needed of the nature of 'demand' for aquatic animals, which are among the most perishable of assets that poor households produce and use. The strategies by which households use wild and cultured fish to manage their food and income security are inadequately understood by institutions promoting their role in rural livelihoods.

Aquaculture may be a particularly valuable intervention to improve the situation of rural women. Many household ponds are close to the homestead, which allows women, often less mobile, to play a key role in husbandry. This is the argument put forward by Setboonsarng in Chapter 18, the only chapter in the volume that deals specifically with women. This chapter addresses the role of women in the wider perspective of the integrated agriculture-aquaculture farming system. Although the fish pond is the core of this system, it appears that fish culture is still dominated by men. Similar findings have also emerged in other recent work, including that of Barman in Bangladesh. Nevertheless, women's groups are the core of the Model Village project under the Northwest Fisheries Extension Project in Bangladesh outlined in Chapter 21, while UNDP projects in Vietnam also emphasise co-operation with the Women's Union at the grassroots level.

Development Models

The volume addresses possible ways to reach the poor and whether they should be explicitly targeted. The Northwest Fisheries Extension Project in Bangladesh outlined in Chapter 21 offers an example of a 'model village' approach in which communities with significant numbers of poor households are targeted rather than individual poor households. Islam and Mardall argue that this approach benefits more poor people than would otherwise be reached, even though many of the beneficiaries are not among the poorest groups. Furthermore the approach is managerially efficient.

The same could be said for the distance extension model described by Turongruang and Demaine in Chapter 22. Clearly the recipients of mass media for dissemination of technical messages cannot be 'targeted' in the sense of restricting distribution, but the messages themselves were developed on the basis of on-farm trials with poorer farmers and the extension materials designed in co-operation with them. Thus, it may be argued that the materials are aimed at poor farmers and, implicitly, that the richer groups looking for a more capital-intensive technology may not find them so relevant. Such approaches recommend themselves in contexts where the availability of conventional extension agents is limited, which is often the case in the aquaculture sector. Governments are increasingly unlikely to offer public funds for separate sectoral extension services. Bangladesh may be the exception in this regard, with one fisheries officer in each *upazila* (average population of around 200,000 people), but elsewhere the presence of trained aquaculturists even at provincial and district level may be very patchy. The paper by Tu and Giang in Chapter 20 offers a clear example from Southeast Vietnam.

A number of projects have recently started in the region such as the UNDP-funded Aquaculture Development in the Northern Uplands project in Vietnam and Phase III of AIT Aqua Outreach, funded by SIDA, which apply participatory rural appraisal techniques to target poorer groups. However, it has still to be established that this approach can be 'up-scaled' and implemented on a widespread basis. There is a danger indeed that the development ideals of the donor community do not fit in with the institutional capacity of the national partners, leading to what the paper by Litdamlong *et al.* in Chapter 12 term 'sunset' projects, in which the impact fades away immediately after the project ends. From the experience of the Lao PDR, the authors recommend strongly that projects should work within the existing system, working in partnership to effect change on a mutually agreed basis. Facilitating the development of operational research and development systems within local institutions may lead to a slower development of appropriate technologies than would be possible through a project team based in-country, but it has a better chance of being sustained.

Turning from wider institutional issues to the specific research-extension framework, many of the chapters in the collection reflect the major change that has been taking place in the wider agricultural development paradigm. This has involved a move from a top-down 'transfer of technology (ToT)' mode to working in partnership with farmers, sometimes called the 'farmer-first' approach.

Several chapters in this volume describe the more traditional ToT models based upon training and demonstration. This is frequently criticised by protagonists of the 'farmer-first' approach. In aquaculture, it is more difficult to observe demonstrations as the fish cannot be readily seen. More fundamentally, however, demonstrations have proved of little relevance for the wider, and particularly the poorer members of a farming community, because those selected for demonstration are usually the better-off who can afford to take risks. Frequently also, demonstration farmers are subsidised, making what they do even less relevant. As a general principle, it may be better in work with small farmers and not provide anything free, except advice and training. In the Lao PDR, the demonstration farm approach has become something of a joke, proving a good way to show poorer farmers what they *cannot* do.

By contrast, the 'farmer-first approach' attempts to address the problem of sustainability in agriculture from the farmers' viewpoint, putting farmers first, and not technology. Such an approach requires changes in roles and in the skills required by the key actors in the research-extension system. Extensionists, for example, have traditionally only a technical background in fisheries; they now need broader skills in participatory approaches to development in how to work with farmers. Although there is still a need for developing technical options for and with farmers, developing household capacity to make decisions or choices in relation to the available information and the risks that they face is more fundamental. Providing farmers with options and ideas and the ability to critically evaluate their choices is the central objective of CARE's involvement in rural aquaculture in Bangladesh. Participatory evaluation is used to develop farmers' potential for meeting their own goals. There is a need to 'spell out' information concerning resource use, technical options and markets in relation to farm livelihoods to avoid risks.

In such an approach, there may even be a need to move away from a commodity such as fish, even a sectoral approach such as fisheries, in consideration of improving living standards for the rural poor, in favour of an even more holistic view in which all aspects of livelihood are considered prior to identifying where key interventions can be made. In this view, rural development projects should not prejudge the issue of which focus might be suitable, including the unsuitability of fish production in some circumstances.

This issue poses a challenge to scientists who have become increasingly specialised in one particular field, particularly those who have built careers and reputations in a narrow technical area. In essence it is part of the debate on the relative importance of strategic and adaptive research. Adaptive research to introduce known generic technologies to specific local contexts is likely to have a greater short-term impact on poverty reduction. However, it would be unwise for donors to fund only adaptive research as strategic research has a potentially higher pay-off in terms of the number of poor who may eventually benefit from the research over a wide area.

Key References

Carney, D. (1999) *Approaches to Sustainable Livelihoods for the Rural Poor.* ODI Poverty Briefing 2, Overseas Development Institute, London.

Chambers, R. (1997) *Whose Reality Counts? Putting the Last First.* Intermediate Technology Publications, London.

Chambers, R., Pacey, A. and Thrupp, L.A. (1989) *Farmer First, Farmer Innovation and Agricultural Research.* Intermediate Technology Publications, London.

Clarke, G. and Mair, G.C. (1998) The Philippines: blue revolution? Who gains and who loses from the introduction of genetics-based technology in Philippine aquaculture. The Rural Extension. Bulletin, October 98, pp. 19-24. Agricultural Extension and Rural Development Department, University of Reading, Reading.

Costa-Pierce, B. (1997) From farmers to fishers, developing reservoir aquaculture for people displaced by dams. World Bank Technical Paper No. 369, Fisheries Series. The World Bank, Washington, DC.

Edwards, P. (2000) Aquaculture, poverty impacts and livelihoods. Natural Resource Perspectives Number 56, June. 4 pp.

Edwards, P. and Demaine, H. (1997) Rural aquaculture: overview and framework for country reviews. RAP Publication 1997/36. RAP/FAO, Bangkok.

Edwards, P., Demaine, H., Innes-Taylor, N. and Turongruang, D. (1996) Sustainable aquaculture for small-scale farmers: need for a balanced model. *Outlook on Agriculture* 25(1), 19-26.

FAO (1999) Report of the *Ad Hoc* Working Group of Experts in Rural Aquaculture, 20-22 October 1999, Bangkok, Asia-Pacific Fishery Commission. FAO Fisheries Report No. 660, FAO, Rome. 22 pp.

FAO (2000) Small ponds make a big difference. Integrating fish with crop and livestock farming. FAO Farm Management and Production Economics Service; FAO Inland Water Resources and Aquaculture Service. FAO, Rome. 30 pp.

Gupta, M.V., Ahmed, M., Bimbao, M.A.P. and Lightfoot, C. (1992) Socioeconomic impact and farmers assessment of Nile tilapia (*Oreochromis niloticus*) culture in Bangladesh. ICLARM Technical Report 35, ICLARM, Manila.

Gupta, M.V. and Rab, M.A. (1994) Adoption and economics of silver barb (*Puntius gonionotus*) culture in seasonal waters in Bangladesh. ICLARM Technical Report 41, ICLARM, Manila.

Gupta, M.V., *et al.* (1998) *Integrating Aquaculture with Rice Farming in Bangladesh: Feasibility & Economic Viability, its Adoption & Impact.* ICLARM Technical Report 55. ICLARM, Manila.

Guttman, H. (1999) Rice field fisheries - a resource for Cambodia. *NAGA* 22 (2), 11-15.

Halwart, M. (1998) Trends in rice-fish farming. *FAO Aquaculture Newsletter* 18, 3-11.

Hambrey, J., Beveridge, M. and McAndrew, K. (2001) Aquaculture and poverty alleviation 1. Cage culture in freshwater in Bangladesh. *World Aquaculture* 32(1), 50-51, 53-55, 67.

Horstkotte-Wesseler, G. (1999) Socioeconomics of rice aquaculture and IPM in the Philippines: synergies, potentials and problems. GTZ, ICLARM and IRRI, Manila.

Kabir, A.K.M.N. and Huque, S.M.Z. (2000) Cage design in CARE-CAGES Project, Bangladesh. In: Liao, I.C. and Lin, C.K. (eds) *Cage Aquaculture in Asia. Proceedings of the First International Symposium on Cage Aquaculture in Asia.* Asian Fisheries Society, Manila and World Aquaculture Society – Southeast Asian Chapter, Bangkok, pp. 125-131.

Lewis, D.J., Wood, G.D. and Gregory, R. (1996) *Trading the Silver Seed, Local Knowledge and Market Moralities in Aquacultural Development.* University Press, Dhaka.

Little, D.C., Innes-Taylor, N.L., Turongruang, D. and Komolmarl, S. (1991) Large seed for small-scale aquaculture. *Aquabyte* 4(2), 2-3.

Martinez-Espinosa, M. (1992) Rural aquaculture, from myth to reality. *FAO Aquaculture Newsletter* 2,13-15.

McAndrew, K.I., Little, D.C. and Beveridge, M.C.M. (2000) Entry points and low risk strategies appropriate for the resource-poor to participate in cage aquaculture: experiences from CARE-CAGES Project, Bangladesh. In: Liao, I.C. and Lin, C.K. (eds) *Cage Aquaculture in Asia.* Proceedings of the First International Symposium on Cage Aquaculture in Asia. Asian Fisheries Society, Manila and World Aquaculture Society – Southeast Asian Chapter, Bangkok, pp. 225-231.

Nandeesha, M.C. and Hanglomong, H. (1997) Women in fisheries in Indo-China countries. Proceedings of the Seminar on Women in Fisheries in Indo-China Countries, Bati Fisheries Station, Bati, Prey Veng, Cambodia.

Scoones, I. and Thompson, J. (1994) *Beyond Farmer First: Rural People's Knowledge, Agricultural Research and Extension Practice.* Intermediate Technology Publications, London.

Surintaraseree, P. and Little, D. (1998) Thailand: diffusing rice-fish culture. The Rural Extension Bulletin, October 98, pp. 30-37.

Van den Berg, F. (1996) The private sector: a potential key element in the development of small-scale aquaculture in Africa. *FAO Aquaculture Newsletter* 12,14-16.

Yap, W.G. (1999a) Rural aquaculture systems in the Philippines. *Aquaculture Asia* 4(2), 45-50.

Index

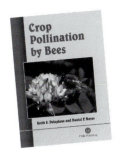
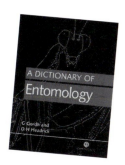
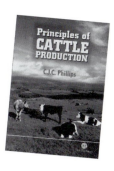